Semigroup Approach to
Nonlinear Diffusion Equations

Semigroup Approach to Nonlinear Diffusion Equations

Viorel Barbu

"Al I Cuza" University of Iaşi, Romania and Romanian Academy

NEW JERSEY • LONDON • SINGAPORE • BEIJING • SHANGHAI • HONG KONG • TAIPEI • CHENNAI • TOKYO

Published by

World Scientific Publishing Co. Pte. Ltd.
5 Toh Tuck Link, Singapore 596224
USA office: 27 Warren Street, Suite 401-402, Hackensack, NJ 07601
UK office: 57 Shelton Street, Covent Garden, London WC2H 9HE

Library of Congress Control Number: 2021045208

British Library Cataloguing-in-Publication Data
A catalogue record for this book is available from the British Library.

SEMIGROUP APPROACH TO NONLINEAR DIFFUSION EQUATIONS

ISBN 978-981-124-651-7 (hardcover)
ISBN 978-981-124-652-4 (ebook for institutions)
ISBN 978-981-124-653-1 (ebook for individuals)

For any available supplementary material, please visit
https://www.worldscientific.com/worldscibooks/10.1142/12534#t=suppl

Printed in Singapore

Preface

A famous and influential book of J.-L. Lions entitled *Quelques méthodes de résolution des problèkes aux limites nonlinéaires*, published with Dunod Gauthier–Villars in 1969, opened a new way in the treatment of nonlinear partial differential equations by emphasizing the role of functional methods arranged in several groups of techniques. From this perspective the methods are more important than the particular results and such a philosophy had a considerable impact on the development of the theory of nonlinear partial differential equations in the last fifty years. Meantime, new functional methods as the theory of nonlinear semigroups of contractions or the variational methods had attained preeminence and undisputed impact in the theory of partial differential equations. This book, which is related and continues some of our previous books, was written in the same idea, emphasizing the role of the theory of m-accretive and maximal monotone operators in Banach spaces in the treatment of well-posedness for some significant classes of nonlinear partial equations with main emphasis on elliptic and parabolic equations. More precisely, it is focusing on the monotonicity method in nonlinear analysis and on nonlinear dynamics described by continuous semigroups of contractions generated by m-accretive operators in Banach spaces and, in particular, in the space of p-Lebesgue integrable functions on $\Omega \subset \mathbb{R}^N$. Roughly speaking, this is the *semigroup approach* to autonomous evolution equations and in a modified version it applies as well to time-varying problems. As regards the existence theory for nonlinear partial differential equations, no generality is claimed whatever while, as mentioned above, the accent is put on the methods which open the applications field to a wider class of equations. In order to keep the thematic unity of this book, we had to omit some important classes of nonlinear equations in mathematical physics such as the wave equations or Navier–Stokes equations though the nonlinear semigroup theory applies in these cases as well.

I am grateful to Dr. Gabriela Marinoschi from Institute of Mathematical Statistics and Applied Mathematics of the Romanian Academy who carefully read the preliminary draft of this book and made helpfully suggestions. Many thanks are due also to Mrs. Elena Mocanu for processing the manuscript of this book.

Iaşi, July 2021 *Viorel Barbu*

Contents

Chapter 1

Preliminaries

Mathematics is the science of skillful operations with concepts and rules invented just for this purpose. The principal emphasis is on the invention of concepts.

Eugen Wigner, *The unreasonable effectiveness of mathematics in the natural sciences* (1960)

The modern theory of partial differential equations is based on the functional analysis and theory of Sobolev spaces. The aim of this chapter is to provide for later use some basic results of functional analysis pertaining to geometric properties of infinite-dimensional normed spaces, convex functions, spaces of distributions, and the variational theory of linear elliptic boundary value problems. Most of these results, which can be easily found in standard textbooks or monographs, are given without proof or with a sketch of proof and suitable references.

1.1 Geometric Properties of Banach Spaces

Throughout this section, X is a real Banach space and X^* denotes its dual. The value of a functional $x^* \in X^*$ at $x \in X$ is denoted either by (x, x^*) or $x^*(x)$, as is convenient. The norm of X is denoted by $\|\cdot\|$, and the norm of X^* is denoted by $\|\cdot\|_*$. If there is no danger of confusion, we omit the asterisk from the notation $\|\cdot\|_*$ and denote both the norms of X and X^* by the symbol $\|\cdot\|$.

We use the symbol lim or $\to$ to indicate *strong convergence* in X and w-lim or $\rightharpoonup$ for *weak convergence* in X. By w^*-$\lim(x, x^*) \rightharpoonup$ we indicate the *weak-star* convergence in X^*. The space X^* endowed with the weak-star topology is denoted by X^*_w.

Define on X the mapping $J : X \to 2^{X^*}$:

$$J(x) = \{x^* \in X^*;\ (x, x^*) = \|x\|^2 = \|x^*\|^2\}, \quad \forall x \in X. \tag{1.1}$$

By the Hahn–Banach theorem we know that, for every $x_0 \in X$, there is some $x_0^* \in X^*$ such that $(x_0^*, x_0) = \|x_0\|$ and $\|x_0^*\| \le 1$, and so $J(x) \ne \emptyset$, $\forall x \in X$. The mapping $J : X \to X^*$ is called the *duality mapping* of the space X and, in general, it is multivalued.

The inverse mapping $J^{-1} : X^* \to X$ defined by

$$J^{-1}(x^*) = \{x \in X;\ x^* \in J(x)\}$$

also satisfies

$$J^{-1}(x^*) = \{x \in X;\ \|x\| = \|x^*\|,\ (x, x^*) = \|x\|^2 = \|x^*\|^2\}.$$

If the space X is reflexive (i.e., $X = X^{**}$), then clearly J^{-1} is just the duality mapping of X^* and so $D(J^{-1}) = X^*$. As a matter of fact, reflexivity plays an important role everywhere in the following and it should be recalled that a normed space is reflexive if and only if its dual X^* is reflexive (see, e.g., Yosida [104], p. 113).

It turns out that the properties of the duality mapping are closely related to the nature of the spaces X and X^*, more precisely, to the convexity and smoothing properties of the closed balls in X and X^*.

Recall that the space X is called *strictly convex* if the unity ball B of X is strictly convex, that is, the boundary ∂B contains no line segments.

The space X is said to be *uniformly convex* if, for each $\varepsilon > 0$, $0 < \varepsilon < 2$, there is $\delta(\varepsilon) > 0$ such that, if $\|x\| = 1$, $\|y\| = 1$ and $\|x - y\| \geq \varepsilon$, then $\|x + y\| \leq 2(1 - \delta(\varepsilon))$.

Obviously, every uniformly convex space X is strictly convex. Hilbert spaces, as well as the spaces $L^p(\Omega)$, $1 < p < \infty$, are uniformly convex spaces. Recall also that, by virtue of the Milman theorem (see, e.g., Yosida [104], p. 127), every uniformly convex Banach space X is reflexive. Conversely, it turns out that every reflexive Banach space X can be renormed such that X and X^* become strictly convex. More precisely, one has the following important result due to Asplund [7].

Theorem 1.1 *Let X be a reflexive Banach space with the norm $\|\cdot\|$. Then there is an equivalent norm $\|\cdot\|_0$ on X such that X is strictly convex in this norm and X^* is strictly convex in the dual norm $\|\cdot\|_0^*$.*

Regarding the properties of the duality mapping associated with strictly or uniformly convex Banach spaces, we have the following.

Theorem 1.2 *Let X be a Banach space. If the dual space X^* is strictly convex, then the duality mapping $J : X \to X^*$ is single-valued and demicontinuous (i.e., it is continuous from X to X^*_w). If the space X^* is uniformly convex, then J is uniformly continuous on every bounded subset of X.*

Let us give a few examples of duality mappings.

(1) $X = H$ is a Hilbert space identified with its own dual. Then $J = I$, the identity operator in H. If H is not identified with its dual H^*, then the duality mapping $J : H \to H^*$ is the canonical isomorphism Λ of H onto H^*. For instance, if $H = H_0^1(\Omega)$ and $H^* = H^{-1}(\Omega)$ and Ω is a bounded and open subset of $\mathbb{R}^N$, then $J = \Lambda$ is defined by

$$(\Lambda u, v) = \int_\Omega \nabla u \cdot \nabla v\, dx, \quad \forall u, v \in H_0^1(\Omega). \tag{1.2}$$

In other words, J is the Laplace operator $-\Delta$ under Dirichlet homogeneous boundary conditions in $\Omega \subset \mathbb{R}^N$. Here $H_0^1(\Omega)$ is the Sobolev space $\{u \in L^2(\Omega);\ \nabla u \in (L^2(\Omega))^N;\ u = 0 \text{ on } \partial\Omega\}$. (See Section 1.3 below.) If $\Omega = \mathbb{R}^N$, that is $X = H^1(\mathbb{R}^N)$, $X^* = H^{-1}(\mathbb{R}^N)$, then the duality mapping $J : X \to X^*$ is given by $J(u) = u - \Delta u$ in $\mathcal{D}'(\mathbb{R}^N)$.

(2) $X = L^p(\Omega)$, where $1 < p < \infty$ and Ω is a measurable subset of $\mathbb{R}^N$. Then, the duality mapping of X is given by

$$J(u)(x) = |u(x)|^{p-2}u(x)\|u\|_{L^p(\Omega)}^{2-p}, \quad \text{a.e. } x \in \Omega,\ \forall u \in L^p(\Omega). \tag{1.3}$$

Indeed, it is readily seen that if Φ_p is the mapping defined by the right-hand side of (1.3), we have

$$\int_\Omega \Phi_p(u)u\,dx = \left(\int_\Omega |u|^p dx\right)^{\frac{2}{p}} = \left(\int_\Omega |\Phi_p(u)|^q dx\right)^{\frac{2}{q}},$$

where $\frac{1}{p} + \frac{1}{q} = 1$. Since the duality mapping J of $L^p(\Omega)$ is single-valued (because L^p is uniformly convex for $p > 1$) and $\Phi_p(u) \in J(u)$, we conclude that $J = \Phi_p$, as claimed. If $X = L^1(\Omega)$, then as we show later (Corollary 2.19)

$$J(u) = \{v \in L^\infty(\Omega);\ v(x) \in \text{sign}(u(x)) \cdot \|u\|_{L^1(\Omega)},\ \text{a.e. } x \in \Omega\}, \tag{1.4}$$

where $\text{sign}(u) = \frac{u}{|u|}$ if $u \neq 0$, $\text{sign}(0) = [-1, 1]$.

(3) Let X be the Sobolev space $W_0^{1,p}(\Omega)$, where $1 < p < \infty$ and Ω is a bounded and open subset of $\mathbb{R}^N$. (See Section 1.3 below.) Then, $X^* = W^{-1,q}(\Omega)$, $\frac{1}{p} + \frac{1}{q} = 1$, and the duality mapping J is given by

$$J(u) = -\|u\|_{W_0^{1,p}(\Omega)}^{2-p} \sum_{i=1}^N \frac{\partial}{\partial x_i}\left(\left|\frac{\partial u}{\partial x_i}\right|^{p-2} \frac{\partial u}{\partial x_i}\right), \quad u \in W_0^{1,p}(\Omega). \tag{1.5}$$

In other words, $J : W_0^{1,p}(\Omega) \to W^{-1,q}(\Omega)$, $\frac{1}{p} + \frac{1}{q} = 1$, is defined by

$$(J(u), v) = \|u\|_{W_0^{1,p}(\Omega)}^{2-p} \sum_{i=1}^N \int_\Omega \left|\frac{\partial u}{\partial x_i}\right|^{p-2} \frac{\partial u}{\partial x_i}\frac{\partial v}{\partial x_i}\,dx, \quad \forall u, v \in W_0^{1,p}(\Omega), \tag{1.6}$$

where $(\cdot, \cdot)$ is the duality pairing on $W_0^{1,p}(\Omega) \times W^{-1,q}(\Omega)$.

1.2 Convex Functions

Let X be a real Banach space with dual X^*. A *proper convex function* on X is a function $\varphi : X \to (-\infty, +\infty] = \overline{\mathbb{R}}$, that is, not identically $+\infty$ and that satisfies the inequality

$$\varphi((1-\lambda)x + \lambda y) \leq (1-\lambda)\varphi(x) + \lambda\varphi(y) \tag{1.7}$$

for all $x, y \in X$ and all $\lambda \in [0, 1]$.

The function φ is called *strictly convex* if the equality in (1.7) for some $\lambda \in (0, 1)$ implies that $x = y$.

The function $\varphi : X \to (-\infty, +\infty]$ is said to be *lower semicontinuous* (l.s.c.) on X if

$$\liminf_{u\to x} \varphi(u) \geq \varphi(x), \quad \forall x \in X,$$

or, equivalently, if every level subset $\{x \in X;\ \varphi(x) \leq \lambda\}$ is closed.

The function $\varphi : X \to]-\infty, +\infty]$ is said to be *weakly lower semicontinuous* if it is lower semicontinuous on the space X endowed with weak topology.

Because every level set of a convex function is convex and every closed convex set is weakly closed, a function φ is l.s.c. if and only if it is weakly lower semicontinuous.

Given a lower semicontinuous convex function $\varphi : X \to (-\infty, +\infty] = \overline{\mathbb{R}}$, $\varphi \not\equiv \infty$, we use the following notations:

$$D(\varphi) = \{x \in X;\ \varphi(x) < \infty\} \quad \text{(the effective domain of } \varphi\text{)}. \tag{1.8}$$

Now, let us briefly describe some elementary properties of l.s.c., convex functions. For proof, we refer to [21], [41], [87], [94].

Proposition 1.3 *Let $\varphi : X \to \overline{\mathbb{R}} =]-\infty, +\infty]$ be a proper, l.s.c., and convex function. Then φ is bounded from below by an affine function; that is there are $x_0^* \in X^*$ and $\beta \in \mathbb{R}$ such that*

$$\varphi(x) \geq (x_0^*, x) + \beta, \quad \forall x \in X. \tag{1.9}$$

Proposition 1.4 *Let $\varphi : X \to \overline{\mathbb{R}}$ be a proper, convex, and l.s.c. function. Then φ is continuous on* $\text{int}(D(\varphi))$.

(Here *int* denotes the interior.)

The function $\varphi^* : X^* \to \overline{\mathbb{R}}$ defined by

$$\varphi^*(p) = \sup\{(p, x) - \varphi(x);\ x \in X\} \tag{1.10}$$

is called the *conjugate* of φ.

Proposition 1.5 *Let $\varphi : X \to \overline{\mathbb{R}}$ be l.s.c., convex, and proper. Then φ^* is l.s.c., convex, and proper on the space X^*.*

Proof. As supremum of a set of affine functions, φ^* is convex and l.s.c. Moreover, by Proposition 1.4 we see that $\varphi^* \not\equiv \infty$. □

Proposition 1.6 *Let $\varphi : X \to \overline{\mathbb{R}}$ be a weakly lower semicontinuous function such that every level set $\{x \in X;\ \varphi(x) \leq \lambda\}$ is weakly compact. Then φ attains its infimum on X. In particular, if X is reflexive and φ is an l.s.c. proper convex function on X such that*

$$\lim_{\|x\|\to\infty} \varphi(x) = \infty, \tag{1.11}$$

then there exists $x_0 \in X$ such that $\varphi(x_0) = \inf\{\varphi(x);\ x \in X\}$.

Given an l.s.c., convex, proper function $\varphi : X \to \overline{\mathbb{R}}$, the mapping $\partial\varphi : X \to X^*$ defined by

$$\partial\varphi(x) = \{x^* \in X^*;\ \varphi(x) \le \varphi(y) + (x^*, x - y),\ \forall y \in X\} \tag{1.12}$$

is called the *subdifferential* of φ.

In general, $\partial\varphi$ is a multivalued operator from X to X^* not everywhere defined and can be seen as a subset of $X \times X^*$. An element $x^* \in \partial\varphi(x)$ (if any) is called a *subgradient* of φ in x. We denote as usual by $D(\partial\varphi)$ the set of all $x \in X$ for which $\partial\varphi(x) \ne \emptyset$.

Let us pause briefly to give some simple examples.

(1) $\varphi(x) = \frac{1}{2}\|x\|^2$. Then, $\partial\varphi = J$ (the duality mapping of the space X).

(2) Let K be a closed convex subset of X. The function $I_K : X \to \overline{\mathbb{R}}$ defined by

$$I_K(x) = \begin{cases} 0, & \text{if } x \in K, \\ +\infty, & \text{if } x \notin K, \end{cases} \tag{1.13}$$

is called the *indicator function* of K, and its dual function H_K,

$$H_K(p) = \sup\{(p, u);\ u \in K\},\quad \forall p \in X^*,$$

is called the *support function* of K. It is readily seen that $D(\partial I_K) = K$, $\partial I_K(x) = 0$ for $x \in \operatorname{int} K$ (if nonempty) and that

$$\partial I_K(x) = N_K(x) = \{x^* \in X^*;\ (x^*, x - u) \ge 0,\ \forall u \in K\},\quad \forall x \in K. \tag{1.14}$$

For every $x \in \partial K$ (the boundary of K), $N_K(x)$ is the *normal cone* at K in x.

(3) Let φ be convex and Gâteaux differentiable at x with the gradient $\nabla\varphi(x)$. Then $\partial\varphi(x) = \nabla\varphi(x)$.

Proposition 1.7 *Let X be a reflexive Banach space and let $\varphi : X \to \overline{\mathbb{R}}$ be an l.s.c., convex, proper function. Then the following conditions are equivalent.*

(i) $x^* \in \partial\varphi(x)$,

(ii) $\varphi(x) + \varphi^*(x^*) = (x, x^*)$,

(iii) $x \in \partial\varphi^*(x^*)$.

In particular, $\partial\varphi^* = (\partial\varphi)^{-1}$ and $(\varphi^*)^* = \varphi$.

We also mention without proof the following results. (See, e.g., [21], [41].)

Proposition 1.8 *Let $\varphi : X \to \overline{\mathbb{R}}$ be an l.s.c., convex, and proper function. Then $D(\partial\varphi)$ is a dense subset of $D(\varphi)$. Moreover, $\operatorname{int} D(\varphi) \subset D(\partial\varphi)$.*

There is a close connection between the range of subdifferential $\partial\varphi$ of a lower semicontinuous convex function $\varphi : X \to \overline{\mathbb{R}}$ and its behavior for $\|x\| \to \infty$. Namely, one has

Proposition 1.9 *The following two conditions are equivalent.*

(j) *$R(\partial\varphi) = X^*$, and $\partial\varphi^* = (\partial\varphi)^{-1}$ is bounded on bounded subsets,*

(jj) $\lim_{\|x\|\to\infty} \frac{\varphi(x)}{\|x\|} = +\infty$.

1.3 Sobolev Spaces

Throughout this section, until further notice, we assume that Ω is an open subset of $\mathbb{R}^N$. Denote by $L^p(\Omega)$, $1 \le p \le \infty$, the space of Lebesgue p-integrable functions on Ω and by $D(\Omega)$ (or $C_0^\infty(\Omega)$) the space of infinitely differentiable functions with compact support in Ω. Denote by $\mathcal{D}'(\Omega)$ the space of Schwartz distributions on Ω. Given $u \in \mathcal{D}'(\Omega)$, by definition, the derivative of order $\alpha = (\alpha_1, ..., \alpha_N)$, $D^\alpha u$, of u, is the distribution

$$(D^\alpha u)(\varphi) = (-1)^{|\alpha|} u(D^\alpha \varphi),\ \forall \varphi \in \mathcal{D}(\Omega), \quad \text{where } |\alpha| = \alpha_1 + \cdots + \alpha_N.$$

Let m be a positive integer. Denote by $H^m(\Omega)$ the set of all real valued functions $u \in L^2(\Omega)$ such that distributional derivatives $D^\alpha u$ of u of order $|\alpha| \le m$ all belong to $L^2(\Omega)$. In other words,

$$H^m(\Omega) = \{u \in L^2(\Omega);\ D^\alpha u \in L^2(\Omega),\ |\alpha| \le m\}.$$

We present below a few basic properties of Sobolev spaces and refer to the books [1, 16, 38] for proofs.

Proposition 1.10 *$H^m(\Omega)$ is a Hilbert space with the scalar product*

$$\langle u, v \rangle_m = \sum_{|\alpha| \le m} \int_\Omega D^\alpha u(x) D^\alpha v(x) dx, \quad \forall u, v \in H^m(\Omega).$$

If $\Omega = (a, b)$, $-\infty < a < b < \infty$, then $H^1(\Omega)$ reduces to the subspace of absolutely continuous functions on the interval $[a, b]$ with derivative in $L^2(a, b)$. More generally, for an integer $m \ge 1$ and $1 \le p \le \infty$, one defines the Sobolev space

$$W^{m,p}(\Omega) = \{u \in L^p(\Omega);\ D^\alpha u \in L^p(\Omega),\ |\alpha| \le m\}$$

with the norm

$$\|u\|_{m,p} = \left(\sum_{|\alpha| \le m} \int_\Omega |D^\alpha u(x)|^p dx \right)^{\frac{1}{p}}. \tag{1.15}$$

For $0 < m < 1$, the space $W^{m,p}(\Omega)$ is defined by (see Adams [1], p. 214)

$$W^{m,p}(\Omega) = \left\{ u \in L^p(\Omega);\ \frac{u(x) - u(y)}{|x - y|^{m + \frac{N}{p}}} \in L^p(\Omega \times \Omega) \right\}$$

with the natural norm. For $m > 1$, $m = s + a$, $s = [m]$, $0 < a < 1$, define

$$W^{m,p}(\Omega) = \{u \in W^{s,p}(\Omega);\ D^\alpha u \in W^{a,p}(\Omega);\ |\alpha| \le s\}.$$

Now, we mention an important property of the space $H^1(\Omega)$ known as the *Sobolev embedding theorem.*

Theorem 1.11 *Let $1 \le p < N$, $N > 2$, and $p^* = \frac{2N}{N-2}$. Then*

$$W^{1,p}(\mathbb{R}^N) \subset L^{p^*}(\mathbb{R}^N), \tag{1.16}$$

$$\|u\|_{L^{p^*}(\mathbb{R}^N)} \le C_{p,N} \|\nabla u\|_{L^p(\mathbb{R}^N)}, \quad \forall u \in W^{1,p}(\mathbb{R}^N). \tag{1.17}$$

Theorem 1.11, known in literature as the *Sobolev embedding theorem*, has a natural extension to the Sobolev space $W^{m,p}(\Omega)$ for any $m > 0$. More precisely, we have (see Adams [1], p. 217, Brezis [38], p. 285).

Theorem 1.12 *Let Ω be a bounded and open subset of $\mathbb{R}^N$. Then we have*

$$\begin{aligned} &W^{m,p}(\Omega) \subset L^{p^*}(\Omega) && \text{if } 1 \le p < \tfrac{N}{m},\ \tfrac{1}{p^*} = \tfrac{1}{p} - \tfrac{m}{N}, \\ &W^{m,p}(\Omega) \subset L^q(\Omega) && \text{for all } q \ge p \text{ if } p = \tfrac{N}{m}, \\ &W^{m,p}(\Omega) \subset L^\infty(\Omega) && \text{if } p > \tfrac{N}{m}, \end{aligned} \tag{1.18}$$

with compact injections.

We also have

$$\|u\|_{W^{1,p}(\Omega)} \le C(\|\nabla u\|_{L^p(\Omega)} + \|u\|_{L^q(\Omega)}) \tag{1.19}$$

for $1 \le q \le Np(N-p)^{-1}$, $1 \le p < N$ and the right-hand side of (1.19) is an equivalent norm on $W^{1,p}(\Omega)$.

In particular, the injection of the space $W^{1,p}(\Omega)$ into $L^p(\Omega)$ is compact for all $p \ge 1$ and N.

If Ω is an open C^1 subset of $\mathbb{R}^N$ with the boundary $\partial\Omega$, then each $u \in C(\overline{\Omega})$ is well defined on $\partial\Omega$. We call the restriction of u to $\partial\Omega$ the *trace* of u to $\partial\Omega$ and it is denoted by $\gamma_0(u)$. If $u \in L^2(\Omega)$, then $\gamma_0(u)$ is no longer well defined. We have, however, the following.

Lemma 1.13 *Let Ω be an open subset of class C^1 with compact boundary $\partial\Omega$ or $\Omega = \mathbb{R}^N_+$. Then, there is $C > 0$ such that*

$$\|\gamma_0(u)\|_{L^2(\partial\Omega)} \le C\|u\|_{H^1(\Omega)}, \quad \forall u \in C_0^\infty(\mathbb{R}^N). \tag{1.20}$$

Definition 1.14 Let Ω be of class C^1 with compact boundary or $\Omega = \mathbb{R}^N_+$. Let $u \in H^1(\Omega)$. Then $\gamma_0(u) = \lim_{j\to\infty} \gamma_0(u_j)$ in $L^2(\partial\Omega)$, where $\{u_j\} \subset C_0^\infty(\mathbb{R}^N)$ is such that $u_j \to u$ in $H^1(\Omega)$.

It turns out that the definition is consistent, that is, $\gamma_0(u)$ is independent of $\{u_j\}$. Indeed, if $\{u_j\}$ and $\{\bar{u}_j\}$ are two sequences in $C_0^\infty(\mathbb{R}^N)$ convergent to u in $H^1(\Omega)$, then, by (1.20),

$$\|\gamma_0(u_j - \bar{u}_j)\|_{L^2(\partial\Omega)} \le C\|u_j - \bar{u}_j\|_{H^1(\Omega)} \to 0 \quad \text{as } j \to \infty.$$

Moreover, it follows by Lemma 1.13 that the map $\gamma_0 : H^1(\Omega) \to L^2(\partial\Omega)$ is continuous. In general (see Adams [1], p. 114), we have for $u \in W^{m,p}(\Omega)$, $\gamma_0(u) \in L^q(\partial\Omega)$ if $mp < N$ and

$$p \le q \le \frac{(N-1)p}{(N-mp)}.$$

A sharper interpolation-trace inequality is in Proposition 1.15 (see [66]).

Proposition 1.15 *Let Ω be bounded with the smooth boundary $\partial\Omega$. Then*

$$\|u\|_{L^2(\partial\Omega)} \le C(\|\nabla u\|_{L^2(\Omega)} + \|u\|_{L^{\sigma+1}(\Omega)})^{\theta}\|u\|^{1-\theta}_{L^{\sigma+1}(\Omega)},$$
$$\forall\, u \in H^1(\Omega) \cap L^{\sigma+1}(\Omega),$$

for all $\sigma \in [0,1]$ and $\theta = (N(1-\sigma)+\sigma+1)(N(1-\sigma)+2(\sigma+1))^{-1}$.

Also the following result will be frequently invoked in the following.

Proposition 1.16 *Let $u \in W^{1,p}(\mathbb{R}^N)$, $1 \le p < \infty$, and let*

$$\Omega_\delta = \{x \in \mathbb{R}^N;\ |u(x)| \le \delta\}.$$

Then

$$\lim_{\delta\to 0}\int_{\Omega_\delta} |\nabla u(x)|^p dx = 0. \tag{1.21}$$

Proof. We set $I_\delta(x) = 1$, a.e. $x \in \Omega_\delta$, $I_\delta(x) = 0$, a.e. $\mathbb{R}^N \setminus \Omega_\delta$. Then, by the Lebesgue theorem, we have

$$\lim_{\delta\to 0}\int_{\Omega_\delta} |\nabla u(x)|^p dx = \lim_{\delta\to 0}\int_{\mathbb{R}^N} I_\delta(x)|\nabla u(x)|^p dx = 0,$$

because $|\nabla u(x)| = 0$, a.e. in $\{x;\ |u(x)| = 0\}$, we have $I_\delta|\nabla u|^2 \to 0$, a.e. in $\mathbb{R}^d$ and $I_\delta|\nabla u|^p \le |\nabla u|^p$, a.e. in $\mathbb{R}^N$. □

Definition 1.17 Let Ω be any open subset of $\mathbb{R}^N$. The space $H_0^1(\Omega)$ is the closure (the completion) of $C_0^\infty(\Omega)$ in the norm of $H^1(\Omega)$.

It follows that $H_0^1(\Omega)$ is a closed subspace of $H^1(\Omega)$ and it is a Hilbert space with the scalar product

$$\langle u, v\rangle_1 = \sum_{i=1}^N \int_\Omega \frac{\partial u}{\partial x_i}\,\frac{\partial v}{\partial x_i}\,dx + \int_\Omega uv\,dx$$

with the corresponding norm

$$\|u\|_1 = \left(\int_\Omega (|\nabla u(x)|^2 + u^2(x))dx\right)^{\frac{1}{2}}.$$

In fact, $H_0^1(\Omega)$ is the subspace of functions $u \in H^1(\Omega)$ that are zero on $\partial\Omega$ in the sense of traces. (Here and everywhere in the following, we shall denote also by $|\cdot|$ the Euclidean norm in $\mathbb{R}^N$.)

Proposition 1.18 *Let Ω be an open and bounded subset of $\mathbb{R}^N$. Then there is $C > 0$ independent of u such that*

$$\|u\|_{L^2(\Omega)} \le C\|\nabla u\|_{L^2(\Omega)},\quad \forall u \in H_0^1(\Omega).$$

In particular, Proposition 1.18 shows that if Ω is bounded, then the scalar product

$$((u,v)) = \int_\Omega \nabla u(x)\cdot\nabla v(x)dx$$

and the corresponding norm

$$\|u\| = \left(\int_\Omega |\nabla u(x)|^2 dx\right)^{\frac{1}{2}}$$

define an equivalent Hilbertian structure on $H_0^1(\Omega)$.

We denote by $H^{-1}(\Omega)$ the dual space of $H_0^1(\Omega)$, that is, the space of all linear continuous functionals on $H_0^1(\Omega)$. Equivalently,

$$H^{-1}(\Omega) = \{u \in \mathcal{D}'(\Omega);\ |u(\varphi)| \le C_u\|\varphi\|_{H^1(\Omega)},\quad \forall\varphi \in C_0^\infty(\Omega)\}.$$

The space $H^{-1}(\Omega)$ is endowed with the dual norm

$$\|u\|_{-1} = \sup\{|u(\varphi)|;\ \|\varphi\| \le 1\},\quad \forall u \in H^{-1}(\Omega).$$

By Riesz's theorem, we know that $H^{-1}(\Omega)$ is isometric to $H_0^1(\Omega)$. Note also that

$$H_0^1(\Omega) \subset L^2(\Omega) \subset H^{-1}(\Omega)$$

in the algebraic and topological sense. In other words, the injections of $L^2(\Omega)$ into $H^{-1}(\Omega)$ and of $H_0^1(\Omega)$ into $L^2(\Omega)$ are continuous. Note also that the above injections are dense.

There is an equivalent definition of $H^{-1}(\Omega)$ given in Theorem 1.19 below.

Theorem 1.19 *The space $H^{-1}(\Omega)$ coincides with the set of all distributions $u \in \mathcal{D}'(\Omega)$ of the form*

$$u = f_0 + \sum_{i=1}^N \frac{\partial f_i}{\partial x_i} \quad \text{in } \mathcal{D}'(\Omega),\ \text{where } f_i \in L^2(\Omega),\ i = 1,\dots,N.$$

The space $W_0^{1,p}(\Omega)$, $p \ge 1$, is similarly defined as the closure of $C_0^1(\Omega)$ into $W^{1,p}(\Omega)$ norm. The dual of $W_0^{1,p}(\Omega)$ is the space $W^{-1,q}(\Omega)$, $\frac{1}{p}+\frac{1}{q}=1$ defined as in Theorem 1.19 with $f_0, f_1, \dots, f_N \in L^q(\Omega)$. As regards the dual space $(H^1(\Omega))'$ of $H^1(\Omega)$, we still have $H^1(\Omega) \subset L^2(\Omega) \subset (H^1(\Omega))'$ with dense and continuous embeddings. In this case, the duality mapping (that is, the canonical isomorphism) of $H^1(\Omega)$ is the operator $J(u) = u - \Delta u$. However, since $C_0^\infty(\Omega)$ is not dense in $H^1(\Omega)$ (as happens for $H_0^1(\Omega)$), the space $(H^1(\Omega))'$ is not a Schwartz-distributions subspace on Ω.

Marcinkievicz spaces

For each $1 < p < \infty$, the Marcinkievicz space of order p, that is, $M^p(\mathbb{R}^N)$, is the set of all Lebesgue measurable functions $u : \mathbb{R}^N \to \mathbb{R}$ such that

$$\|u\|_{M^p} = \min\left\{\alpha > 0;\ \int_K |u(x)|dx \le \alpha(\text{meas}\, K)^{\frac{1}{p'}} \text{ for all Lebesgue measurable sets } K \subset \mathbb{R}^N\right\}. \tag{1.22}$$

Here, $\frac{1}{p} + \frac{1}{p'} = 1$. It is easily seen that, for $1 \le q < p < \infty$, $M^p(\mathbb{R}^N) \subset L^q_{\text{loc}}(\mathbb{R}^N)$ with continuous injection.

Elliptic boundary value problems

Let V be a real Hilbert space and let V^* be the topological dual space of V. For each $v^* \in V^*$ and $v \in V$ we denote by (v^*, v) the value $v^*(v)$ of the functional v^* at v. The functional $a : V \times V \to \mathbb{R}$ is said to be *bilinear* if, for each $u \in V$, $v \to a(u,v)$ is linear on V. The function a is said to be

- *continuous* if $|a(u,v)| \le M\|u\|_V\|v\|_V, \quad \forall u, v \in V$,
- *coercive* if $a(u,u) \ge \omega\|u\|_V^2, \quad \forall u \in V$, for some $\omega > 0$, and
- *symmetric* if $a(u,v) = a(v,u), \quad \forall u, v \in V$.

Lemma 1.20 (Lax–Milgram) *Let $a : V \times V \to \mathbb{R}$ be a bilinear, continuous and coercive functional. Then, for each $f \in V^*$, there is a unique $u^* \in V$ such that*

$$a(u^*, v) = (f, v), \quad \forall v \in V. \tag{1.23}$$

Moreover, the map $f \to u^*$ is Lipschitzian from V^* to V with a Lipschitz constant $\le \omega^{-1}$. If a is symmetric, then u^* minimizes the function $u \to \frac{1}{2}\,a(u,u) - (f,u)$ on V, that is,

$$\frac{1}{2}\,a(u^*, u^*) - (f, u^*) = \min\left\{\frac{1}{2}\,a(u,u) - (f,u);\ u \in V\right\}.$$

If a is symmetric, then the Lax–Milgram lemma is a simple consequence of Riesz's representation theorem because, in this case, $(u,v) \to a(u,v)$ is an equivalent scalar product on V and so, by the Riesz theorem, the functional $v \to (f,v)$ can be represented as (1.23) for some $u^* \in V$. In the general case, we note that the operator $A : V \to V^*$, defined by $(v, Au) = a(u,v)$, $\forall v \in V$, is linear, continuous and coercive and so, its range is all of V^*.

The Lax–Milgram lemma is the abstract formulation of a large class of elliptic boundary value problems and a few of them are recalled below.

The Dirichlet problem

Consider the Dirichlet problem

$$\begin{cases} -\Delta u + c(x)u = f & \text{in } \Omega, \\ u = 0 & \text{on } \partial\Omega, \end{cases} \tag{1.24}$$

where Ω is an open set of $\mathbb{R}^N$, $c \in L^\infty(\Omega)$, and $f \in H^{-1}(\Omega)$ is given.

Definition 1.21 The function u is said to be a *weak* or *variational solution* to the Dirichlet problem (1.24) if $u \in H^1_0(\Omega)$ and

$$\int_\Omega \nabla u(x) \cdot \nabla\varphi(x)dx + \int_\Omega c(x)u(x)\varphi(x)dx = (f, \varphi) \tag{1.25}$$

for all $\varphi \in H^1_0(\Omega)$ (equivalently, for all $\varphi \in C^\infty_0(\Omega)$).

In (1.25), ∇u is taken in the sense of distributions and (f, φ) is the value of the functional $f \in H^{-1}(\Omega)$ into $\varphi \in H_0^1(\Omega)$. If $f \in L^2(\Omega) \subset H^{-1}(\Omega)$, then

$$(f, \varphi) = \int_\Omega f(x)\varphi(x)dx.$$

Then, by the Lax–Milgram Lemma 1.20 applied to the functional $a = a(u, \varphi)$, defined by the left-hand side of (1.25), we get

Theorem 1.22 *Let Ω be a bounded open set of $\mathbb{R}^N$ and let $c \in L^\infty(\Omega)$ be such that $c(x) \geq 0$, a.e. $x \in \Omega$. Then, for each $f \in H^{-1}(\Omega)$ the Dirichlet problem (1.24) has a unique weak solution $u^* \in H_0^1(\Omega)$. Moreover, u^* minimizes on $H_0^1(\Omega)$ the functional*

$$\frac{1}{2}\int_\Omega (|\nabla u(x)|^2 + c(x)u^2(x))dx - (f, u) \tag{1.26}$$

and the map $f \to u^*$ is Lipschitzian from $H^{-1}(\Omega)$ to $H_0^1(\Omega)$.

The Neumann problem

Consider the boundary value problem

$$\begin{cases} -\Delta u + cu = f & \text{in } \Omega, \\ \dfrac{\partial u}{\partial n} = g & \text{on } \partial\Omega, \end{cases} \tag{1.27}$$

where $\frac{\partial}{\partial n}$ is the normal derivative, that is, $\frac{\partial u}{\partial n} = \nabla u \cdot n$, where $n : \partial\Omega \to \mathbb{R}^N$ is the outward normal to $\partial\Omega$, $c \in L^\infty(\Omega)$ is such that $c(x) \geq \rho > 0$, a.e. $x \in \Omega$, and $f \in L^2(\Omega)$, $g \in L^2(\partial\Omega)$.

Definition 1.23 The function $u \in H^1(\Omega)$ is said to be a *weak solution* to the Neumann problem (1.27) if

$$\int_\Omega \nabla u \cdot \nabla v\, dx + \int_\Omega cuv\, dx = \int_\Omega fv\, dx + \int_{\partial\Omega} gv\, d\sigma, \quad \forall v \in H^1(\Omega). \tag{1.28}$$

Because, as seen earlier, for each $v \in H^1(\Omega)$ the trace $\gamma_0(v)$ of v on $\partial\Omega$ is in $L^2(\partial\Omega)$, the integral $\int_{\partial\Omega} gv\, d\sigma$ is well defined and so (1.28) makes sense.

By the Lax–Milgram lemma applied to $V = H^1(\Omega)$ and $a(u, v)$ defined by the left-side hand of (1.28), we obtain as above

Theorem 1.24 *Let Ω be an open subset of $\mathbb{R}^N$. Then, for each $f \in L^2(\Omega)$ and $g \in L^2(\partial\Omega)$, problem (1.27) has a unique weak solution $u \in H^1(\Omega)$ that minimizes the functional*

$$u \to \frac{1}{2}\int_\Omega (|\nabla u(x)|^2 + c(x)u^2(x))dx - \int_\Omega f(x)u(x)dx - \int_{\partial\Omega} gu\, d\sigma \quad \textit{on } H^1(\Omega).$$

Now, we briefly recall the regularity of the weak solutions to the Dirichlet problem

$$\begin{cases} -\Delta u = f & \text{in } \Omega, \\ u = 0 & \text{on } \partial\Omega. \end{cases} \tag{1.29}$$

By Theorem 1.22 we know that if Ω is a bounded and open subset of $\mathbb{R}^N$ and $f \in L^2(\Omega)$, then problem (1.29) has a unique solution $u \in H_0^1(\Omega)$. It turns out that if $\partial\Omega$ is smooth enough, then this solution is actually in $H^2(\Omega) \cap H_0^1(\Omega)$.

Theorem 1.25 *Let Ω be a bounded and open subset of $\mathbb{R}^N$ of class C^2. Let $f \in L^2(\Omega)$ and let $u \in H_0^1(\Omega)$ be the weak solution to* (1.29). *Then, $u \in H^2(\Omega)$ and*

$$\|u\|_{H^2(\Omega)} \leq C\|f\|_{L^2(\Omega)}, \tag{1.30}$$

where C is independent of f.

We also have

Theorem 1.26 *Under the assumptions of Theorem* 1.24, *if $g \in H^{\frac{1}{2}}(\partial\Omega)$, the weak solution $u \in H^1(\Omega)$ to problem* (1.27) *belongs to $H^2(\Omega)$ and*

$$\|u\|_{H^2(\Omega)} \leq C\left(\|f\|_{L^2(\Omega)} + \|g\|_{H^{\frac{1}{2}}(\partial\Omega)}\right). \tag{1.31}$$

Theorem 1.25 remains true in $L^p(\Omega)$ for $p > 1$. Namely, we have (Agmon, Douglas and Nirenberg [2])

Theorem 1.27 *Let Ω be a bounded open subset of $\mathbb{R}^N$ with smooth boundary $\partial\Omega$ and let $1 < p < \infty$. Then, for each $f \in L^p(\Omega)$, the boundary value problem*

$$-\Delta u = f \quad \text{in } \Omega, \quad u = 0 \quad \text{on } \partial\Omega \tag{1.32}$$

has a unique weak solution $u \in W_0^{1,p}(\Omega) \cap W^{2,p}(\Omega)$. Moreover, one has

$$\|u\|_{W^{2,p}(\Omega)} \leq C\|f\|_{L^p(\Omega)},$$

where C is independent of f.

In the case $p = 1$, it follows that equation (1.32) has a unique solution $u \in W^{1,q}(\Omega)$, where $1 \leq q < \frac{N}{N-1}$.

Equation $-\Delta u = f$ in $L^1(\mathbb{R}^N)$

Let E_N be the fundamental solution of the Laplace operator $-\Delta$, that is,

$$\begin{aligned} E_N(x) &= \frac{1}{(N-2)\omega_N}|x|^{2-N}, && \forall x \neq 0, \text{ for } N \geq 3, \\ E_N(x) &= \frac{1}{2\pi}\log\frac{1}{|x|}, && \forall x \neq 0, \text{ for } N = 2, \end{aligned} \tag{1.33}$$

where ω_N is the volume of unit N-ball.

For each $f \in L^1(\mathbb{R}^N)$,

$$u(x) = (E_N * f)(x) = \int_{\mathbb{R}^N} E_N(x-y)f(y)dy, \ x \in \mathbb{R}^N, \tag{1.34}$$

is a solution in $L^1_{\text{loc}}(\mathbb{R}^N)$ to the equation

$$-\Delta u = f \ \text{ in } \mathcal{D}'(\mathbb{R}^N). \tag{1.35}$$

On the other hand, we have, by Fubini's theorem,

$$\int_K |E_N * f(x)|dx \leq \|E_N\|_{M^p} \|f\|_{L^1(\mathbb{R}^N)} (\text{meas } K)^{\frac{1}{p'}}, \tag{1.36}$$

for each measurable set $K \subset \mathbb{R}^N$ and $1 < p < \infty$, $\frac{1}{p'} = 1 - \frac{1}{p}$.

Moreover, if $N \geq 3$, one has

$$\|E_N * f\|_{M^{\frac{N}{N-2}}} \leq C^1_N \|f\|_{L^1(\mathbb{R}^N)}, \tag{1.37}$$

$$\|\nabla(E_N * f)\|_{M^{\frac{N}{N-1}}} \leq C^2_N \|f\|_{L^1(\mathbb{R}^N)}. \tag{1.38}$$

As regards the uniqueness in equation (1.35), one has (see [32], Lemma A.10)

Proposition 1.28 *If $N \geq 3$ and $f \in L^1(\mathbb{R}^N)$, then $u = E_N * f \in M^{\frac{N}{N-2}}$ with $\nabla u \in M^{\frac{N}{N-1}}$ is the unique solution to equation (1.35). If $N = 2$, then $u = E_N * f \in W^{1,1}_{\text{loc}}(\mathbb{R}^2)$ and $|\nabla u| \in M^2(\mathbb{R}^2)$.*

This result extends to the second order elliptic operators with constant coefficients

$$L(D)u = \sum_{i,j=1}^N a_{ij} \frac{\partial^2 u}{\partial x_i \partial x_j},$$

where

$$a_{ij} \in \mathbb{R}, \ a_{ij} = a_{ji}, \ i,j = 1,...,N, \ \sum_{i,j=1}^N \xi_i \xi_j \geq \alpha |\xi|^2, \ \forall \xi \in \mathbb{R}^N, \ \alpha > 0.$$

The space $BV(\Omega)$

Let Ω be an open subset of $\mathbb{R}^N$ with smooth boundary $\partial\Omega$ (of class C^1, for instance).

A function $f \in L^1(\Omega)$ is said to be of bounded variation on Ω if its gradient Df in the sense of distributions is an $\mathbb{R}^N$-valued measure on Ω, that is,

$$|Df|(\Omega) := \sup \left\{ \int_\Omega f \operatorname{div} \psi \, d\xi : \psi \in C_0^\infty(\Omega; \mathbb{R}^N), \ \|\psi\|_{L^\infty(\Omega)} \leq 1 \right\} < +\infty.$$

Thus, $u \in BV(\Omega)$ if and only if there are Radon measures μ_j on Ω such that $D_j u = \mu_j$, $j = 1,...,N$, in the sense of distributions.

The space of all functions of bounded variation on Ω is denoted by $BV(\Omega)$. It is a Banach space with the norm

$$\|f\|_{BV(\Omega)} = \|f\|_{L^1(\Omega)} + |Df|(\Omega).$$

Let $f \in BV(\Omega)$. Then there is a Radon measure μ_f on $\overline{\Omega}$ and a μ_f-measurable function $\sigma_f : \Omega \to \mathbb{R}^N$ such that $|\sigma_f(x)| = 1$, μ_f, a.e., and

$$\int_\Omega f \operatorname{div} \psi \, d\xi = -\int_\Omega \psi \cdot \sigma_f \, d\mu_f, \quad \forall \psi \in C_0^1(\Omega; \mathbb{R}^N). \tag{1.39}$$

For each $f \in BV(\Omega)$ there is the trace $\gamma(f)$ on $\partial\Omega$ (assumed sufficiently smooth) defined by

$$\int_\Omega f \operatorname{div} \psi \, d\xi = -\int_\Omega \psi{\cdot}\sigma_f \, d\mu_f + \int_{\partial\Omega} \gamma(f)\psi \cdot n \, dH^{N-1}, \ \forall \psi \in C^1(\overline{\Omega}; \mathbb{R}^N), \tag{1.40}$$

where $n = n(x) \in \mathbb{R}^N$ is the outward normal to $\partial\Omega$ and dH^{N-1} is the Hausdorff measure on $\partial\Omega$. We have that $|\gamma(f)| \in L^1(\partial\Omega; dH^{N-1})$.

Theorem 1.29 *Let $1 \le p \le \frac{N}{N-1}$ and let Ω be a bounded open subset of $\mathbb{R}^N$. Then, we have $BV(\Omega) \subset L^p(\Omega)$ with continuous embedding and the function $u \to \|Du\|$ is lower semicontinuous in $L^p(\Omega)$. Such embedding is compact if $1 \le p < \frac{N}{N-1}$.*

We refer the reader to [4] for proofs and other basic results on functions with bounded variations.

Compactness in $L^p(\Omega)$

Let Ω be a Lebesgue measurable subset of $\mathbb{R}^N$. Contrary to what happens in $L^p(\Omega)$ spaces with $1 < p < \infty$ that are reflexive, a bounded subset $\mathcal{M}$ of $L^1(\Omega)$ is not necessarily weakly compact. This happens, however, under some additional conditions on $\mathcal{M}$.

Theorem 1.30 (Dunford–Pettis) *Let $\mathcal{M}$ be a bounded subset of $L^1(\Omega)$ having the property that the family of integrals $\left\{\int_E u(x)dx;\ E \subset \Omega \text{ measurable, } u \in \mathcal{M}\right\}$ is uniformly absolutely continuous, that is, for every $\varepsilon > 0$ there is $\delta(\varepsilon) > 0$ independent of u, such that $\int_E |u(x)|dx \le \varepsilon$ for $m(E) < \delta(\varepsilon)$ (m is the Lebesgue measure). Then the set $\mathcal{M}$ is weakly sequentially compact in $L^1(\Omega)$.*

For the proof, we refer the reader to Edwards [69], p. 270. Theorem 1.30 remains true, of course, in $(L^1(\Omega))^m$, $\forall m \in \mathbb{N}$.

As regard the strong compactness in $L^p(\Omega)$, $\Omega \subset \mathbb{R}^N$, we have the following theorem (Kolmogorov–Riesz–Fréchet).

Theorem 1.31 *Let $1 \le p < \infty$, and let $\mathcal{M}$ be a bounded subset of $L^p(\mathbb{R}^N)$ such that*

$$\lim_{h\to\infty} \int_{\mathbb{R}^N} |u(x+h) - u(x)|^p dx = 0$$

uniformly in $u \in \mathcal{M}$. Then the set $\{u|_\Omega;\ u \in \mathcal{M}\}$ is relatively compact in $L^p(\Omega)$ for any measurable set Ω with bounded (Lebesgue) measure.

For the proof, we refer to H. Brezis [38], p. 111.

1.4 Infinite-dimensional Sobolev Spaces

Let X be a real (or complex) Banach space with the norm denoted by $\|\cdot\|$ and let $[a,b]$ be a fixed interval on the real axis. A function $x:[a,b]\to X$ is said to be *finitely valued* if it is constant on each of a finite number of disjoint measurable sets $A_k\subset[a,b]$ and equal to zero on $[a,b]\setminus\cup_k A_k$. The function x is said to be *strongly measurable* on $[a,b]$ if there is a sequence $\{x_n\}$ of finite-valued functions that converges strongly in X and almost everywhere on $[a,b]$ to x. The function x is said to be *Bochner integrable* if there exists a sequence $\{x_n\}$ of finitely valued functions on $[a,b]$ to X that converges almost everywhere to x such that

$$\lim_{n\to\infty}\int_a^b\|x_n(t)-x(t)\|dt=0.$$

A necessary and sufficient condition guaranteeing that $x:[a,b]\to X$ is Bochner integrable is that x is strongly measurable and that $\int_a^b\|x(t)\|dt<\infty$. The space of all Bochner integrable functions $x:[a,b]\to X$ is a Banach space with the norm

$$\|x\|_1=\int_a^b\|x(t)\|dt,$$

and is denoted by $L^1(a,b;X)$.

More generally, the space of all (classes of) strongly measurable functions x from $[a,b]$ to X such that

$$\|x\|_p=\left(\int_a^b\|x(t)\|^p dt\right)^{\frac{1}{p}}<\infty$$

for $1\le p<\infty$ and $\|x\|_\infty=\operatorname{ess\,sup}_{t\in[a,b]}\|x(t)\|<\infty$, is denoted by $L^p(a,b;X)$. This is a Banach space in the norm $\|\cdot\|_p$.

If X is reflexive, then the dual of $L^p(a,b;X)$ is the space $L^q(a,b;X^*)$, where $p<\infty$, $\frac{1}{p}+\frac{1}{q}=1$ (see [69]). Recall also that a function $x:[a,b]\to X$ is said to be *weakly measurable* if for any $x^*\in X^*$, the function $t\to(x^*,x(t))$ is measurable. According to the Pettis theorem, if X is separable then every weakly measurable function is strongly measurable, and so these two notions coincide.

An X-valued function x defined on $[a,b]$ is said to be *absolutely continuous* on $[a,b]$ if for each $\varepsilon>0$ there exists $\delta(\varepsilon)$ such that $\sum_{n=1}^N\|x(t_n)-x(s_n)\|\le\varepsilon$, whenever $\sum_{n=1}^N|t_n-s_n|\le\delta(\varepsilon)$ and $(t_n,s_n)\cap(t_m,s_m)=\emptyset$ for $m\ne n$. Here, (t_n,s_n) is an arbitrary subinterval of (a,b).

A classical result in real analysis says that any real-valued absolutely continuous function is almost everywhere differentiable and it is expressed as the indefinite integral of its derivative. It should be mentioned that this result fails for X-valued absolutely continuous functions if X is a general Banach space.

However, if the space X is reflexive, we have (see, e.g., [82]):

Theorem 1.32 *Let X be a reflexive Banach space. Then every X-valued absolutely continuous function x on $[a,b]$ is almost everywhere differentiable on $[a,b]$ and*

$$x(t) = x(a) + \int_a^t \frac{d}{ds}x(s)ds, \quad \forall t \in [a,b], \tag{1.41}$$

where $\left(\frac{dx}{dt}\right) : [a,b] \to X$ is the derivative of x, that is,

$$\frac{d}{dt}x(t) = \lim_{\varepsilon \to 0} \frac{x(t+\varepsilon) - x(t)}{\varepsilon}.$$

Let us denote, as above, by $\mathcal{D}(a,b)$ the space of all infinitely differentiable real-valued functions on $[a,b]$ with compact support in (a,b), and by $\mathcal{D}'(a,b;X)$ the space of all continuous operators from $\mathcal{D}(a,b)$ to X. An element u of $\mathcal{D}'(a,b;X)$ is called an X-valued distribution on (a,b). If $u \in \mathcal{D}'(a,b;X)$ and j is a natural number, then

$$u^{(j)}(\varphi) = (-1)^j u(\varphi^{(j)}), \quad \forall \varphi \in \mathcal{D}(a,b),$$

defines another distribution $u^{(j)}$, which is called the derivative of order j of u.

We note that every element $u \in L^1(a,b;X)$ defines uniquely the distribution (again denoted u)

$$u(\varphi) = \int_a^b u(t)\varphi(t)dt, \quad \forall \varphi \in \mathcal{D}(a,b), \tag{1.42}$$

and so $L^1(a,b;X)$ can be regarded as a subspace of $\mathcal{D}'(a,b;X)$. In all that follows, we identify a function $u \in L^1(a,b;X)$ with the distribution u defined by (1.42).

Let k be a natural number and $1 \le p \le \infty$. We denote by $W^{k,p}([a,b];X)$ the space of all X-valued distributions $u \in \mathcal{D}'(a,b;X)$ such that

$$u^{(j)} \in L^p(a,b;X) \quad \text{for } j = 0,1,...,k. \tag{1.43}$$

By $W^{k,p}((a,b);X)$ we denote the space of all $u \in \mathcal{D}'(a,b;X)$ such that $u \in W^{k,p}([a+\delta, b-\delta];X)$ for all $0 < \delta < b-a$. Similarly, there are defined the spaces $W^{k,p}((a,b);X)$ and $W^{k,p}([a,b]);X)$.

We denote by $A^{1,p}([a,b];X)$, $1 \le p \le \infty$, the space of all absolutely continuous functions u from $[a,b]$ to X having the property that they are a.e. differentiable on (a,b) and $\frac{du}{dt} \in L^p(a,b;X)$. If the space X is reflexive, it follows by Theorem 1.32 that $u \in A^{1,p}([a,b];X)$ if and only if u is absolutely continuous on $[a,b]$ and $\frac{du}{dt} \in L^p(a,b;X)$.

It turns out that the space $W^{1,p}$ can be identified with $A^{1,p}$. More precisely, we have (see Brezis [41])

Theorem 1.33 *Let X be a Banach space and let $u \in L^p(a,b;X)$, $1 \le p \le \infty$. Then the following conditions are equivalent.*

(i) $u \in W^{1,p}([a,b];X)$.

(ii) *There is* $u^0 \in A^{1,p}([a,b];X)$ *such that*

$$u(t) = u^0(t), \ \ a.e., \ t \in (a,b).$$

Moreover, $u' = \frac{du^0}{dt}$, *a.e. in* (a,b).

Theorem 1.34 *Let* X *be a reflexive Banach space and let* $u \in L^p(a,b;X)$, $1 < p \leq \infty$. *Then the following two conditions are equivalent.*

(i) $u \in W^{1,p}([a,b];X)$.
(ii) *There is* $C > 0$ *such that*

$$\int_a^{b-h} \|u(t+h) - u(t)\|^p dt \leq C|h|^p, \ \ \forall h \in [0, b-a],$$

with the usual modification in the case $p = \infty$.

Let V be a reflexive Banach space and H be a real Hilbert space such that $V \subset H \subset V'$ in the algebraic and topological senses. Here, V' is the dual space of V and H is identified with its own dual. Denote by $|\cdot|$ and $\|\cdot\|$ the norms of H and V, respectively, and by $(\cdot,\cdot)$ the duality between V and V'. If $v_1, v_2 \in H$, then (v_1, v_2) is the scalar product in H of v_1 and v_2.

Denote by $W_p([a,b];V)$, $1 < p < \infty$, the space

$$W_p([a,b];V) = \{u \in L^p(a,b;V); \ u' \in L^q(a,b;V')\}, \ \ \frac{1}{p} + \frac{1}{q} = 1, \tag{1.44}$$

where u' is the derivative of u in the sense of $\mathcal{D}'(a,b;V)$. By Theorem 1.33, we know that every $u \in W_p([a,b];V)$ can be identified with an absolutely continuous function $u^0 : [a,b] \to V'$. However, we have a more precise result (see, e.g., [15], p. 25).

Theorem 1.35 *Let* $u \in W_p([a,b];V)$. *Then there is a continuous function* $u^0 : [a,b] \to H$ *such that* $u(t) = u^0(t)$, *a.e.,* $t \in (a,b)$.

Moreover, if $u, v \in W_p([a,b];V)$, *then the function* $t \to (u(t), v(t))$ *is absolutely continuous on* $[a,b]$ *and*

$$\frac{d}{dt}(u(t), v(t)) = (u'(t), v(t)) + (u(t), v'(t)), \quad a.e. \ t \in (a,b). \tag{1.45}$$

We also note the following compactness result.

Theorem 1.36 (Aubin–Lions–Simon) *Let* X_0, X_1, X_2 *be Banach spaces such that* $X_0 \subset X_1 \subset X_2$, *and the injection of* X_0 *into* X_1 *is compact. Let* $-\infty < a < b < \infty$ *and* $1 \leq p \leq \infty$. *Then the space*

$$W = L^p(a,b;X_0) \cap W^{1,1}([a,b];X_2)$$

is compactly embedded in $L^p(a,b;X_1)$.

Moreover, for $r > 1$, *the space* $L^\infty(a,b;X_0) \cap W^{1,r}([a,b];X_2)$ *is compactly embedded in* $C([a,b];X_1)$.

The proof relies on the following property of the spaces X_i (see Lions [78], p. 58). For every $\varepsilon > 0$ there exists $C_\varepsilon > 0$ such that

$$\|u\|_{X_1} \leq \varepsilon \|u\|_{X_0} + C_\varepsilon \|u\|_{X_2}, \quad \forall u \in X_0.$$

Theorem 1.36 is due to J.P. Aubin [8] and to J.L. Lions [78] for $1 < p < \infty$ and reflexive Banach spaces X_i, and to J. Simon [100] in the general case of Banach spaces X_i.

Chapter 2

Monotone and Accretive Operators in Banach Spaces

Before the mathematician demonstrates he must invent. But nobody has ever invented anything by pure deduction. Pure logic cannot create anything; there is only one way to discovery, namely induction; for the mathematicians as well for the physicists. Henri Poincaré

The intuition is the first source of evidence and that the immediate and intermediate relations derived from it are the only absolute truth.
Arthur Schopenhauer

In this chapter, we present the basic results of the theory of maximal monotone and m-accretive operators in Banach spaces. Moreover, the fundamentals of the existence theory for the Cauchy problem associated with nonlinear m-accretive operators in Banach spaces is briefly treated. The presentation is confined to the essential results necessary for the treatment of nonlinear partial differential equations, and so most of them are given here without proof and refer the reader to the book [15] for the complete treatment.

2.1 Maximal Monotone Operators

Let X be a real Banach space with the dual X^* and let $J : X \to 2^{X^*}$ be the *duality mapping*

$$J(x) = \{x^* \in X^*;\ (x, x^*) = \|x\|^2 = \|x^*\|^2\}, \quad \forall x \in X. \tag{2.1}$$

(See Section 1.1.) For the sake of simplicity, we shall denote by the same symbol $\|\cdot\|$ the norm of X and of X^*. If X is a Hilbert space unless otherwise stated we implicitly assume that it is identified with its own dual.

If X and Y are two linear spaces, we denote by $X \times Y$ their Cartesian product. The elements of $X \times Y$ are written as $[x, y]$, where $x \in X$ and $y \in Y$.

If A is a multivalued operator from X to Y, we may identify it with its graph in $X \times Y$:

$$\{[x, y] \in X \times Y;\ y \in Ax\}, \tag{2.2}$$

and, therefore, view it as a subset of $X \times Y$.

Conversely, if $A \subset X \times Y$, then we define

$$Ax = \{y \in Y;\ [x,y] \in A\}, \quad D(A) = \{x \in X;\ Ax \neq \emptyset\}, \tag{2.3}$$

$$R(A) = \bigcup_{x \in D(A)} Ax, \qquad A^{-1} = \{[y,x];\ [x,y] \in A\}. \tag{2.4}$$

In this way, here and everywhere in the following we identify the operators from X to Y with their graphs in $X \times Y$ and so we equivalently speak of subsets of $X \times Y$ instead of operators from X to Y.

If $A, B \subset X \times Y$ and λ is a real number, we set:

$$\lambda A = \{[x, \lambda y];\ [x,y] \in A\}; \tag{2.5}$$

$$A + B = \{[x, y+z];\ [x,y] \in A,\ [x,z] \in B\}; \tag{2.6}$$

$$AB = \{[x,z];\ [x,y] \in B,\ [y,z] \in A \text{ for some } y \in Y\}. \tag{2.7}$$

Definition 2.1 The set $A \subset X \times X^*$ (equivalently the operator $A : X \to X^*$) is said to be *monotone* if

$$(x_1 - x_2, y_1 - y_2) \geq 0, \quad \forall [x_i, y_i] \in A,\ i = 1, 2. \tag{2.8}$$

A monotone set $A \subset X \times X^*$ is said to be *maximal monotone* if it is not properly contained in any other monotone subset of $X \times X^*$.

Note that if A is a single-valued operator from X to X^*, then A is monotone if

$$(x_1 - x_2, Ax_1 - Ax_2) \geq 0, \quad \forall x_1, x_2 \in D(A). \tag{2.9}$$

Definition 2.2 Let A be a single-valued operator from X to X^* with $D(A) = X$. The operator A is said to be *hemicontinuous* if, for all $x, y \in X$,

$$w^*\text{-}\lim_{\lambda \to 0} A(x + \lambda y) = Ax.$$

A is said to be *demicontinuous* if it is continuous from X to X^*_w, that is,

$$w^*\text{-}\lim_{x_n \to x} Ax_n = Ax.$$

A is said to be *coercive* if

$$\lim_{n \to \infty} (x_n - x^0, y_n) \|x_n\|^{-1} = \infty \tag{2.10}$$

for some $x^0 \in X$ and all $[x_n, y_n] \in A$ such that $\lim_{n \to \infty} \|x_n\| = \infty$.

A is said to be *bounded* if it is bounded on each bounded subset.

In the following, we give without proof some basic results pertaining maximal monotone operators and refer to [12, 14] for proof.

Proposition 2.3 *Let $A \subset X \times X^*$ be maximal monotone. Then:*

(i) *A is weakly-strongly closed in $X \times X^*$, that is, if $y_n = Ax_n$, $x_n \rightharpoonup x$ in X, and $y_n \to y$ in X^*, then $[x, y] \in A$,*
(ii) *A^{-1} is maximal monotone in $X^* \times X$,*
(iii) *For each $x \in D(A)$, Ax is a closed convex subset of X^*.*

The next theorem, due to G. Minty [85] and F. Browder [50], is the principal result of the theory of maximal monotone operators. It relates the maximal monotonicity which is an algebraic property to that of the surjectivity of the operator $A+\lambda J$, and so opens the way to the existence theory for nonlinear operator equations of monotone type.

Theorem 2.4 *Let X be a reflexive Banach space. Let $A \subset X \times X^*$ be a monotone subset of $X \times X^*$ and let $J : X \to X^*$ be the duality mapping of X. Then A is maximal monotone if and only if, for any $\lambda > 0$ (equivalently, for some $\lambda > 0$), $R(A + \lambda J) = X^*$.*

The main ingredient of the proof of Theorem 2.4 which will be sketched in the following is the existence result below.

Proposition 2.5 *Let X be a reflexive Banach space and let A and B be two monotone sets of $X \times X^*$ such that $0 \in D(A)$, B is single-valued, hemicontinuous, and coercive, that is,*

$$\lim_{\|x\| \to \infty} \frac{(x, Bx)}{\|x\|} = +\infty. \tag{2.11}$$

Then there exists $x \in K = \overline{\text{conv}\, D(A)}$ such that

$$(u - x, Bx + v) \geq 0 \qquad \forall [u, v] \in A. \tag{2.12}$$

Here, $\overline{\text{conv}\, D(A)}$ is the convex hull of the set $\overline{D(A)}$, that is, the set

$$\left\{ \sum_{i=1}^{m} \lambda_i x_i,\ x_i \in \overline{D(A)},\ 0 \leq \lambda_i \leq 1,\ \sum_{i=1}^{m} \lambda_i = 1,\ m \in \mathbb{N} \right\}.$$

Proof. We note first that, if X is a finite-dimensional Banach space and B is a hemicontinuous monotone operator from X to X^*, then B is continuous. Indeed, as easily seen, by monotonicity B is bounded on bounded sets of X.

Now, let $\{x_n\}$ be convergent to x_0 and let y_0 be a cluster point of $\{Bx_n\}$. Again by the monotonicity of B, we have

$$(x_0 - x, y_0 - Bx) \geq 0, \quad \forall x \in X.$$

If in this inequality we take $x = tu + (1 - t)x_0$, $0 \leq t \leq 1$, u arbitrary in X, we get

$$(x_0 - u, y_0 - B(tu + (1 - t)x_0)) \geq 0, \quad \forall t \in [0, 1],\ u \in X.$$

Then, letting t tend to zero and using the hemicontinuity of B, we get

$$(x_0 - u, y_0 - Bx_0) \geq 0, \quad \forall u \in X,$$

which clearly implies that $y_0 = Bx_0$, as claimed.

The next step in the proof of Proposition 2.5 is the case where the space X is finite-dimensional.

Lemma 2.6 *Let X be a finite-dimensional Banach space and let A and B be two monotone subsets of $X \times X^*$ such that $0 \in D(A)$, and B is single-valued, continuous, and satisfies* (2.11). *Then there exists $x \in \overline{\text{conv}\, D(A)}$ such that*

$$(u - x, Bx + v) \geq 0, \quad \forall [u, v] \in A. \tag{2.13}$$

Proof. Redefining A if necessary, we may assume that the set $K = \overline{\text{conv}\, D(A)}$ is bounded. Indeed, if Lemma 2.1 is true in this case, then replacing A by $A_n = \{[x, y] \in A; \|x\| \leq n\}$, we infer that for every n there exists $x_n \in K_n = K \cap \{x; \|x\| \leq n\}$ such that

$$(u - x_n, Bx_n + v) \geq 0, \quad \forall [u, v] \in A_n. \tag{2.14}$$

This yields

$$(x_n, Bx_n)\|x_n\|^{-1} \leq \|\xi\|, \quad \text{for some } \xi \in A0,$$

and, by the coercivity condition (2.11), we see that there is $M > 0$ such that $\|x_n\| \leq M$ for all n. Now, on a subsequence, for simplicity again denoted n, we have $x_n \to x$. By (2.14) and the continuity of B, it is clear that x is a solution to (2.13), as claimed.

Let $T : K \to K$ be the multivalued operator defined by

$$Tx = \{y \in K; \ (u - y, Bx + v) \geq 0, \quad \forall [u, v] \in A\}.$$

Let us show first that $Tx \neq \emptyset$, $\forall x \in K$. To this end, define the sets

$$K_{uv} = \{y \in K; \ (u - y, Bx + v) \geq 0\},$$

and notice that

$$Tx = \bigcap_{[u,v] \in A} K_{uv}.$$

Inasmuch as K_{uv} are closed subsets (if nonempty) of the compact set K, to show that $\bigcap_{[u,v] \in A} K_{uv} \neq \emptyset$ it suffices to prove that every finite collection $\{K_{u_i, v_i}; i = 1, ..., m\}$ has a nonempty intersection. Equivalently, it suffices to show that the system

$$(u_i - y, Bx + v_i) \geq 0, \qquad i = 1, ..., m, \tag{2.15}$$

has a solution $y \in K$ for any set of pairs $[u_i, v_i] \in A$, $i = 1, ..., m$.

Consider the function $H : U \times U \to \mathbb{R}$,

$$H(\lambda, \mu) = \sum_{i=1}^{m} \mu_i \left(\sum_{j=1}^{m} \lambda_j u_j - u_i, Bx + v_i \right), \quad \forall \lambda, \mu \in U, \tag{2.16}$$

where $U = \{\lambda \in \mathbb{R}^m; \ \lambda = (\lambda_1, ..., \lambda_m), \ \lambda_i \geq 0, \ \sum_{i=1}^{m} \lambda_i = 1\}$.

The function H is continuous, convex in λ, and concave in μ. Then, according to the classical Von Neumann min–max theorem (see, e.g., [65], p. 125), H has a saddle point $(\lambda_0, \mu_0) \in U \times U$, that is,

$$H(\lambda_0, \mu) \leq H(\lambda_0, \mu_0) \leq H(\lambda, \mu_0), \quad \forall \lambda, \mu \in U. \tag{2.17}$$

On the other hand, we have

$$H(\lambda,\lambda) = \sum_{i=1}^{m} \lambda_i \left(\sum_{j=1}^{m} \lambda_j u_j - u_i, Bx + v_i \right)$$
$$= \sum_{i=1}^{m}\sum_{j=1}^{m} \lambda_i\lambda_j(v_i, u_j - u_i) + \sum_{i=1}^{m}\sum_{j=1}^{m} \lambda_i\lambda_j(u_j - u_i, Bx) \le 0, \quad \forall \lambda \in U,$$

because, by monotonicity of B, $(v_i - v_j, u_i - u_j) \ge 0$ for all i, j.

Then, by (2.17) we see that $H(\lambda_0, \mu) \le 0, \quad \forall \mu \in U$, that is,

$$\sum_{i=1}^{m} \mu_i \left(\sum_{j=1}^{m} (\lambda_0)_j u_j - u_i, Bx + v_i \right) \le 0, \quad \forall \mu \in U.$$

In particular, it follows that

$$\left(\sum_{j=1}^{m} (\lambda_0)_j u_j - u_i, Bx + v_i \right) \le 0, \quad \forall i = 1, ..., m.$$

Hence, $y = \sum_{j=1}^{m} (\lambda_0)_j u_j \in K$ is a solution to (2.15). We have, therefore, proved that T is well defined on K and that $T(K) \subset K$. It is also clear that for every $x \in K$, Tx is a closed convex subset of X and T is upper semicontinuous on K. Indeed, because the range of T belongs to a compact set, to verify that T is upper-semicontinuous it suffices to show that T is closed in $K \times K$, that is, if $[x_n, y_n] \in T$, $x_n \to x$, and $y_n \to y$, then $y \in Tx$. But the last property is obvious if one takes into account the definition of T. Then, applying the Kakutani fixed point theorem (see, e.g., [65], p. 128), we conclude that there exists $x \in K$ such that $x \in Tx$, thereby completing the proof of Lemma 2.6. □

Proof of Proposition 2.5. Let Λ be the family of all finite dimensional subspaces X_α of X ordered by the inclusion relation. For every $X_\alpha \in \Lambda$, denote by $j_\alpha : X_\alpha \to X$ the injection mapping of X_α into X and by $j_\alpha^* : X^* \to X_\alpha^*$ the dual mapping, that is, the projection of X^* onto X_α^*. The operators $A_\alpha = j_\alpha^* A j_\alpha$ and $B_\alpha = j_\alpha^* B j_\alpha$ map X_α into X_α^* and are monotone in $X_\alpha \times X_\alpha^*$. Because B is hemicontinuous from X to X^* and j_α^* are continuous from X^* to X_α^*, it follows that B_α is continuous from X_α to X_α^*.

We may, therefore, apply Lemma 2.6, where $X = X_\alpha$, $A = A_\alpha$, $B = B_\alpha$, and $K = K_\alpha = \overline{\text{conv}\, D(A_\alpha)}$. Hence, for each $X_\alpha \in \Lambda$, there exists $x_\alpha \in K_\alpha$ such that

$$(u - x_\alpha, B_\alpha x_\alpha + v) \ge 0, \quad \forall [u, v] \in A,$$

or, equivalently,

$$(u - x_\alpha, Bx_\alpha + v) \ge 0, \quad \forall [u, v] \in A_\alpha. \tag{2.18}$$

By using the coercivity condition (2.11), we deduce from (2.18) that $\{x_\alpha\}$ remain in a bounded subset of X. The space X is reflexive, thus every bounded subset of X is sequentially weakly compact and so there exists a sequence $\{x_{\alpha_n}\} \subset \{x_\alpha\}$ such that

$$x_{\alpha_n} \rightharpoonup x \quad \text{in } X \text{ as } n \to \infty. \tag{2.19}$$

Moreover, because the operator B is bounded on bounded subsets, we may assume that $Bx_{\alpha_n} \rightharpoonup y$ in X^* as $n \to \infty$. Because the closed convex subsets are weakly closed, we infer that $x \in K$. Moreover, by (2.18) we see that

$$\limsup_{n\to\infty}(x_{\alpha_n}, Bx_{\alpha_n}) \le (u - x, v) + (u, y), \quad \forall [u, v] \in A. \tag{2.20}$$

Without loss of generality, we may assume that A is maximal in the class of all monotone subsets $\widetilde{A} \subset X \times X^*$ such that $D(\widetilde{A}) \subset K = \overline{\text{conv}\, D(A)}$. (If not, we may extend A by Zorn's lemma to a maximal element of this class.) To complete the proof, let us show first that $\limsup_{n\to\infty}(x_{\alpha_n} - x, Bx_{\alpha_n}) \le 0$. Indeed, if this is not the case, it follows from (2.20) that

$$(u - x, v + y) \ge 0, \quad \forall [u, x] \in A,$$

and because $x \in K$ and A is maximal in the class of all monotone operators $\widetilde{A}$ with domain in K, it follows that $[x, -y] \in A$. Then, putting $u = x$ in (2.20), we obtain an inequality which contradicts the working hypothesis.

Now, for u arbitrary but fixed in $D(A)$ consider $u_\lambda = \lambda x + (1 - \lambda)u$, $0 \le \lambda \le 1$, and notice that, by virtue of the monotonicity of B, we have

$$(x_{\alpha_n} - u_\lambda, Bx_{\alpha_n}) \ge (x_{\alpha_n} - u_\lambda, Bu_\lambda).$$

This yields

$$(1-\lambda)(x_{\alpha_n} - u, Bx_{\alpha_n}) + \lambda(x_{\alpha_n} - x, Bx_{\alpha_n}) \ge (1-\lambda)(x_{\alpha_n} - u, Bu_\lambda) + \lambda(x_{\alpha_n} - x, Bu_\lambda)$$

and so,

$$(x - u, Bu_\lambda) \le \limsup_{n\to\infty}(x_{\alpha_n} - u, Bx_{\alpha_n}) \le (u - x, v), \quad \forall [u, v] \in A.$$

Inasmuch as B is hemicontinuous, the latter inequality yields for $\lambda \to 1$,

$$(u - x, v + Bx) \ge 0, \quad \forall [u, v] \in A,$$

thereby completing the proof of Proposition 2.5. □

Proof of Theorem 2.4. *"If" part.* Assume that $R(A + \lambda J) = X^*$ for some $\lambda > 0$. We suppose that A is not maximal monotone, and argue from this to a contradiction. Indeed, in this case there exists $[x_0, y_0] \in X \times X^*$ such that $[x_0, y_0] \notin A$ and

$$(x - x_0, y - y_0) \ge 0, \quad \forall [x, y] \in A. \tag{2.21}$$

On the other hand, by hypothesis, there exists $[x_1, y_1] \in A$ such that

$$\lambda J(x_1) + y_1 = \lambda J(x_0) + y_0.$$

Substituting $[x_1, y_1]$ in place of $[x, y]$ in (2.21), this yields

$$(x_1 - x_0, J(x_1) - J(x_0)) \le 0.$$

Taking into account the definition of J, we get

$$(x_1, J(x_0)) = (x_0, J(x_1)) = \|x_1\|^2 = \|x_0\|^2.$$

Hence $J(x_0) = J(x_1)$ and, because the duality mapping J^{-1} of X^* is single-valued (because X is strictly convex), we infer that $x_0 = x_1$. Hence $[x_0, y_0] = [x_1, y_1] \in A$, which contradicts the hypothesis.

"Only if" part. By the Asplund theorem (Theorem 1.1) renorming the space X^*, we may assume that X^* is strictly convex, J is single-valued and demicontinuous on X. Let y_0 be an arbitrary element of X^* and let $\lambda > 0$. Applying Proposition 2.5, where $Bu = \lambda J(u) - y_0$, $\forall u \in X$, we conclude that there is $x \in X$ such that

$$(u - x, \lambda J(x) - y_0 + v) \geq 0, \quad \forall [u, v] \in A.$$

Since A is maximal monotone, this implies that $[x, -\lambda J(x) + y_0] \in A$, that is, $y_0 \in \lambda J(x) + Ax$. □

By Proposition 2.5, part (2.12), it follows also that

Theorem 2.7 *Let X be reflexive and let B be a hemicontinuous monotone and bounded operator from X to X^*. Let $A \subset X \times X^*$ be maximal monotone. Then $A + B$ is maximal monotone.*

In particular, it follows by Theorem 2.7 that every monotone, hemicontinuous, and bounded operator from X to X^* is maximal monotone. In fact, it follows that the boundedness assumption is redundant. Namely,

Theorem 2.8 *Let X be a reflexive Banach space and let $B : X \to X^*$ be a monotone and hemicontinuous operator. Then B is maximal monotone in $X \times X^*$.*

Let $f \in X^*$. Then, if A is maximal monotone, the equation $Ax \ni f$ can be approximated by

$$\lambda J(x_\lambda) + Ax_\lambda \ni f, \ \lambda > 0,$$

and, as easily seen, if A is coercive, then $x_\lambda \to x \in A^{-1}(f)$ as $\lambda \to 0$. Therefore, we have

Corollary 2.9 *Let X be a reflexive Banach space and let A be a coercive maximal monotone subset of $X \times X^*$. Then A is surjective, that is, $R(A) = X^*$.*

In the following, we define for each $x \in X$ and $\lambda > 0$

$$J_\lambda(x) = x_\lambda; \ J(x_\lambda - x) + \lambda A x_\lambda \ni 0, \tag{2.22}$$

$$A_\lambda(x) = \frac{1}{\lambda} J(x - x_\lambda). \tag{2.23}$$

The operator $A_\lambda : X \to X^*$ is called the *Yosida approximation* of A and plays an important role in the smooth approximation of A. We collect in Proposition 2.10 several basic properties of the operators A_λ and J_λ. For the proof, see [14] or [15], p. 38.

Proposition 2.10 *Let X and X^* be strictly convex and reflexive. Then:*

(i) A_λ *is single-valued, monotone, bounded, and demicontinuous from X to X^*.*
(ii) $\|A_\lambda x\| \le |Ax| = \inf\{\|y\|;\ y \in Ax\}$ *for every* $x \in D(A)$, $\lambda > 0$.
(iii) $J_\lambda : X \to X$ *is bounded on bounded subsets and*

$$\lim_{\lambda\to 0} J_\lambda x = x, \quad \forall x \in \overline{\operatorname{conv} D(A)}. \tag{2.24}$$

(iv) *If* $\lambda_n \to 0$, $x_n \to x$, $A_{\lambda_n} x_n \rightharpoonup y$ *and*

$$\limsup_{n,m\to\infty} (x_n - x_m, A_{\lambda_n} x_n - A_{\lambda_m} x_m) \le 0, \tag{2.25}$$

then $[x, y] \in A$ *and*

$$\lim_{m,n\to\infty} (x_n - x_m, A_{\lambda_n} x_n - A_{\lambda_m} x_m) = 0.$$

(v) *For* $\lambda \to 0$, $A_\lambda x \rightharpoonup A^0 x$, $\forall x \in D(A)$, *where* $A^0 x$ *is the element of minimum norm in* Ax, *that is,* $\|A^0 x\| = |Ax|$. *If X^* is uniformly convex, then* $A_\lambda x \to A^0 x$, $\forall x \in D(A)$.

We also note the following convergence result, which is similar to Proposition 2.10, part (iv).

Proposition 2.11 *Let X be a reflexive Banach space and let $A \subset X \times X^*$ be a maximal monotone subset. Let $[u_n, v_n] \in A$ be such that $u_n \rightharpoonup u$ in X, $v_n \rightharpoonup v$ in X^*, and*

$$\limsup_{n\to\infty} (u_n, v_n) \le (u, v).$$

Then, $[u, v] \in A$.

Proof. For each $(x, y) \in A$, we have

$$(u_n - x, v_n - y) \ge 0, \quad \forall\, n \in \mathbb{N},$$

which yields

$$(u - x, v - y) \ge 0, \quad \forall (x, y) \in A,$$

and since A is maximal monotone, we have $(u, v) \in A$. □

In particular, Proposition 2.11 is important to pass to limit in an equation of the form $Au_n \ni f_n$, where $u_n \rightharpoonup u$ in X and $f_n \rightharpoonup f$ in X^*. It implies that, if $\limsup_{n\to\infty}(f_n, u_n) \le (f, u)$, then $f \in Au$.

Now, we mention an important property of monotone operators with nonempty interior domain.

Theorem 2.12 *Let A be a monotone subset of $X \times X^*$. Then A is locally bounded at any interior point of $D(A)$, that is, on* $\operatorname{int}(D(A))$.

We also have (Rockafellar [96]).

Theorem 2.13 *Let X be a reflexive Banach space and let A and B be maximal monotone subsets of $X \times X^*$ such that*

$$(\text{int}(D(A))) \cap D(B) \neq \emptyset. \tag{2.26}$$

Then $A + B$ is maximal monotone in $X \times X^*$.

Corollary 2.14 *Let X be a reflexive Banach space, $A \subset X \times X^*$ a maximal monotone operator, and let $B : X \to X^*$ be a demicontinuous (equivalently, hemicontinuous) monotone operator. Then $A + B$ is maximal monotone.*

The subdifferential of a lower semicontinuous convex function is an important example of maximal monotone operator that closes the bridge between the theory of nonlinear maximal monotone operators and convex analysis. Such an operator is also called a *subpotential maximal monotone operator.* (We refer to Section 1.2 for definitions and basic property of convex, lower-semicontinuous functions.) We have

Theorem 2.15 *Let X be a real Banach space and let $\varphi : X \to \overline{\mathbb{R}}$ be an l.s.c. proper convex function. Then $\partial\varphi$ is a maximal monotone subset of $X \times X^*$.*

It is readily seen that $\partial\varphi$ is monotone in $X \times X^*$. If X is reflexive, then it is easily seen that, for each $y \in X^*$, the equation

$$Jx + \partial\varphi(x) \ni y$$

or, equivalently,

$$x = \arg\min\left\{\varphi(x) + \frac{1}{2}\|x\|^2 - (x, y)\right\}$$

has a solution, which implies that $\partial\varphi$ is maximal monotone. In the nonreflexive case, the proof is however more complicated (see R.T. Rockafellar [96]).

For every $\lambda > 0$, define the function (the Hille–Moreau regularization)

$$\varphi_\lambda(x) = \inf\left\{\frac{\|x - u\|^2}{2\lambda} + \varphi(u);\ u \in X\right\},\quad \forall x \in X, \tag{2.27}$$

where $\varphi : X \to \overline{\mathbb{R}}$ is an l.s.c. proper convex function. By Propositions 1.3 and 1.6 it follows that $\varphi_\lambda(x)$ is well defined for all $x \in X$ and the infimum defining it is attained (if the space X is reflexive). This implies by a straightforward argument that φ_λ is convex and l.s.c. on X. (Because φ_λ is everywhere defined, we conclude by Proposition 1.4, that φ_λ is continuous.) We have (see [16])

Theorem 2.16 *Let X be a reflexive and strictly convex Banach space with strictly convex dual. Let $\varphi : X \to \overline{\mathbb{R}}$ be an l.s.c. convex, proper function and let $A = \partial\varphi \subset X \times X^*$. Then the function φ_λ is convex, continuous, Gâteaux differentiable, and $\nabla\varphi_\lambda = A_\lambda$ for all $\lambda > 0$. Moreover:*

$$\varphi_\lambda(x) = \frac{\|x - J_\lambda x\|^2}{2\lambda} + \varphi(J_\lambda x),\quad \forall \lambda > 0,\ x \in X; \tag{2.28}$$

$$\lim_{\lambda \to 0} \varphi_\lambda(x) = \varphi(x),\quad \forall x \in X; \tag{2.29}$$

$$\varphi(J_\lambda x) \le \varphi_\lambda(x) \le \varphi(x),\quad \forall \lambda > 0,\ x \in X. \tag{2.30}$$

If X is a Hilbert space (not necessarily identified with its dual), then φ_λ is Fréchet differentiable on X.

Examples of subpotential operators

1. *Maximal monotone sets (graphs) in* $\mathbb{R}\times\mathbb{R}$. Every maximal monotone set (graph) of $\mathbb{R}\times\mathbb{R}$ is the subdifferential of an l.s.c., convex, proper function on $\mathbb{R}$. Indeed, let β be a maximal monotone set in $\mathbb{R}\times\mathbb{R}$ and let $\beta^0:\mathbb{R}\to\mathbb{R}$ be the function defined by

$$\beta^0(r)=\{y\in\beta(r);\ |y|=\inf\{|z|;\ z\in\beta(r)\}\},\quad \forall r\in\mathbb{R}.$$

We know that $\overline{D(\beta)}=[a,b]$, where $-\infty\le a\le b\le\infty$. The function β^0 is monotonically increasing and so the integral

$$j(r)=\int_{r_0}^{r}\beta^0(u)du,\quad \forall r\in\mathbb{R},\tag{2.31}$$

where $r_0\in D(\beta)$, is well defined (unambiguously a real number or $+\infty$). Clearly, the function j is continuous on (a,b) and convex on $\mathbb{R}$. Moreover,

$$\liminf_{r\to b} j(r)\ge j(b)\quad\text{and}\quad \liminf_{r\to a} j(r)\ge j(a).$$

Finally,

$$j(r)-j(t)=\int_t^r\beta^0(u)du\le v(r-t),\quad \forall[r,v]\in\beta,\ t\in\mathbb{R}.$$

Hence $\beta=\partial j$, where j is the l.s.c. convex function defined by (2.31).

It is easily seen that if $\beta:\mathbb{R}\to\mathbb{R}$ is a continuous and monotonically increasing function, then β is a maximal monotone graph in $\mathbb{R}\times\mathbb{R}$ in the sense of general definition, that is, the range of $u\to u+\beta(u)$ is all of $\mathbb{R}$. (By a monotonically increasing function we mean, here and everywhere in the following, a monotonically nondecreasing function.) If β is a monotonically increasing function discontinuous in $\{r_j\}_{j=1}^\infty$, then as seen earlier one gets from β a maximal monotone graph $\widetilde{\beta}\subset\mathbb{R}\times\mathbb{R}$ by "filling" the jumps of β in r_j, that is,

$$\widetilde{\beta}(r)=\begin{cases}\beta(r), & \text{for } r\ne r_j,\\ [\beta(r_j-0),\beta(r_j+0)], & \text{for } r=r_j.\end{cases}$$

2. *Self-adjoint operators.* Let H be a real Hilbert space (identified with its own dual) with scalar product $(\cdot,\cdot)$ and norm $|\cdot|$, and let A be a linear self-adjoint positive operator on H. Then, $A=\partial\varphi$, where

$$\varphi(x)=\begin{cases}\dfrac{1}{2}\,|A^{\frac12}x|^2, & x\in D(A^{\frac12}),\\ +\infty, & \text{otherwise.}\end{cases}\tag{2.32}$$

(Here, $A^{\frac12}$ is the square root of the operator A.)

Conversely, any linear, densely defined operator, that is, the subdifferential of an l.s.c. convex function on H is self-adjoint.

To prove these assertions, we note first that any self-adjoint positive operator A in a Hilbert space is maximal monotone. Indeed, it is readily seen that the range of the operator $I + A$ is simultaneously closed and dense in H. On the other hand, if $\varphi : H \to \overline{\mathbb{R}}$ is the function defined by (2.32), then clearly it is convex, l.s.c., and

$$\varphi(x) - \varphi(u) = \frac{1}{2}\,(|A^{\frac{1}{2}}x|^2 - |A^{\frac{1}{2}}u|^2) \le (Ax, x - u), \quad \forall x \in D(A),\ u \in D(A^{\frac{1}{2}}).$$

Hence $A \subset \partial\varphi$, and, because A is maximal monotone, we conclude that $A = \partial\varphi$.

Now, let A be a linear, densely defined operator on H of the form $A = \partial\psi$, where $\psi : H \to \overline{\mathbb{R}}$ is an l.s.c. convex function. By Theorem 2.16, we know that $A_\lambda = \nabla\psi_\lambda$, where $A_\lambda = \lambda^{-1}(I - \lambda A)^{-1}$. This yields

$$\frac{d}{dt}\,\psi_\lambda(tu) = t(A_\lambda u, u), \quad \forall u \in H,\ t \in [0,1],$$

and therefore $\psi_\lambda(u) = \frac{(A_\lambda u, u)}{2}$ for all $u \in H$ and $\lambda > 0$. Calculating the Fréchet derivative of ψ_λ, we see that

$$\nabla\psi_\lambda = A_\lambda = \frac{1}{2}\,(A_\lambda + A_\lambda^*).$$

Hence $A_\lambda = A_\lambda^*$, and letting $\lambda \to 0$, this implies that $A = A^*$, as claimed.

More generally, if A is a linear continuous, symmetric operator from a Hilbert space V to its dual V^* (not identified with V), then $A = \partial\varphi$, where $\varphi : V \to \mathbb{R}$ is the function

$$\varphi(u) = \frac{1}{2}\,(Au, u), \quad \forall u \in V.$$

Conversely, every linear continuous operator $A : V \to V'$ of the form $\partial\varphi$ is symmetric.

In particular, by virtue of Theorem 1.25, if Ω is a bounded and open domain of $\mathbb{R}^N$ with sufficiently smooth boundary (of class C^2, for instance), then the operator $A : D(A) \subset L^2(\Omega) \to L^2(\Omega)$ defined by

$$Ay = -\Delta y, \quad \forall y \in D(A), \quad D(A) = H_0^1(\Omega) \cap H^2(\Omega),$$

is self-adjoint and $A = \partial\varphi$, where $\varphi : L^2(\Omega) \to \overline{\mathbb{R}}$, is given by

$$\varphi(y) = \begin{cases} \dfrac{1}{2}\displaystyle\int_\Omega |\nabla y|^2 dx & \text{if } y \in H_0^1(\Omega), \\ +\infty & \text{otherwise.} \end{cases}$$

Similarly, if $\phi : L^2(\Omega) \to \overline{\mathbb{R}}$ is defined by

$$\phi(y) = \begin{cases} \dfrac{1}{2}\displaystyle\int_\Omega |\nabla y|^2 dx & \text{if } y \in H^1(\Omega), \\ +\infty, & \text{otherwise} \end{cases}$$

then $\partial\phi = \widetilde{A}$, where

$$\widetilde{A}y = -\Delta y,\ D(\widetilde{A}) = \left\{ y \in H^2(\Omega);\ \frac{\partial y}{\partial n} = 0 \text{ on } \partial\Omega \right\}.$$

This result remains true for a nonsmooth bounded open domain if it is convex.

More generally, one could take $A : L^2(\Omega) \to L^2(\Omega)$, the elliptic operator

$$Ay = -\sum_{i=1}^{N} \frac{\partial}{\partial x_i}\left(a_{ij}(x)\,\frac{\partial y}{\partial x_j}\right) + a_0(x)y, \quad x \in \Omega, \tag{2.33}$$

with the domain $D(A) = H_0^1(\Omega) \cap H^2(\Omega)$ or, in the case of Neumann boundary conditions,

$$D(A) = \left\{ y \in H^2(\Omega);\ \sum_{i,j=1}^{N} a_{ij}(x)\,\frac{\partial y}{\partial x_j} \cdot n_i = 0 \text{ on } \partial\Omega \right\},$$

where $n = \{n_i\}_{i=1}^N$ is the outward normal to $\partial\Omega$, and

$$a_{ij} = a_{ji},\ \forall i,j = 1,...,N;\ a_0, a_{ij} \in C(\overline{\Omega}),$$
$$\sum_{i,j=1}^{N} a_{ij}(x)\xi_i\xi_j \geq \omega|\xi|^2,\ \forall \xi \in \mathbb{R}^N,\ x \in \Omega,$$

and

$$\phi(y) = \begin{cases} \displaystyle\int_\Omega \left(\frac{1}{2} \sum_{i,j=1}^{N} a_{ij}(x)\,\frac{\partial y}{\partial x_i}\,\frac{\partial y}{\partial x_j} + a_0(x)y^2 \right) dx & \text{if } y \in H_0^1(\Omega), \\ +\infty & \text{otherwise.} \end{cases}$$

3. *Convex integrands.* Let Ω be a measurable subset of the Euclidean space $\mathbb{R}^N$ and let $L^p(\Omega)$, $1 \leq p < \infty$, be the space of all p summable functions on Ω. We set $L^p_m(\Omega) = (L^p(\Omega))^m$.

The function $g : \Omega \times \mathbb{R}^m \to \overline{\mathbb{R}}$ is said to be a *normal convex integrand* if the following conditions hold.

(i) For almost all $x \in \Omega$, the function $g(x,\cdot) : \mathbb{R}^m \to \overline{\mathbb{R}}$ is convex, l.s.c., and not identically $+\infty$.

(ii) g is $\mathcal{L} \times \mathcal{B}$ measurable on $\Omega \times \mathbb{R}^m$, that is, it is measurable with respect to the σ-algebra of subsets of $\Omega \times \mathbb{R}^m$ generated by products of Lebesgue measurable subsets of Ω and Borel subsets of $\mathbb{R}^m$.

We note that if g is convex in y and $\operatorname{int} D(g(x,\cdot)) \neq \emptyset$ for every $x \in \Omega$, then condition (ii) holds if and only if $g = g(x,y)$ is measurable in x for every $y \in \mathbb{R}^m$.

A special case of an $\mathcal{L} \times \mathcal{B}$ measurable integrand is the *Carathéodory integrand.* Namely, one has the following.

Lemma 2.17 *Let $g = g(x,y) : \Omega \times \mathbb{R}^m \to \mathbb{R}$ be continuous in y for every $x \in \Omega$ and measurable in x for every y. Then g is $\mathcal{L} \times \mathcal{B}$ measurable.*

Proof. Let $\{z_i^n\}_{i=1}^\infty$ be a dense subset of $\mathbb{R}^m$ and let $\lambda \in \mathbb{R}$ arbitrary but fixed. Inasmuch as g is continuous in y, it is clear that $g(x,y) \leq \lambda$ if and only if for every n

there exists z_i^n such that $\|z_i^n - y\| \leq \frac{1}{n}$ and $g(x, z_i^n) \leq \lambda + \frac{1}{n}$. Denote by Ω_{in} the set $\{x \in \Omega;\ g(x, z_i^n) \leq \lambda + \frac{1}{n}\}$ and put $Y_{in} = \{y \in \mathbb{R}^m;\ \|y - z_i^n\| \leq \frac{1}{n}\}$. Inasmuch as

$$\{(x, y) \in \Omega \times \mathbb{R}^m;\ g(x, y) \leq \lambda\} = \bigcap_{n=1}^{\infty} \bigcup_{i=1}^{\infty} \Omega_{in} \times Y_{in},$$

we infer that g is $\mathcal{L} \times \mathcal{B}$ measurable, as desired. □

Let us assume, in addition to conditions (i) and (ii), the following.

(iii) There are $\alpha \in L^q_m(\Omega)$, $\frac{1}{p} + \frac{1}{q} = 1$, and $\beta \in L^1(\Omega)$ such that

$$g(x, y) \geq (\alpha(x), y) + \beta(x), \quad \text{a.e. } x \in \Omega,\ y \in \mathbb{R}^m, \tag{2.34}$$

where $(\cdot, \cdot)$ is the usual scalar product in $\mathbb{R}^m$.

(iv) There is $y_0 \in L^p_m$ such that $g(x, y_0) \in L^1(\Omega)$.

Let us remark that if g is independent of x, then conditions (iii) and (iv) automatically hold by virtue of Proposition 1.3.

Define on the space $X = L^p_m(\Omega)$ the function $I_g : X \to \overline{\mathbb{R}}$,

$$I_g(y) = \begin{cases} \displaystyle\int_\Omega g(x, y(x))dx & \text{if } g(x, y) \in L^1(\Omega), \\ +\infty & \text{otherwise.} \end{cases} \tag{2.35}$$

Proposition 2.18 *Let g satisfy assumptions* (i)–(iv). *Then the function I_g is convex, lower semicontinuous, and proper. Moreover,*

$$\partial I_g(y) = \{w \in L^q_m(\Omega);\ w(x) \in \partial g(x, y(x)),\ a.e.\ x \in \Omega\}. \tag{2.36}$$

Here, ∂g is the subdifferential of the function $y \to g(x, y)$.

Proof. Let us show that I_g is well defined (unambiguously a real number or $+\infty$) for every $y \in L^q_m(\Omega)$. Note first that for every Lebesgue measurable function $y : \Omega \to \mathbb{R}^m$ the function $x \to g(x, y(x))$ is Lebesgue measurable on Ω. For a fixed $\lambda \in \mathbb{R}$, we set

$$E = \{(x, y) \in \Omega \times \mathbb{R}^m;\ g(x, y) \leq \lambda\}.$$

Let us denote by $\mathcal{S}$ the class of all sets $S \subset \Omega \times \mathbb{R}^m$ having the property that the set $\{x \in \Omega;\ (x, y(x)) \in S\}$ is Lebesgue measurable. Obviously, S contains every set of the form $T \times D$, where T is a measurable subset of Ω and D is an open subset of $\mathbb{R}^m$. Because $\mathcal{S}$ is a σ-algebra, it follows that it contains the σ-algebra generated by the products of Lebesgue measurable subsets of Ω and Borel subsets of $\mathbb{R}^m$. Hence, $E \in \mathcal{S}$, and therefore $g(x, y(x))$ is Lebesgue measurable, that is, I_g is well defined. By assumption (i), it follows that I_g is convex, whereas by (iv) we see that $I_g \not\equiv +\infty$. Let $\{y_n\} \subset L^p_m(\Omega)$ be strongly convergent to y. Then there is $\{y_{n_k}\} \subset \{y_n\}$ such that

$$y_{n_k}(x) \to y(x), \quad \text{a.e. } x \in \Omega \text{ for } n_k \to \infty.$$

Then, by assumption (iii) and by Fatou's lemma, it follows that

$$\liminf_{n_k\to\infty}\int_\Omega (g(x,y_{n_k}(x)) - (\alpha(x),y_{n_k}(x)) - \beta(x))dx \geq \int_\Omega (g(x,y(x)) - (\alpha(x),y(x)) - \beta(x))dx,$$

and, therefore,

$$\liminf_{n_k\to\infty} I_g(y_{n_k}) \geq I_g(y).$$

Clearly, this implies that $\liminf_{n\to\infty} I_g(y_n) \geq I_g(y)$, that is, I_g is l.s.c. on X.

Let us now prove (2.36). It is easily seen that every $w \in L^q_m(\Omega)$ such that $w(x) \in \partial g(x,y(x))$ belongs to $\partial I_g(y)$. Now, let $w \in \partial I_g$, that is,

$$\int_\Omega (g(x,y(x)) - g(x,u(x)))dx \leq \int_\Omega (w(x),y(x) - u(x))dx, \quad \forall u \in L^p_m(\Omega).$$

Let D be an arbitrary measurable subset of Ω and let $u \in L^p_m(\Omega)$ be defined by

$$u(x) = \begin{cases} y_0 & \text{for } x \in D, \\ y(x) & \text{for } x \in \Omega \setminus \overline{D}, \end{cases}$$

where y_0 is arbitrary in $\mathbb{R}^m$. Substituting in the previous inequality, we get

$$\int_D (g(x,y(x)) - g(x,y_0) - (w(x),y(x) - y_0))dx \leq 0.$$

D is arbitrary, therefore this implies, a.e. $x \in \Omega$,

$$g(x,y(x)) \leq g(x,y_0) + (w(x),y(x) - y_0), \quad \forall y_0 \in \mathbb{R}^m.$$

Hence, $w(x) \in \partial g(x,y(x))$, a.e. $x \in \Omega$, as claimed. □

The case $p = \infty$ is more subtle, because the elements of $\partial I_g(y) \subset (L^\infty_m(\Omega))^*$ are no longer Lebesgue integrable functions on Ω. It turns out, however, that in this case $\partial I_g(y)$ is of the form $\mu_a + \mu_s$, where $\mu_a \in L^1_m(\Omega)$, $\mu_a(x) \in \partial g(y(x))$, a.e., $x \in \Omega$, and μ_s is a singular element of $(L^\infty_m(\Omega)))^*$. Of course, if g is convex and continuous, then $\mu_s = 0$. We refer the reader to Rockafellar [98] for the complete description of ∂I_g in this case.

Now, let us consider the special case where

$$g(x,y) = I_K(y) = \begin{cases} 0 & \text{if } y \in K, \\ +\infty & \text{if } y \notin K, \end{cases}$$

K being a closed convex subset of $\mathbb{R}^m$. Then, I_g is the indicator function of the closed convex subset $\mathcal{K}$ of $L^p_m(\Omega)$ defined by

$$\mathcal{K} = \{y \in L^p_m(\Omega);\ y(x) \in K, \quad \text{a.e. } x \in \Omega\},$$

and so by formula (2.36) we see that the normal cone $N_\mathcal{K} \subset L^q_m(\Omega)$ to $\mathcal{K}$ is defined by

$$N_\mathcal{K}(y) = \{w \in L^q_m(\Omega);\ w(x) \in N_K(y(x)), \quad \text{a.e. } x \in \Omega\}, \tag{2.37}$$

where $N_K(y) = \{z \in \mathbb{R}^m;\ (z, y - u) \geq 0,\ \forall u \in K\}$ is the normal cone at K in $y \in K$.

In particular, if $m = 1$ and $K = [a, b]$, then

$$N_K(y) = \{w \in L^q(\Omega);\ w(x) = 0,\ \text{a.e. in } [x \in \Omega;\ \ a < y(x) < b], \\ w(x) \geq 0,\ \text{a.e. in } [x \in \Omega;\ y(x) = b],\ w(x) \leq 0,\ \text{a.e. in } [x \in \Omega;\ y(x) = a]\}. \tag{2.38}$$

Let us take now $K = \{y \in \mathbb{R}^m;\ \|y\| \leq \rho\}$. Then,

$$N_K(y) = \begin{cases} 0 & \text{if } \ \|y\| < \rho, \\ \displaystyle\bigcup_{\lambda > 0} \lambda y & \text{if } \ \|y\| = \rho, \end{cases}$$

and so $N_{\mathcal{K}}$ is given by

$$N_{\mathcal{K}}(y) = \{w \in L^q_m(\Omega);\ w(x) = 0,\ \text{a.e. in } [x \in \Omega;\ \|y(x)\| < \rho],\ w(x) = \lambda(x)y(x), \\ \text{a.e. in } [x \in \Omega;\ \|y(x)\| = \rho],\ \text{where } \lambda \in L^q_m(\Omega),\ \lambda(x) \geq 0,\ \text{a.e. } x \in \Omega\}.$$

Elliptic nonlinear operators on bounded open domains of $\mathbb{R}^N$ with appropriate boundary value conditions represent another source of maximal monotone operators and, in particular, of subpotential operators. We give a few examples here.

Corollary 2.19 *The mapping $\phi_1 : L^1(\Omega) \to L^\infty(\Omega)$ defined by*

$$\phi_1(u) = \{\|u\|_{L^1(\Omega)} w;\ w(x) \in L^\infty(\Omega),\ w(x) \in \operatorname{sign} u(x) \ \textit{a.e. } x \in \Omega\}$$

is the duality mapping J of the space $X = L^1(\Omega)$. Here, $\operatorname{sign} r = \frac{r}{|r|}$ *for $r \neq 0$,* $\operatorname{sign} 0 = [-1.1]$.

Proof. It is easily seen that $\phi_1(u) \in J(u)$, $\forall u \in L^1(\Omega)$. On the other hand, by Proposition 2.18 we have

$$\partial\|u\|_{L^1(\Omega)} = \{w \in L^\infty(\Omega);\ w(x) \in \operatorname{sign} u(x),\ \ \text{a.e. } x \in \Omega\}.$$

This implies that

$$\partial\left(\frac{1}{2}\ \|u\|^2_{L^1(\Omega)}\right) = \phi_1(u),\ \ \forall u \in L^1(\Omega)$$

and, because by Theorem 2.15 the mapping $\partial\left(\frac{1}{2}\ \|u\|^2_{L^1(\Omega)}\right)$ is maximal monotone in $L^1(\Omega) \times L^\infty(\Omega)$, we conclude that so is ϕ_1 and, because $\phi_1 \subset J$, we have $\phi_1 = J$ as claimed. □

4. *Semilinear elliptic operators in $L^2(\Omega)$.* Let Ω be an open bounded subset of $\mathbb{R}^N$, and let $g : \mathbb{R} \to \overline{\mathbb{R}}$ be a lower semicontinuous, convex, proper function such that $0 \in D(\partial g)$.

Define the function $\varphi : L^2(\Omega) \to \overline{\mathbb{R}}$ by

$$\varphi(y) = \begin{cases} \displaystyle\int_\Omega \left(\frac{1}{2}\,|\nabla y|^2 + g(y)\right) dx & \text{if } \ y \in H^1_0(\Omega) \text{ and } g(y) \in L^1(\Omega), \\ +\infty & \text{otherwise.} \end{cases} \tag{2.39}$$

Proposition 2.20 *The function φ is convex, l.s.c., and $\not\equiv +\infty$. Moreover, if the boundary $\partial\Omega$ is sufficiently smooth (for instance, of class C^2) or if Ω is convex, then $\partial\varphi \subset L^2(\Omega) \times L^2(\Omega)$ is given by*

$$\begin{aligned} \partial\varphi = \{[y,w];\ w \in L^2(\Omega);\ y \in H_0^1(\Omega) \cap H^2(\Omega), \\ w(x) + \Delta y(x) \in \partial g(y(x)),\ a.e.\ x \in \Omega\}. \end{aligned} \tag{2.40}$$

Proof. It is readily seen that φ is convex and $\not\equiv +\infty$. Let $\{y_n\} \subset L^2(\Omega)$ be strongly convergent to y as $n \to \infty$. As seen earlier,

$$\liminf_{n\to\infty} \int_\Omega g(y_n)dx \geq \int_\Omega g(y)dx,$$

and it is also clear, by weak lower semicontinuity of the $L^2(\Omega)$-norm, that

$$\liminf_{n\to\infty} \int_\Omega |\nabla y_n|^2 dx \geq \int_\Omega |\nabla y|^2 dx.$$

Hence, $\liminf_{n\to\infty} \varphi(y_n) \geq \varphi(y)$.

Let us denote by $\Gamma \subset L^2(\Omega) \times L^2(\Omega)$ the operator defined by the second part of (2.40), that is,

$$\begin{aligned} \Gamma = \{\ [y,w] \in (H_0^1(\Omega) \cap H^2(\Omega)) \times L^2(\Omega); \\ w(x) \in -\Delta y(x) + \partial g(y(x)),\ \text{a.e. } x \in \Omega\}. \end{aligned}$$

The inclusion $\Gamma \subset \partial\varphi$ is obvious, thus it suffices to show that Γ is maximal monotone in $L^2(\Omega)$. To this end, observe that $\Gamma = A_2 + B$, where $A_2 y = -\Delta y$, $\forall y \in D(A_2) = H_0^1(\Omega) \cap H^2(\Omega)$, and $By = \{v \in L^2(\Omega);\ v(x) \in \partial g(y(x)), \text{a.e. } x \in \Omega\}$. As seen earlier, the operators A_2 and B are maximal monotone in $L^2(\Omega) \times L^2(\Omega)$. Replacing B by $y \to By - y_0$, where $y_0 \in B(0)$, we may assume without loss of generality that $0 \in B(0)$. On the other hand, it is readily seen that $(B_\lambda u)(x) = \beta_\lambda(u(x))$, a.e. $x \in \Omega$ for all $u \in L^2(\Omega)$, where $\beta = \partial g$, and $\beta_\lambda = \lambda^{-1}(1 - (1 + \lambda\beta)^{-1})$ is the Yosida approximation of β. We have

$$(A_2 u, B_\lambda u) = -\int_\Omega \Delta u \beta_\lambda(u) dx \geq 0, \qquad \forall u \in H_0^1(\Omega) \cap H^2(\Omega), \tag{2.41}$$

or, equivalently,

$$\int_\Omega g(1 + \lambda A_2)^{-1} y(x) dx \leq \int_\Omega g(y(x)) dx, \quad \forall y \in L^2(\Omega),$$

which results from the following simple argument. We set $z = (I + \lambda A_2)^{-1} y$:

$$z - \lambda\Delta z = y \quad \text{in } \Omega;\ z \in H_0^1(\Omega) \cap H^2(\Omega).$$

If we multiply the latter by $\beta_\mu(z) = \frac{1}{\mu}(z - (1 + \mu\beta)^{-1} z)$, $\mu > 0$, and integrate on Ω, we obtain that

$$\int_\Omega \beta_\mu(z)(z - y) dx \leq 0, \quad \forall \mu > 0,$$

because (inasmuch as $\beta'_\mu \geq 0$) we have

$$\int_\Omega \Delta z \beta_\mu(z)dx = -\int_\Omega \beta'_\mu(z)|\nabla z|^2 dx \leq 0, \quad \forall \mu > 0.$$

This yields

$$\int_\Omega g_\mu(z)dx \leq \int_\Omega g_\mu(y)dx, \quad \forall \mu > 0,$$

where $g_\mu = \beta_\mu$. Then, letting $\mu \to 0$, and recalling Theorem 2.16, we get the desired inequality. (As a matter of fact, this calculation works if $\beta_\lambda \in C^1(\mathbb{R})$ but, in a general situation, we replace β_λ by a C^1 mollifier regularization $(\beta_\lambda)_\varepsilon$ and let ε tend to zero.)

By Corollary 2.14, we know that A_2+B_λ is maximal monotone in $L^2(\Omega)\times L^2(\Omega)$. This means that, for each $f \in L^2(\Omega)$, the equation

$$u + A_2 u + B_\lambda u = f \tag{2.42}$$

has a unique solution $u_\lambda \in D(A_2)$. Taking into account (2.41), it follows by (2.42) that $\{A_2 u_\lambda\}$ and $\{B_\lambda u_\lambda\}$ are bounded in $L^2(\Omega)$. Hence, $\{u_\lambda\}$ is bounded in $H^1_0(\Omega)\cap H^2(\Omega)$ and, on a subsequence $\{\lambda\} \to 0$ we have

$$\begin{aligned} u_\lambda &\to u && \text{strongly in } H^1_0(\Omega) \text{ weakly in } H^2(\Omega)\\ A_2 u_\lambda &\to A_2 u && \text{weakly in } L^2(\Omega),\\ B_\lambda u_\lambda &\to \eta && \text{weakly in } L^2(\Omega)\\ (I+\lambda B)^{-1} u_\lambda &\to u && \text{strongly in } L^2(\Omega). \end{aligned}$$

Then, by Proposition 2.3, part (i), $\eta \in Bu$ and so, $f \in u + \Gamma u$, as claimed. □

Remark 2.21 Because $A_2 + B$ is coercive, it follows from Corollary 2.9 that $R(A_2 + B) = L^2(\Omega)$. Hence, for every $f \in L^2(\Omega)$, the Dirichlet problem

$$\begin{cases} -\Delta y + \beta(y) \ni f, & \text{a.e. in } \Omega,\\ y = 0, & \text{on } \partial\Omega, \end{cases} \tag{2.43}$$

has a unique solution $y \in H^1_0(\Omega) \cap H^2(\Omega)$.

In the special case, where $\beta \subset \mathbb{R} \times \mathbb{R}$ is given by

$$\beta(r) = \begin{cases} 0 & \text{if } r > 0,\\ \mathbb{R}^- & \text{if } r = 0, \end{cases}$$

and so $g(y) = 0$ for $y \geq 0$, $g(y) = \emptyset$ for $y < 0$, problem (2.43) reduces to the *obstacle problem*

$$\begin{cases} -\Delta y = f, & \text{a.e. in } [y > 0],\\ -\Delta y \geq f,\ y \geq 0, & \text{a.e. in } \Omega,\\ y = 0, & \text{on } \partial\Omega. \end{cases} \tag{2.44}$$

This is an *elliptic variational inequality* describing a free boundary problem arising in fluid dynamics (the *dam problem*) as well in the theory of nonlinear diffusion phenomena and of phase transition (see, e.g., [78]).

We also note that the solution y to (2.43) is the limit in $H_0^1(\Omega)$ of the solutions y_ε to the approximating problem

$$\begin{cases} -\Delta y + \beta_\varepsilon(y) = f, & \text{in } \Omega, \\ y = 0, & \text{on } \partial\Omega, \end{cases} \tag{2.45}$$

where β_ε is the Yosida approximation of β. Indeed, multiplying (2.45) by y_ε, we get

$$\|y_\varepsilon\|^2_{H_0^1(\Omega)} + \|\Delta y_\varepsilon\|^2_{L^2(\Omega)} \le C, \quad \forall \varepsilon > 0,$$

and therefore $\{y_\varepsilon\}$ is bounded in $H_0^1(\Omega) \cap H^2(\Omega)$. By (2.45), it follows also that

$$\int_\Omega |\nabla(y_\varepsilon - y_\lambda)|^2 dx + \int_\Omega (\beta_\varepsilon(y_\varepsilon) - \beta_\lambda(y_\lambda))(y_\varepsilon - y_\lambda)dx = 0,$$

and, therefore,

$$\int_\Omega |\nabla(y_\varepsilon - y_\lambda)|^2 dx + \int_\Omega (\beta_\varepsilon(y_\varepsilon) - \beta_\lambda(y_\lambda))(\varepsilon\beta_\varepsilon(y_\varepsilon) - \lambda\beta_\lambda(y_\lambda))dx \le 0,$$

because $\beta_\varepsilon(y) \in \beta((1+\varepsilon\beta)^{-1}y)$ and β is monotone. Hence, $\{y_\varepsilon\}$ is Cauchy in $H_0^1(\Omega)$, and so $y = \lim_{\varepsilon\to 0} y_\varepsilon$ exists in $H_0^1(\Omega)$. Clearly, this also implies that, for $\varepsilon \to 0$,

$$\begin{aligned} \Delta y_\varepsilon &\to \Delta y && \text{weakly in } L^2(\Omega), \\ y_\varepsilon &\to y && \text{weakly in } H^2(\Omega), \\ \beta_\varepsilon(y_\varepsilon) &\to g && \text{weakly in } L^2(\Omega). \end{aligned}$$

Now, by Proposition 2.3, part (i), we see that $g(x) \in \beta(y(x))$, a.e. $x \in \Omega$, and so y is the solution to problem (2.43).

5. *Neumann nonlinear boundary conditions.* Let Ω be a bounded and open subset of $\mathbb{R}^N$ with the boundary $\partial\Omega$ of class C^2. Let $j : \mathbb{R} \to \overline{\mathbb{R}}$ be an l.s.c., proper, convex function and let $\beta = \partial j$. Define the function $\varphi : L^2(\Omega) \to \overline{\mathbb{R}}$ by

$$\varphi(u) = \begin{cases} \dfrac{1}{2}\displaystyle\int_\Omega |\nabla u|^2 dx + \int_{\partial\Omega} j(u)dx & \text{if } u \in H^1(\Omega),\ j(u) \in L^1(\partial\Omega), \\ +\infty & \text{otherwise.} \end{cases} \tag{2.46}$$

Because for every $u \in H^1(\Omega)$ the trace of u on $\partial\Omega$ is well defined and belongs to $L^2(\partial\Omega)$ (see Definition 1.14), formula (2.46) makes sense. Moreover, arguing as in the previous example, it follows that φ is convex and l.s.c. on $L^2(\Omega)$. Regarding its subdifferential $\partial\varphi \subset L^2(\Omega) \times L^2(\Omega)$, it is completely described in Proposition 2.22, due to Brezis [42].

Proposition 2.22 *We have*

$$\partial\varphi(u) = -\Delta u, \quad \forall u \in D(\partial\varphi), \tag{2.47}$$

where

$$D(\partial\varphi) = \left\{ u \in H^2(\Omega);\ -\frac{\partial u}{\partial n} \in \beta(u),\ \text{a.e. on } \partial\Omega \right\}$$

and $\frac{\partial}{\partial n}$ *is the normal derivative to* $\partial\Omega$. *Moreover, there are some positive constants* C_1, C_2 *such that*

$$\|u\|_{H^2(\Omega)} \le C_1\|u - \Delta u\|_{L^2(\Omega)} + C_2, \quad \forall u \in D(\partial\varphi). \tag{2.48}$$

6. *The nonlinear diffusion operator.* Let Ω be a bounded and open subset of $\mathbb{R}^N$ with a sufficiently smooth boundary $\partial\Omega$. Denote as usual by $H_0^1(\Omega)$ the Sobolev space of all $u \in H^1(\Omega)$ having null trace on $\partial\Omega$ and by $H^{-1}(\Omega)$ the dual of $H_0^1(\Omega)$. Note that $H^{-1}(\Omega)$ is a Hilbert space with the scalar product

$$\langle u, v\rangle = (J^{-1}u, v) \qquad \forall u, v \in H^{-1}(\Omega),$$

where $J = -\Delta$ is the canonical isomorphism (duality mapping) of $H_0^1(\Omega)$ onto $H^{-1}(\Omega)$ and $(\cdot,\cdot)$ is the pairing between $H_0^1(\Omega)$ and $H^{-1}(\Omega)$.

Let $j : \mathbb{R} \to \overline{\mathbb{R}}$ be an l.s.c., convex, proper function and let $\beta = \partial j$. Define the function $\varphi : H^{-1}(\Omega) \to \overline{\mathbb{R}}$ by

$$\varphi(u) = \begin{cases} \displaystyle\int_\Omega j(u(x))dx & \text{if } u \in L^1(\Omega) \text{ and } j(u) \in L^1(\Omega), \\ +\infty & \text{otherwise.} \end{cases} \tag{2.49}$$

It turns out (see Proposition 2.23 below) that the subdifferential $\partial\varphi : H^{-1}(\Omega) \to H^{-1}(\Omega)$ of φ is just the operator $u \to -\Delta\beta(u)$ with appropriate boundary conditions.

The equation $\lambda u - \Delta\beta(u) = f$, where $\lambda \geq 0$, is known in the literature as the *nonlinear diffusion equation* or the *porous media equation.*

Proposition 2.23 *Let us assume that*

$$\lim_{|r|\to\infty} \frac{j(r)}{|r|} = +\infty. \tag{2.50}$$

Then the function φ is convex and lower semicontinuous on $H^{-1}(\Omega)$. Moreover, $\partial\varphi \subset H^{-1}(\Omega) \times H^{-1}(\Omega)$ is given by

$$\begin{aligned} \partial\varphi = \{[u, w] \in (H^{-1}(\Omega) \cap L^1(\Omega)) \times H^{-1}(\Omega);\ w = -\Delta v, \\ v \in H_0^1(\Omega),\ v(x) \in \beta(u(x)),\ a.e.\ x \in \Omega\}. \end{aligned} \tag{2.51}$$

Proof. Obviously, φ is convex. To prove that φ is l.s.c., consider a sequence $\{u_\lambda\} \subset H^{-1}(\Omega) \cap L^1(\Omega)$ such that $u_n \to u$ in $H^{-1}(\Omega)$ and $\varphi(u_n) \leq \lambda$, that is, $\int_\Omega j(u_n)dx \leq \lambda$, $\forall n$. We must prove that $\int_\Omega j(u)dx \leq \lambda$. We have already seen in the proof of Proposition 2.18 that the function $u \to \int_\Omega j(u)dx$ is lower semicontinuous on $L^1(\Omega)$. Because this function is convex, it is weakly lower semicontinuous in $L^1(\Omega)$ and so it suffices to show that $\{u_n\}$ is weakly compact in $L^1(\Omega)$. According to the Dunford–Pettis criterion (see Theorem 1.30), we must prove that the integrals $\int_E |u_n|dx$ are uniformly absolutely continuous, that is, for every $\varepsilon > 0$ there is $\delta(\varepsilon)$ such that $\int_E |u_n(x)|dx \leq \varepsilon$ if $m(E) \leq \delta(\varepsilon)$ (E is a measurable set of Ω and m is the Lebesgue measure). By condition (2.50), for every $p > 0$ there exists $R(p) > 0$ such that $j(r) \geq p|r|$ if $|r| \geq R(p)$. In particular, this clearly implies that $\int_\Omega |u_n(x)|dx \leq C$, $\forall n$.

Moreover, for every measurable subset E of Ω, we have

$$\begin{aligned}\int_E |u_n(x)|dx &\le \int_{E\cap\{|u_n|\ge R(p)\}} |u_n(x)|dx + \int_{E\cap\{|u_n|<R(p)\}} |u_n(x)|dx \\ &\le \frac{1}{p}\int_\Omega |u_n(x)|dx + R(p)m(E) \le \varepsilon,\end{aligned}$$

if we choose $p > (2\varepsilon)^{-1} \sup \int_\Omega |u_n(x)|dx$ and $m(E) \le \frac{\varepsilon}{2R(p)}$. Hence, $\{u_n\}$ is weakly compact in $L^1(\Omega)$.

To prove (2.51), consider the operator $A \subset H^{-1}(\Omega) \times H^{-1}(\Omega)$ defined by

$$Au = \{-\Delta v;\ v \in H_0^1(\Omega),\ v(x) \in \beta(u(x)),\ \text{a.e. } x \in \Omega\},$$

where $D(A) = \{u \in H^{-1}(\Omega) \cap L^1(\Omega);\ \exists v \in H_0^1(\Omega),\ v(x) \in \beta(u(x)),\ \text{a.e. } x \in \Omega\}$. To prove that $A = \partial\varphi$, proceeding as in the previous case, we show separately that $A \subset \partial\varphi$ and that A is maximal monotone. Let us show first that $R(I + A) = H^{-1}(\Omega)$. Let f be arbitrary but fixed in $H^{-1}(\Omega)$. We must show that there exist $u \in H^{-1}(\Omega) \cap L^1(\Omega)$ and $v \in H_0^1(\Omega)$ such that

$$u - \Delta v = f \quad \text{in } \Omega,\quad v(x) \in \beta(u(x)), \quad \text{a.e. } x \in \Omega;$$

or, equivalently,

$$\begin{aligned} u - \Delta v = f \quad &\text{in } \Omega,\ u(x) \in \gamma(v(x)), \quad \text{a.e. } x \in \Omega, \\ &u \in H^{-1}(\Omega) \cap L^1(\Omega),\ v \in H_0^1(\Omega), \end{aligned} \tag{2.52}$$

where $\gamma = \beta^{-1}$.

Consider the approximating equation

$$\gamma_\lambda(v) - \Delta v = f \quad \text{in } \Omega, \quad v = 0 \quad \text{on } \partial\Omega, \tag{2.53}$$

where $\gamma_\lambda = \lambda^{-1}(1-\lambda\gamma)^{-1}$, $\lambda > 0$. It is readily seen that (2.53) has a unique solution $v_\lambda \in H_0^1(\Omega)$. Indeed, because $-\Delta$ is maximal monotone from $H_0^1(\Omega)$ to $H^{-1}(\Omega)$ and $v \to \gamma_\lambda(v)$ is monotone and continuous from $H_0^1(\Omega)$ to $H^{-1}(\Omega)$ (in fact, from $L^2(\Omega)$ to itself), we infer by Theorem 2.7 that $v \to \gamma_\lambda(v) - \Delta v$ is maximal monotone in $H_0^1(\Omega) \times H^{-1}(\Omega)$, and by Corollary 2.9 that it is surjective. Let $v_0 \in D(\gamma)$. Multiplying equation (2.53) by $v_\lambda - v_0$, we get

$$\int_\Omega |\nabla v_\lambda|^2 dx + \int_\Omega \gamma(v_0)(v_\lambda - v_0)dx \le (v_\lambda - v_0, f) \le \|v_\lambda - v_0\|_{H_0^1(\Omega)} \|f\|_{H^{-1}(\Omega)}.$$

(Here, $(\cdot,\cdot)$ is the duality pairing in $H_0^1(\Omega) \times H^{-1}(\Omega)$ which coincides with the scalar product of $L^2(\Omega)$ on $L^2(\Omega) \times L^2(\Omega)$.) Hence, $\{v_\lambda\}$ is bounded in $H_0^1(\Omega)$ and compact in $L^2(\Omega)$. Then, on a subsequence, again denoted by λ, we have

$$v_\lambda \rightharpoonup v \quad \text{in } H_0^1(\Omega), \quad v_\lambda \to v \quad \text{in } L^2(\Omega).$$

Thus, extracting further subsequences, we may assume that

$$\begin{aligned} v_\lambda(x) &\to v(x), \quad \text{a.e. } x \in \Omega, \\ (1 + \lambda\gamma)^{-1} v_\lambda(x) &\to v(x), \quad \text{a.e. } x \in \Omega, \end{aligned} \tag{2.54}$$

because, by condition (2.50) and Proposition 1.9, it follows that $D(\gamma) = R(\beta) = \mathbb{R}$ (β is coercive) and so $\lim_{\lambda\to 0}(1+\lambda\gamma)^{-1}r = r$ for all $r \in \mathbb{R}$ (Proposition 2.10).

We get $g_\lambda = \gamma_\lambda(v_\lambda)$. Then, letting λ tend to zero in (2.53), we see that $g_\lambda \to u$ in $H^{-1}(\Omega)$ and

$$u - \Delta v = f \quad \text{in } \Omega, \quad v \in H_0^1(\Omega).$$

It remains to be shown that $u \in L^1(\Omega)$ and $u(x) \in \gamma(v(x))$, a.e. $x \in \Omega$.

Multiplying equation (2.53) by v_λ, we see that

$$\int_\Omega g_\lambda v_\lambda dx \le C, \quad \forall \lambda > 0.$$

On the other hand, for some $u_0 \in D(j)$ we have $j(g_\lambda(x)) \le j(u_0) + (g_\lambda(x) - u_0)v$, $\forall v \in \beta(g_\lambda(x))$. This yields

$$\int_\Omega j(g_\lambda(x))dx \le C, \quad \forall \lambda > 0,$$

because $(1+\lambda\gamma)^{-1}v_\lambda \in \beta(g_\lambda)$.

As seen before, this implies that $\{g_\lambda\}$ is weakly compact in $L^1(\Omega)$. Hence, $u \in L^1(\Omega)$ and

$$g_\lambda \rightharpoonup u \quad \text{in } L^1(\Omega) \text{ for } \lambda \to 0. \tag{2.55}$$

On the other hand, by (2.54) it follows by virtue of the Egorov theorem that for every $\varepsilon > 0$ there exists a measurable subset $E_\varepsilon \subset \Omega$ such that $m(\Omega \setminus E_\varepsilon) \le \varepsilon$, $\{(1+\lambda\gamma)^{-1}v_\lambda\}$ is bounded in $L^\infty(E_\varepsilon)$, and

$$(1+\lambda\gamma)^{-1}v_\lambda \to v \quad \text{uniformly in } E_\varepsilon \text{ as } \lambda \to 0. \tag{2.56}$$

Recalling that $g_\lambda(x) \in \gamma((1+\lambda\gamma)^{-1}v_\lambda(x))$ and that the operator

$$\widetilde{\gamma} = \{[u,v] \in L^1(E_\varepsilon) \times L^\infty(E_\varepsilon);\ u(x) \in \gamma(v(x)), \text{ a.e. } x \in E_\varepsilon\},$$

is maximal monotone in $L^1(E_\varepsilon)\times L^\infty(E_\varepsilon)$, we infer, by (2.55) and (2.56), that $[u,v]\in\widetilde{\gamma}$, that is, $v(x) \in \beta(u(x))$, a.e. $x \in E_\varepsilon$. Because ε is arbitrary, we infer that $v(x) \in \beta(u(x))$, a.e. $x \in \Omega$, as desired. □

To prove that $A \subset \partial\varphi$, we must use the definition of A. However, in order to avoid a formal calculus with the symbol (v,u) in the case where $v \in H^{-1}(\Omega)\cap L^1(\Omega)$ and $u \in H_0^1(\Omega)$, we need the following lemma, which is a special case of a general result due to Brezis and Browder [44]. (See [15], p. 71, for the proof.)

Lemma 2.24 *Let Ω be an open subset of $\mathbb{R}^N$. If $v \in H^{-1}(\Omega) \cap L^1(\Omega)$ and $u \in H_0^1(\Omega)$ are such that*

$$v(x)u(x) \ge -|h(x)|, \quad \text{a.e. } x \in \Omega, \tag{2.57}$$

for some $h \in L^1(\Omega)$, then $vu \in L^1(\Omega)$ and

$$v(u) = \int_\Omega v(x)u(x)dx. \tag{2.58}$$

(Here, $v(u) = (v, u)$ is the value of functional $v \in H^{-1}(\Omega)$ at $u \in H_0^1(\Omega)$.) We note that, in our case, that is of v, u found above, condition (2.57) holds and so $v \in Au$, as claimed. □

Remark 2.25 As seen in Proposition 1.9, condition (2.50) is equivalent to $R(\beta)=\mathbb{R}$ and β^{-1} is bounded on bounded sets.

Remark 2.26 Proposition 2.23 extends with the same proof to the functions $j : \Omega \times \mathbb{R} \to \overline{\mathbb{R}}$, $\beta(x, \cdot) = \partial j(x, \cdot)$, and $\varphi : H^{-1}(\Omega) \to \overline{\mathbb{R}}$,

$$\varphi(u) = \begin{cases} \int_\Omega j(x, u(x))dx & \text{if } u \in L^1(\Omega),\ j(x, u) \in L^1(\Omega), \\ +\infty & \text{otherwise,} \end{cases}$$

where $j = j(x, lr)$ is convex and lower-semicontinuous in $r \in \mathbb{R}$, Lebesgue measurable in x, $j(x, u_0(x)) \in L^1(\Omega)$ for some $u_0 \in L^1(\Omega)$ and

$$\lim_{|r|\to\infty} \frac{j(x, r)}{|r|} = +\infty \text{ uniformly in } x \in \mathbb{R}. \tag{2.59}$$

Let X be a reflexive Banach space with the dual X^* and let $A : X \to X^*$ be a monotone operator (linear or nonlinear). Let $\varphi : X \to \overline{\mathbb{R}}$ be a lower semicontinuous convex function on X, $\varphi \not\equiv +\infty$. If f is a given element of X, consider the following problem.

Find $y \in X$ such that

$$(y - z, Ay) + \varphi(y) - \varphi(z) \le (y - z, f), \quad \forall z \in X. \tag{2.60}$$

This is an *abstract elliptic variational* inequality associated with the operator A and the convex function φ, and it can be equivalently expressed as

$$Ay + \partial\varphi(y) \ni f, \tag{2.61}$$

where $\partial\varphi \subset X \times X^*$ is the subdifferential of φ. In the special case where $\varphi = I_K$ is the indicator function of a closed convex set K, that is,

$$I_K(x) = \begin{cases} 0 & \text{if } x \in K, \\ +\infty & \text{otherwise,} \end{cases}$$

problem (2.60) becomes:

Find $y \in K$ such that

$$(y - z, Ay) \le (y - z, f), \quad \forall z \in K. \tag{2.62}$$

If the operator A is itself a subdifferential $\partial\psi$ of a continuous convex function $\psi : X \to \mathbb{R}$, then the variational inequality (2.60) is equivalent to the minimization problem (the Dirichlet principle)

$$\min\{\psi(z) + \varphi(z) - (z, f);\ z \in X\} \tag{2.63}$$

or, in the case of problem (2.62),

$$\min\{\psi(z) - (z, f);\ z \in K\}. \tag{2.64}$$

We have the following result.

Theorem 2.27 *Let $A : X \to X^*$ be a monotone demicontinuous operator and let $\varphi : X \to \overline{\mathbb{R}}$ be a lower semicontinuous, proper, convex function. Assume that there exists $y_0 \in D(\varphi)$ such that*

$$\lim_{\|y\|\to\infty} \frac{(y - y_0, Ay) + \varphi(y)}{\|y\|} = +\infty. \tag{2.65}$$

Then, problem (2.60) has at least one solution. Moreover, the set of solutions is bounded, convex, and closed in X and if the operator A is strictly monotone (i.e., $(Au - Av, u - v) = 0 \iff u = v$,) then the solution is unique.

Proof. By Theorem 2.8, the operator $A + \partial\varphi$ is maximal monotone in $X \times X^*$. By condition (2.65) it is also coercive, therefore we conclude (see Corollary 2.9) that it is surjective. Hence, equation (2.61) (equivalently, (2.60)) has at least one solution.

The set of all solutions y to (2.60) is $(A + \partial\varphi)^{-1}(f)$, thus we infer that this set is closed and convex (see Proposition 2.3). By the coercivity condition (2.65), it is also bounded. Finally, if A (or, more generally, if $A + \partial\varphi$) is strictly monotone, then $(A + \partial\varphi)^{-1}f$ consists of a single element. □

In the special case $\varphi = I_K$, we have the following.

Corollary 2.28 *Let $A : X \to X^*$ be a monotone demicontinuous operator and let K be a closed convex subset of X. Assume either that there is $y_0 \in K$ such that*

$$\lim_{\|y\|\to\infty} \frac{(y - y_0, Ay)}{\|y\|} = +\infty, \tag{2.66}$$

or that K is bounded. Then problem (2.60) has at least one solution. The set of all solutions is bounded, convex, and closed. If A is strictly monotone, then the solution to (2.60) is unique.

As mentioned earlier, most of the elliptic free boundary problems can be represented as a variational inequality (2.62) and this is the principal motivation for study the multi-valued maximal monotone operators. In fact, the remarkable discovery of Guido Stampacchia at the end of the sixties — that several elliptic problems with free boundary (for instance, the obstacle problem and Signorini's problem) can be represented as a nonlinear equation of this type — opened the way to a simple and complete mathematical treatment of this important class of problems.

2.2 Accretive Operators

Throughout this section, X is a real Banach space with the norm $\|\cdot\|$, X^* is its dual space, and $(\cdot,\cdot)$ the pairing between X and X^*. We denote as usual by $J : X \to X^*$ the duality mapping of the space X.

Definition 2.29 A subset A of $X \times X$ (equivalently, a multivalued operator from X to X) is called *accretive* if for every pair $[x_1, y_1], [x_2, y_2] \in A$, there is $w \in J(x_1 - x_2)$ such that

$$(y_1 - y_2, w) \geq 0. \tag{2.67}$$

An accretive set is said to be *maximal accretive* if it is not properly contained in any accretive subset of $X \times X$. An accretive set A is said to be *m-accretive* if

$$R(I + A) = X. \tag{2.68}$$

Here we have denoted I the unity operator in X, but when there is no danger of confusion, we simply write 1 instead of I.

We denote by $D(A) = \{x \in X; Ax \neq \emptyset\}$ the domain of A and by $R(A) = \{y \in Ax;\ [x, y] \in A\}$ the range of A. As in the case of operators from X to X^*, we identify an operator (eventually multivalued) $A : D(A) \subset X \to X$ with its graph $\{[x, y];\ y \in Ax\}$ and so view A as a subset of $X \times X$.

A subset A is called *dissipative* (*maximal dissipative*, *m-dissipative*, respectively) if $-A$ is accretive (maximal accretive, m-accretive, respectively).

Given $\omega \in \mathbb{R}$, the operator A is said to be *ω-accretive* (*ω-m-accretive*), if $A + \omega I$ is accretive (m-accretive, respectively). A subset $A \subset X \times X$, that is, ω-accretive or ω-m-accretive for some $\omega \in \mathbb{R}$ is also called *quasi-accretive*, respectively, *quasi-m-accretive.*

As we show below, the accretiveness of A is, in fact, a metric geometric property that can be equivalently expressed as

$$\|x_1 - x_2\| \le \|x_1 - x_2 + \lambda(y_1 - y_2)\|, \quad \forall \lambda > 0,\ [x_i, y_i] \in A,\ i = 1, 2, \tag{2.69}$$

using the following lemma (Kato's lemma).

Lemma 2.30 *Let $x, y \in X$. Then there exists $w \in J(x)$ such that $(y, w) \ge 0$ if and only if*

$$\|x\| \le \|x + \lambda y\|, \quad \forall \lambda > 0 \tag{2.70}$$

holds.

Proof. Let x and y in X be such that $(y, w) \ge 0$ for some $w \in J(x)$. Then, by definition of J, we have

$$\|x\|^2 = (x, w) \le (x + \lambda y, w) \le \|x + \lambda y\| \cdot \|w\| = \|x + \lambda y\| \cdot \|x\|, \quad \forall \lambda > 0,$$

and (2.70) follows.

Suppose now that (2.70) holds. For $\lambda > 0$, let w_λ be an arbitrary element of $J(x + \lambda y)$. Without loss of generality, we may assume that $x \neq 0$. Then, $w_\lambda \neq 0$ for λ small. We set $f_\lambda = w_\lambda \|w_\lambda\|^{-1}$. Because $\{f_\lambda\}_{\lambda > 0}$ is weak-star compact in X^*, there exists a generalized sequence, again denoted λ, such that $f_\lambda \rightharpoonup f$ in X^* as $\lambda \to 0$. On the other hand, from the inequality

$$\|x\| \le \|x + \lambda y\| = (x + \lambda y, f_\lambda) \le \|x\| + \lambda(y, f_\lambda)$$

it follows that

$$(y, f_\lambda) \ge 0, \quad \forall \lambda > 0.$$

Hence, $(y, f) \ge 0$ and $\|x\| \le (x, f)$. Because $\|f\| \le 1$, this implies that $\|x\| = (x, f)$, $\|f\| = 1$, and therefore $w = f\|x\| \in J(x)$, $(y, w) \ge 0$, as claimed. $\square$

Proposition 2.31 *A subset A of $X \times X$ is accretive if and only if inequality* (2.69) *holds for all $\lambda > 0$ and all $[x_i, y_i] \in A$, $i = 1, 2$.*

Proposition 2.31 is an immediate consequence of Lemma 2.30. In particular, it follows that A is ω-accretive iff

$$\begin{aligned} &\|x_1 - x_2 + \lambda(y_1 - y_2)\| \geq (1 - \lambda\omega)\|x_1 - x_2\| \\ &\text{for } 0 < \lambda < \frac{1}{\omega} \quad \text{and} \quad [x_i, y_i] \in A, \ i = 1, 2. \end{aligned} \tag{2.71}$$

Hence, if A is accretive, then the operator $(I + \lambda A)^{-1}$ is single-valued and non-expansive on $R(I + \lambda A)$, that is,

$$\|(I + \lambda A)^{-1}x - (I + \lambda A)^{-1}y\| \leq \|x - y\|, \ \ \forall \lambda > 0, \ x, y \in R(I + \lambda A). \tag{2.72}$$

If A is ω-accretive, then it follows by (2.71) that the operator $(I + \lambda A)^{-1}$ is single-valued and Lipschitzian with Lipschitz constant not greater than $\frac{1}{1-\lambda\omega}$ on $R(I+\lambda A)$, $0 < \lambda < \frac{1}{\omega}$.

It should be mentioned, however, that condition (2.72) is not sufficient for the accretivity of the operator. In fact, we have

Proposition 2.32 *The operator A is accretive if and only if, for each $\lambda > 0$ and $y \in \mathbb{R}(I + \lambda A)$, the equation $(I + \lambda A)x \ni y$ has at most one solution and* (2.72) *holds.*

Let us define for $\lambda > 0$ the operators J_λ, $A_\lambda : X \to X$,

$$J_\lambda x = (I + \lambda A)^{-1}x, \quad x \in R(I + \lambda A); \tag{2.73}$$

$$A_\lambda x = \lambda^{-1}(x - J_\lambda x), \quad x \in R(I + \lambda A). \tag{2.74}$$

As in the case of maximal monotone operators in $X \times X^*$, the operator A_λ is called the *Yosida approximation of A*.

In Proposition 2.33 below, we collect some elementary properties of J_λ and A_λ.

Proposition 2.33 *Let A be ω-accretive in $X \times X$. Then:*

(a) $\|J_\lambda x - J_\lambda y\| \leq (1 - \lambda\omega)^{-1}\|x - y\|$, $\forall \lambda \in \left(0, \frac{1}{\omega}\right)$, $\forall x, y \in R(I + \lambda A)$.
(b) *A_λ is ω-accretive and Lipschitz continuous with Lipschitz constant not greater than $\frac{2}{1-\lambda\omega}$ in $R(I + \lambda A)$, $0 < \lambda < \frac{1}{\omega}$.*
(c) $A_\lambda x \in AJ_\lambda x$, $\forall x \in R(I + \lambda A)$, $0 < \lambda < \frac{1}{\omega}$.
(d) $(1 - \lambda\omega)\|A_\lambda x\| \leq |Ax| = \inf\{\|y\|;\ y \in Ax\}$, $\forall x \in D(A) \cap R(I + \lambda A)$.
(e) $\lim_{\lambda \to 0} J_\lambda x = x$, $\forall x \in \overline{D(A)} \cap_{0<\lambda<\frac{1}{\omega}} R(I + \lambda A)$.
(f) *If $\lambda \geq \mu > 0$, then* $(1-\lambda\omega)\|A_\lambda x\| \leq (1-\mu\omega)\|A_\mu x\|$, $\forall x \in R(I+\lambda A) \cap R(I+\mu A)$.
(g) *If $x \in \mathcal{D} = \bigcup_{\nu>0} \bigcap_{0<\lambda<\nu} R(I + \lambda A)$, then $\lim_{\lambda \downarrow 0} \|A_\lambda x\| = L(x)$ and $L(x) \leq |Ax|$, $\forall x \in D(A)$ if $D(A) \subset \mathcal{D}$.*

Proof. (a) and (b) are immediate consequences of inequality (2.71).

(c) Let $x \in R(I+\lambda A)$. Then, $A_\lambda x \in \lambda^{-1}((I+\lambda A)J_\lambda x - J_\lambda x) \in AJ_\lambda x$.

(d) For $x \in D(A) \cap R(I+\lambda A)$, we have $A_\lambda x = \lambda^{-1}(J_\lambda(I+\lambda A)x - J_\lambda x)$ and, therefore, $\|A_\lambda x\| \le |Ax|(1-\lambda\omega)^{-1}$, $\forall x \in D(A)$.

(e) For every $x \in D(A) \cap R(I+\lambda A)$, we have

$$\|J_\lambda x - x\| = \lambda\|A_\lambda x\| \le \frac{\lambda}{1-\lambda\omega}\,|Ax|, \quad \forall \lambda \in \left(0, \frac{1}{\omega}\right).$$

Hence, $\lim_{\lambda\to 0} J_\lambda x = x$. Clearly, this extends to all of $\overline{D(A)} \bigcap_{0<\lambda<\frac{1}{\omega}} R(I+\lambda A)$, as claimed.

As regards (f), it follows by a simple computation from the resolvent formula (see (2.79) below)

$$J_\lambda x = J_\mu\left(\frac{\mu}{\lambda}x + \left(1-\frac{\mu}{\lambda}\right)J_\lambda x\right), \ \forall x \in R(I+\lambda A) \cap R(I+\mu A). \tag{2.75}$$

Finally, (g) follows by (f), but we omit the details. □

In the following we confine ourselves to the study of accretive subsets, the extensions to the quasi-accretive sets being immediate.

Proposition 2.34 *An accretive set $A \subset X \times X$ is m-accretive if and only if $R(I+\lambda A) = X$ for all (equivalently, for some $\lambda > 0$.)*

Proof. Let A be m-accretive and let $y \in X$, $\lambda > 0$, be arbitrary but fixed. Then, the equation

$$x + \lambda Ax \ni y \tag{2.76}$$

may be written as

$$x = J_1\left(\frac{y}{\lambda} + \left(1-\frac{1}{\lambda}\right)x\right).$$

Then, by the contraction principle, we infer that equation (2.76) has a unique solution $x = (I+\lambda A)^{-1}y$ for $\frac{1}{2} < \lambda < +\infty$.

Now, fix $\lambda_0 > \frac{1}{2}$ and write equation (2.76) as

$$x = (I+\lambda_0 A)^{-1}\left(\left(1-\frac{\lambda_0}{\lambda}\right)x + \frac{\lambda_0}{\lambda}\,y\right). \tag{2.77}$$

Because $J_{\lambda_0} = (I+\lambda_0 A)^{-1}$ is nonexpansive, this equation has a solution for $\lambda \in \left(\lambda_0 - \frac{1}{2}, \lambda_0\right)$. Repeating this argument, we conclude that $R(I+\lambda A) = X$ for all $\lambda > 0$. Assume now that $R(I+\lambda_0 A) = X$ for some $\lambda_0 > 0$. Then, if we set equation (2.76) into the form (2.77), we conclude as before that $R(I+\lambda A) = X$ for all $\lambda \in \left(\frac{\lambda_0}{2}, \infty\right)$ and so, iterating the argument we see that $R(I+\lambda A) = X$ for all $\lambda > 0$, as claimed. □

Combining Propositions 2.33 and 2.34, we conclude that $A \subset X \times X$ is m-accretive if and only if for all $\lambda > 0$ the operator $(I+\lambda A)^{-1}$ is single valued and nonexpansive on all of X.

Similarly, A is ω-m-accretive if and only if, for all $0 < \lambda < \frac{1}{\omega}$, or equivalently for some $0 < \lambda < \frac{1}{\omega}$,

$$\|(I + \lambda A)^{-1}x - (I + \lambda A)^{-1}y\| \leq \frac{1}{1 - \lambda\omega} \|x - y\|, \quad \forall x, y \in X. \tag{2.78}$$

By Theorem 2.4, if $X = H$ is a Hilbert space, then A is m-accretive if and only if it is maximal accretive.

In particular, by Proposition 2.34 we get the following resolvent relation for the operator $J_\lambda = (I + \lambda A)^{-1}$

$$J_{\lambda_2}x = J_{\lambda_1}\left(\frac{\lambda_1}{\lambda_2}x + \left(1 - \frac{\lambda_1}{\lambda_2}\right)J_{\lambda_2}x\right), \quad \forall \lambda_1, \lambda_2 > 0. \tag{2.79}$$

A subset $A \subset X \times X$ is said to be *closed* if $x_n \to x$, $y_n \to y$, and $[x_n, y_n] \in A$ for all $n \in \mathbb{N}$ imply that $[x, y] \in A$. Taking into account that inequality (2.78) is invariant to the strong convergence, we have

Proposition 2.35 *Let A be an m-accretive set of $X \times X$. Then A is closed and if $\lambda_n \in \mathbb{R}$, $x_n \in X$ are such that $\lambda_n \to 0$ and*

$$x_n \to x, \; A_{\lambda_n}x_n \to y \quad \text{for } n \to \infty, \tag{2.80}$$

then $[x, y] \in A$. If X^* is uniformly convex, then A is demiclosed, and if

$$x_n \to x, \; A_{\lambda_n}x \rightharpoonup y \quad \text{for } n \to \infty, \tag{2.81}$$

then $[x, y] \in A$.

If X^* is uniformly convex, then it follows that, if A is accretive, then for every $x \in D(A)$ we have the following algebraic description of Ax

$$Ax = \{y \in X; \; (y - v, J(x - u)) \geq 0, \; \forall [u, v] \in A\},$$

which follows by Definition 2.29 taking into account that, in this case, J is single-valued and demicontinuous. In particular, it follows that Ax is a closed convex subset of X. Denote by A^0x the element of minimum norm on Ax (i.e., the projection of the origin into Ax). Because the space X is reflexive, by Proposition 1.6 it follows that $A^0x \neq \emptyset$ for every $x \in D(A)$. The set $A^0 \subset A$ is called the *minimal section* of A. If the space X is strictly convex, then, as easily seen, A^0 is single-valued.

Proposition 2.36 *Let X and X^* be uniformly convex and let A be an m-accretive set of $X \times X$. Then:*

(i) *$A_\lambda x \to A^0x$, $\forall x \in D(A)$ for $\lambda \to 0$.*
(ii) *$\overline{D(A)}$ is a convex set of X.*

Proof. (i) Let $x \in D(A)$. As seen in Proposition 2.33, $\|A_\lambda x\| \le |Ax| = \|A^0 x\|$, $\forall \lambda > 0$. Now, let $\lambda_n \to 0$ be such that $A_{\lambda_n} x \rightharpoonup y$. By Proposition 2.31, we know that $y \in Ax$, and thus

$$\lim_{n\to\infty} \|A_{\lambda_n} x\| = \|y\| = \|A^0 x\|.$$

The space X is uniformly convex; therefore this implies that $A_{\lambda_n} x \to y = A^0 x$. Hence, $A_\lambda x \to A^0 x$ for $\lambda \to 0$.

(ii) Let $x_1, x_2 \in D(A)$, and $0 \le \alpha \le 1$. We set $x_\alpha = \alpha x_1 + (1-\alpha) x_2$. Then, as it is readily verified,

$$\|J_\lambda(x_\alpha) - x_1\| \le \|x_\alpha - x_1\| + \lambda |Ax_1|, \qquad \forall \lambda > 0,$$

$$\|J_\lambda(x_\alpha) - x_2\| \le \|x_\alpha - x_2\| + \lambda |Ax_2|, \qquad \forall \lambda > 0,$$

and, because the space X is uniformly convex, these estimates imply, by a standard geometrical device we omit here, that

$$\|J_\lambda(x_\alpha) - x_\alpha\| \le \delta(\lambda), \quad \forall \lambda > 0,$$

where $\lim_{\lambda\to 0} \delta(\lambda) = 0$. Hence, $x_\alpha \in \overline{D(A)}$. □

Regarding the single-valued linear m-accretive (equivalently, m-dissipative) operators, it is useful to note the following density result.

Proposition 2.37 *Let X be a reflexive Banach space. Then any m-accretive linear operator $A : X \to X$ is densely defined (i.e., $\overline{D(A)} = X$).*

Proof. Let $y \in X$ be arbitrary but fixed. For every $\lambda > 0$, the equation $x_\lambda + \lambda A x_\lambda = y$ has a unique solution $x_\lambda \in D(A)$. We know that $\|x_\lambda\| \le \|y\|$ for all $\lambda > 0$ and so, on a subsequence $\lambda_n \to 0$,

$$x_{\lambda_n} \rightharpoonup x, \ \lambda_n A x_{\lambda_n} \rightharpoonup y - x \quad \text{in } X.$$

Because A is closed, its graph in $X \times X$ is weakly closed (it is a linear subspace of $X \times X$) and so $\lambda_n x_{\lambda_n} \to 0$, $A(\lambda_n x_{\lambda_n}) \rightharpoonup y - x$ imply that $y - x = 0$. Hence,

$$(1 + \lambda_n A)^{-1} y \rightharpoonup y.$$

We have, therefore, proven that $y \in \overline{D(A)}$ (recall that the weak closure of $D(A)$ coincides with the strong closure). □

Proposition 2.37 is not true for a general Banach space X. Such an example is (see A. Pazy [92]), $X = C([0,1])$ and $Ay = y'$ with $D(A) = \{y \in C^1[0,1];\ y(0) = 0\}$.

We conclude this section by introducing another convenient way to define the accretiveness. Toward this aim, denote by $[\cdot,\cdot]_s$ the directional derivative of the function $x \to \|x\|$, that is,

$$[x,y]_s = \lim_{\lambda \downarrow 0} \frac{\|x + \lambda y\| - \|x\|}{\lambda}, \quad x, y \in X. \tag{2.82}$$

The function $\lambda \to \|x + \lambda y\|$ is convex, thus we may define, equivalently, $[\cdot,\cdot]_s$ as

$$[x,y]_s = \inf_{\lambda > 0} \frac{\|x + \lambda y\| - \|x\|}{\lambda}, \qquad \forall x, y \in X. \tag{2.83}$$

Roughly speaking, $[\cdot,\cdot]_s$ can be viewed as a "scalar product" on $X \times X$.

Let us now briefly list some properties of the bracket $[\cdot,\cdot]_s$.

Proposition 2.38 *Let X be a Banach space. We have the following.*

(i) $[\cdot,\cdot]_s : X \times X \to \mathbb{R}$ *is upper semicontinuous.*
(ii) $[\alpha x, \beta y]_s = \beta [x,y]_s$, *for all* $\beta \geq 0$, $\alpha \in \mathbb{R}$, $x, y \in X$.
(iii) $[x, \alpha x + y]_s = \alpha \|x\| + [x,y]_s$ *if* $\alpha \in \mathbb{R}^+$, $x \in X$.
(iv) $|[x,y]_s| \leq \|y\|$, $[x, y+z]_s \leq [x,y]_s + [x,z]_s$, $\forall x, y, z \in X$.
(v) $[x,y]_s = \max\{(y, x^*);\ x^* \in \Phi(x)\}$, $\forall x, y \in X$, *where*

$$\Phi(x) = \{x^* \in X^*;\ (x, x^*) = \|x\|,\ \|x^*\| = 1\},\ \text{if}\ x \neq 0,$$
$$\Phi(0) = \{x^* \in X^*;\ \|x^*\| \leq 1\}.$$

Proof. (i) Let $x_n \to x$ and $y_n \to y$ as $n \to \infty$. For every n there exist $h_n \in X$ and $\lambda_n \in (0,1)$ such that $\|h_n\| + \lambda_n \leq \frac{1}{n}$ and

$$[x_n, y_n]_s \leq (\|x_n + h_n + \lambda_n y_n\| - \|x_n + y_n\|)\lambda_n^{-1} + \frac{1}{n}.$$

This yields

$$\limsup_{n\to\infty} [x_n, y_n]_s \leq [x,y]_s,$$

as claimed.

Note that (ii)–(iv) are immediate consequences of the definition. To prove (v), we note first that

$$\Phi(x) = \partial(\|x\|),\quad \forall x \in X,$$

and apply formula (1.12). □

Now, coming back to the definition of accretiveness, we see that, by virtue of part (v) of Proposition 2.38, condition (2.69) can be equivalently written as

$$[x_1 - x_2, y_1 - y_2]_s \geq 0,\quad \forall [x_i, y_i] \in A,\ i = 1,2. \tag{2.84}$$

Similarly, condition (2.71) is equivalent to

$$[x_1 - x_2, y_1 - y_2]_s \geq -\omega \|x_1 - x_2\|,\quad \forall [x_i, y_i] \in A,\ i = 1,2. \tag{2.85}$$

Summarizing, we may see that a subset A of $X \times X$ is ω-accretive if one of the following equivalent conditions holds.

(i) If $[x_1, y_1], [x_2, y_2] \in A$, then there is $w \in J(x_1 - x_2)$ such that $(y_1 - y_2, w) \geq -\omega \|x_1 - x_2\|$.
(ii) $\|x_1 - x_2 + \lambda(y_1 - y_2)\| \geq (1 - \lambda\omega)\|x_1 - x_2\|$ for $0 < \lambda < \frac{1}{\omega}$ and all $[x_i, y_i] \in A$, $i = 1,2$.
(iii) $[x_1 - x_2, y_1 - y_2]_s \geq -\omega \|x_1 - x_2\|$, $\forall [x_i, y_i] \in A$, $i = 1,2$.

In applications, however, it is more convenient to use condition (i) to verify the ω-accretiveness.

As seen earlier, if A is m-accretive in $X \times X$, then the operator $(I + A)^{-1}$ is nonexpansive from X to itself (that is, Lipschitz with Lipschitz constant 1) and this property characterizes the m-accretive operators. Conversely, we have

Proposition 2.39 *Let $T : X \to X$ be a nonexpansive operator. Then there is a unique m-accretive operator $A \subset X \times X$ such that $(I + A)^{-1} = T$. Moreover, $D(A) = T(X) = \{Tu;\ u \in X\}$.*

Proof. We define the operator $A \subset X \times X$ by

$$Ay = -y + u;\ y = Tu,\ u \in X.$$

Clearly, $(I + A)^{-1} = T$ and $D(A) = T(X)$, as claimed. □

Definition 2.40 The operator $A_0 : D(A_0) \subset X \to X$ is said to be pseudo-m-accretive if $R(I + \lambda A_0) = X$ for each $\lambda > 0$ and there is $J_\lambda : X \to X$ such that

$$\|J_\lambda(x) - J_\lambda(y)\| \le \|x - y\|,\ \forall x, y \in X, \tag{2.86}$$

$$J_\lambda(x) \in (I + \lambda A_0)^{-1}x,\ \forall x \in X, \tag{2.87}$$

$$J_{\lambda_2}(x) = J_{\lambda_1}\left(\frac{\lambda_1}{\lambda_2}\,x + \left(1 - \frac{\lambda_1}{\lambda_2}\right) J_{\lambda_2}(x)\right),\ \forall \lambda_1, \lambda_2 > 0,\ x \in X. \tag{2.88}$$

It should be emphasized that an pseudo-m-accretive operator A_0 is not m-accretive but this happens, however, if it is also accretive, that is, $(I+\lambda A_0)^{-1}$ is single-valued (see Proposition 2.32).

However, it turns out that a pseudo-m-accretive operator has a section $A \subset A_0$ which is m-accretive. Namely,

Proposition 2.41 *Let A_0 be pseudo-m-accretive and let $\{J_\lambda\}_{\lambda>0}$ a family of nonexpansive operators on X satisfying* (2.87)–(2.88). *Then the operator $A : D(A) \subset X \to X$ defined by*

$$Au = A_0 u,\ \forall u \in J_{\lambda_0}(X),\quad D(A) = J_{\lambda_0}(X), \tag{2.89}$$

for some $\lambda_0 > 0$, is m-accretive and

$$(I + \lambda A)^{-1} = J_\lambda,\ \forall \lambda > 0. \tag{2.90}$$

Moreover, A is independent of λ.

Proof. For each $f \in X$ we have for $u = J_{\lambda_0}(f)$ that $u + \lambda_0 A_0 u \ni f$ and so, by (2.89) it follows that $u + \lambda_0 Au = f$ and, therefore,

$$R(I + \lambda_0 A) = X,\quad J_{\lambda_0}(f) = (I + \lambda_0 A)^{-1} f.$$

On the other hand, for $u_i = J_{\lambda_0}(v_i)$, $i = 1, 2$, we have

$$Au_i = A_0 u_i,\quad u_i + \lambda_0 A_0 u_i \ni v_i,\ i = 1, 2,$$

and this yields

$$\frac{1}{\lambda_0}(v_i - u_i) = w_i \in Au_i,\quad i = 1, 2.$$

Then, by (2.86) we have

$$\|u_1 - u_2\| \le \|v_1 - v_2\| \le \|u_1 - u_2 + \lambda_0(w_1 - w_2)\|$$

and so A is accretive. Since $R(I+\lambda A_0) = R(I+\lambda_0 A) = X$, we infer by Proposition 2.34 that A is m-accretive.

Let us show now that A constructed in (2.89) is independent of λ and that (2.90) holds. To this end, we fix $f \in X$ and take $u_\lambda = (I+\lambda A)^{-1}f$, which exists for all $\lambda > 0$ because A is m-accretive. Moreover, by (2.79) we have, for all $\lambda > 0$,

$$(I+\lambda A)^{-1}f = (I+\lambda_0 A)^{-1}\left(\frac{\lambda_0}{\lambda}f + \left(1-\frac{\lambda_0}{\lambda}\right)(I+\lambda A)^{-1}f\right) = J_{\lambda_0}(g), \quad (2.91)$$

where

$$g = \frac{\lambda_0}{\lambda}f + \left(1-\frac{\lambda_0}{\lambda}\right)(I+\lambda A)^{-1}f.$$

On the other hand, by (2.88) we have

$$J_{\lambda_0}(g) = J_\lambda\left(\frac{\lambda}{\lambda_0}g + \left(1-\frac{\lambda}{\lambda_0}\right)J_{\lambda_0}(g)\right)$$

and so, combined the latter with (2.91), we get

$$(I+\lambda A)^{-1}f = J_\lambda\left(f + \left(\frac{\lambda}{\lambda_0}-1\right)(I+\lambda A)^{-1}f - \left(\frac{\lambda}{\lambda_0}-1\right)J_{\lambda_0}(g)\right),$$

which implies the desired equation (2.90). □

Very often, a pseudo-m-accretive operator A_0 arises as a limit of a family $(A_0)_\varepsilon$ of m-accretive operators $(A_0)_\varepsilon : D((A_0)_\varepsilon) \subset X \to X$. Then, for each f and $\lambda > 0$, the equation

$$u_\varepsilon + \lambda(A_0)_\varepsilon(u_\varepsilon) = f$$

has a unique solution u_ε and the mapping $f \to u_\varepsilon$ is nonexpansive. If, for $\varepsilon \to 0$, $u_\varepsilon \to u = u(\lambda, f)$ in the strong or weak topology of X, and $(A_0)_\varepsilon(u_\varepsilon) \to \eta \in A_0(u)$, then $u + \lambda A_0(u) \ni f$ and $u(\lambda, f) = J_\lambda(f)$ satisfies conditions (2.87)–(2.88).

A simple example, which will be treated in a more general setting in Section 3.4, is the operator $A_0 : D(A_0) \subset L^1(\mathbb{R}) \to L^1(\mathbb{R})$,

$$A_0(u) = -(\beta(u))'' + (b(u)u)', \ D(A_0) = \{u \in L^1(\mathbb{R}); -(\beta(u))'' + (b(u)u)' \in L^1(\mathbb{R})\},$$

where $\beta : \mathbb{R} \to \mathbb{R}$ is continuous and monotonically nondecreasing, while $b : \mathbb{R} \to \mathbb{R}$ is continuous and bounded. Then, for each $\lambda > 0$ and $f \in L^1(\mathbb{R})$, the equation $u + \lambda A_0 u = f$ has a solution $u(\lambda, f)$ (in general, not unique) given by $u(\lambda, f) = \lim_{\varepsilon\to 0} u_\varepsilon$ in $L^1(\mathbb{R})$, where u_ε is the solution to the equation

$$u_\varepsilon - \lambda(\varepsilon u_\varepsilon'' + (\beta(u_\varepsilon))'') + \lambda(b_\varepsilon(u_\varepsilon)u_\varepsilon)' = f,$$

where b_ε is a smooth approximation of b. It turns out under suitable conditions on b that the mapping $f \to u = u(\lambda, f)$ is Lipschitzian and $\|u(\lambda, f_1) - u(\lambda, f_2)\|_{L^1(\mathbb{R})} \leq \|f_1 - f_2\|_{L^1(\mathbb{R})}$. This implies that A_0 is pseudo-m-accretive but not m-accretive.

We know from Section 2.1 that, if X is a Hilbert space, then a continuous accretive operator is m-accretive. More generally, we have the following result established in [12]. (See also [14].)

Theorem 2.42 *Let X be a real Banach space, A be an m-accretive set of $X \times X$, and let $B : X \to X$ be a continuous, m-accretive operator with $D(B) = X$. Then $A + B$ is m-accretive.*

This result (which can be compared with Corollary 2.14) is, in particular, useful to treat continuous nonlinear accretive perturbations of equations involving m-accretive operators.

Other m-accretive criteria for the sum $A+B$ of two m-accretive operators $A, B \in X \times X$ can be obtained as in Proposition 2.43 below by approximating the equation $x + Ax + Bx \ni y$ by

$$x + Ax + B_\lambda x \ni y,$$

where B_λ is the Yosida approximation of B.

Proposition 2.43 *Let X be a Banach space with uniformly convex dual X^* and let A and B be two m-accretive sets in $X \times X$ such that $D(A) \cap D(B) \neq \emptyset$ and*

$$(Au, J(B_\lambda u)) \geq -C(\|u\| + \|B_\lambda(u)\|), \quad \forall \lambda > 0, \ u \in D(A), \tag{2.92}$$

where C is independent of λ. Then $A + B$ is m-accretive.

Proof. Since this result will be frequently used in the following, we shall give the proof. Let $f \in X$ and $\lambda > 0$. We approximate the equation

$$u + Au + Bu \ni f \tag{2.93}$$

by

$$u + Au + B_\lambda u \ni f, \quad \lambda > 0, \tag{2.94}$$

where B_λ is the Yosida approximation B, that is, $B_\lambda = \lambda^{-1}(I - (I + \lambda B)^{-1})$. We may write equation (2.94) as

$$u = \left(1 + \frac{\lambda}{1+\lambda} A\right)^{-1} \left(\frac{\lambda f}{1+\lambda} + \frac{(I+\lambda B)^{-1} u}{1+\lambda}\right),$$

which, by the Banach fixed point theorem, has a unique solution $u_\lambda \in D(A)$ (because $(I+\lambda B)^{-1}$ and $(I+\lambda A)^{-1}$ are nonexpansive). Now, we multiply the equation

$$u_\lambda + Au_\lambda + B_\lambda u_\lambda \ni f \tag{2.95}$$

by $J(B_\lambda u_\lambda)$ to get by (2.92) that

$$\|B_\lambda u_\lambda\| \leq C(\|f\| + \|u_\lambda\|), \quad \forall \lambda > 0.$$

(In the following, we shall denote by C several positive constants independent of λ.) On the other hand, multiplying (2.92) by $J(u_\lambda - u_0)$, where $u_0 \in D(A) \cap D(B)$, we get

$$\|u_\lambda - u_0\| \leq \|u_0\| + \|f\| + \|\xi_0\| + \|B_\lambda u_0\| \leq \|u_0\| + \|f\| + \|\xi_0\| + |Bu_0|, \quad \forall \lambda > 0,$$

where $\xi_0 \in Au_0$. Hence, $\|u_\lambda\| + \|B_\lambda u_\lambda\| \leq C$, $\forall \lambda > 0$.

Now, we multiply the equation (in the sense of the duality between X and X^*)

$$u_\lambda - u_\mu + Au_\lambda - Au_\mu + B_\lambda u_\lambda - B_\mu u_\mu \ni 0$$

by $J(u_\lambda - u_\mu)$. Because A is accretive, we have

$$\|u_\lambda - u_\mu\|^2 + (B_\lambda u_\lambda - B_\mu u_\mu, J(u_\lambda - u_\mu)) \le 0, \quad \forall \lambda, \mu > 0,$$

while

$$\begin{aligned}(B_\lambda u_\lambda &- B_\mu u_\mu, J(u_\lambda - u_\mu)) \\ &\ge \left(B_\lambda u_\lambda - B_\mu u_\mu, J(u_\lambda - u_\mu) - J\left((I+\lambda B)^{-1}u_\lambda - (I+\mu B)^{-1}u_\mu\right)\right)\end{aligned}$$

because B is accretive and $B_\lambda u \in B((I+\lambda B)^{-1}u)$. Because J is uniformly continuous on bounded subsets (Theorem 1.2) and $\{\|B_\lambda u_\lambda\|\}$ is bounded, we have

$$\|u_\lambda - (I+\lambda B)^{-1}u_\lambda\| + \|u_\mu - (I+\mu B)^{-1}u_\mu\| \le C(\lambda + \mu),$$

this implies that $\{u_\lambda\}$ is a Cauchy sequence and so $u = \lim_{\lambda \to 0} u_\lambda$ exists. Extracting further subsequences, we may assume that

$$B_\lambda u_\lambda \rightharpoonup y, \quad f - B_\lambda u_\lambda - u_\lambda \rightharpoonup z, \quad \text{as } \lambda \to 0.$$

Then, by Proposition 2.35, we see that $y \in Bu$, $z \in Au$, and so u is a solution (obviously unique) to equation (2.93). □

The accretivity property of an operator A defined in a Banach space X should not be mixed up with that of monotonicity. The first is defined for operators A from X to itself and it is a metric geometric property, whereas the second is defined for operators A from X to dual space X^* and is a variational property. As mentioned earlier, these two notions coincide if X is a Hilbert space identified with its own dual.

As regards the field of applications, the maximal monotonicity is used to treat variational solutions to partial differential equations while the m-accretivity is a property specific to generators of nonlinear semigroups of contractions which will be treated in the next section.

2.3 Existence Theory for the Cauchy Problem

Let X be a real Banach space with the norm $\|\cdot\|$ and dual X^* and let $A \subset X \times X$ be a quasi-accretive set of $X \times X$. As mentioned earlier, this means that A is ω-accretive, that is, that $A + \omega I$ is accretive for some $\omega \in \mathbb{R}$. Similarly, we say that A is quasi-m-accretive if A is ω-m-accretive, i.e. $A + \omega I$ is m-accretive for some $\omega \in \mathbb{R}$.

Consider the Cauchy problem

$$\begin{cases} \dfrac{dy}{dt}(t) + Ay(t) \ni f(t), \quad t \in [0,T], \\ y(0) = y_0, \end{cases} \tag{2.96}$$

where $y_0 \in X$ and $f \in L^1(0,T;X)$.

If A is Lipschitzian, then, as is well known the initial value, problem (2.96) has a unique absolutely continuous solution $y : [0,T] \to X$ which satisfies the equation almost everywhere on $(0,T)$. However, if A is only continuous, such a result fails in an infinite dimensional Banach space X and, as a matter of facts, this is a characteristic property of finite dimensional Banach spaces X.

The main result of this chapter is that, if A is quasi-m-accretive and $u_0 \in \overline{D(A)}$, then problem (2.96) has a unique generalized continuous solution $y : [0,T] \to X$ obtained as limit of the finite difference scheme associated to (2.96). If X is reflexive, then this generalized solution satisfies a.e. equation (2.96), that is, it is a *strong solution.* Let us first recall the definition of strong solutions.

Definition 2.44 A strong solution to (2.96) is a function

$$y \in W^{1,1}((0,T];X) \cap C([0,T];X)$$

such that

$$f(t) - \frac{dy}{dt}(t) \in Ay(t), \quad \text{a.e. } t \in (0,T),\ y(0) = y_0.$$

Here, $W^{1,1}((0,T];X) = \{y \in L^1(0,T;X);\ y' \in L^1(\delta,T;X),\ \forall \delta \in (0,T)\}$.

Proposition 2.45 *Let A be ω-accretive, $f_i \in L^1(0,T;X)$, $y_0^i \in \overline{D(A)}$, $i = 1,2$, and let $y_i \in W^{1,1}((0,T];X)$, $i = 1,2$, be corresponding strong solutions to problem* (2.96)*. Then,*

$$\begin{aligned}\|y_1(t)-y_2(t)\| &\le e^{\omega t}\|y_0^1-y_0^2\|+\int_0^t e^{\omega(t-s)}[y_1(s)-y_2(s), f_1(s)-f_2(s)]_s ds\\ &\le e^{\omega t}\|y_0^1-y_0^2\|+\int_0^t e^{\omega(t-s)}\|f_1(s)-f_2(s)\|ds,\ \forall t \in [0,T].\end{aligned} \tag{2.97}$$

Here (see Proposition 2.38)

$$[x,y]_s = \inf_{\lambda>0} \lambda^{-1}(\|x+\lambda y\| - \|x\|) = \max\{(y,x^*);\ x^* \in \Phi(x)\}, \tag{2.98}$$

$\|x\|\Phi(x) = J(x)$ is the duality mapping of X, that is, $\Phi(x) = \partial\|x\|$ and $\overline{D(A)}$ is the closure of the domain $D(A)$ in X.

The main ingredient of the proof is the following chain differentiation rule lemma.

Lemma 2.46 *Let $y = y(t)$ be an X-valued function on $[0,T]$. Assume that $y(t)$ and $\|y(t)\|$ are differentiable at $t = s$. Then,*

$$\|y(s)\| \frac{d}{ds}\|y(s)\| = \left(\frac{dy}{ds}(s), w\right), \quad \forall w \in J(y(s)). \tag{2.99}$$

Here, $J : X \to X^*$ is the duality mapping of X.

Proof. Let $\varepsilon > 0$. We have

$$(y(s+\varepsilon) - y(s), w) \le (\|y(s+\varepsilon)\| - \|y(s)\|)\|w\|, \quad \forall w \in J(y(s)),$$

and this yields

$$\left(\frac{dy}{ds}(s), w\right) \leq \frac{d}{ds}\|y(s)\|\,\|y(s)\|.$$

Similarly, from the inequality

$$(y(s-\varepsilon) - y(s), w) \leq (\|y(s-\varepsilon)\| - \|y(s)\|)\|w\|,$$

we get

$$\left(\frac{d}{ds}y(s), w\right) \geq \frac{d}{ds}\|y(s)\|\,\|y(s)\|,$$

as claimed. □

In particular, it follows by (2.99) that

$$\frac{d}{ds}\|y(s)\| = \left[y(s), \frac{dy}{ds}(s)\right]_s. \tag{2.100}$$

Proof of Proposition 2.45. We have

$$\frac{d}{ds}(y_1(s) - y_2(s)) + Ay_1(s) - Ay_2(s) \ni f_1(s) - f_2(s), \quad \text{a.e. } s \in (0,T). \tag{2.101}$$

On the other hand, because A is ω-accretive, we have (see (2.85))

$$[y_1(s) - y_2(s), Ay_1(s) - Ay_2(s)]_s \geq -\omega\|y_1(s) - y_2(s)\|$$

and so, by (2.100) and (2.101), we see that

$$\frac{d}{ds}\|y_1(s) - y_2(s)\| \leq [y_1(s) - y_2(s), f_1(s) - f_2(s)]_s + \omega\|y_1(s) - y_2(s)\|, \quad \text{a.e. } s \in (0,T).$$

Then, integrating on $[0,t]$, we get (2.97), as claimed. □

The above proposition suggests that, as far as the uniqueness and continuous dependence of solution of data are concerned, the class of quasi-accretive operators A offers a suitable framework for the Cauchy problem. However, for the existence we must extend the notion of the solution for the Cauchy problem (2.96) from differentiable to continuous functions.

Definition 2.47 Let $f \in L^1(0,T;X)$ and $\varepsilon > 0$ be given. An *ε-discretization* on $[0,T]$ of the equation $y' + Ay \ni f$ consists of a partition $0 = t_0 \leq t_1 \leq t_2 \leq \cdots \leq t_N$ of the interval $[0,t_N]$ and a finite sequence $\{f_i\}_{i=1}^N \subset X$ such that

$$t_i - t_{i-1} < \varepsilon \quad \text{for } i = 1,\dots,N,\ T - \varepsilon < t_N \leq T, \tag{2.102}$$

$$\sum_{i=1}^N \int_{t_{i-1}}^{t_i} \|f(s) - f_i\|ds < \varepsilon. \tag{2.103}$$

In the following, we denote by $D_A^\varepsilon(0 = t_0, t_1, ..., t_N; f_1, ..., f_N)$ such an ε-discretization.

By $D_A^\varepsilon(0 = t_0, t_1, ..., t_N; f_1, ..., f_N)$ *solution* to (2.96) we call a piecewise constant function $z : [0, t_N] \to X$ whose constant values z_i on $(t_{i-1}, t_i]$ satisfy the finite difference equation

$$\frac{z_i - z_{i-1}}{t_i - t_{i-1}} + Az_i \ni f_i, \quad i = 1, ..., N. \tag{2.104}$$

Such a function $z = \{z_i\}_{i=1}^N$ is called an *ε-approximate solution* to the Cauchy problem (2.96) if it further satisfies

$$\|z(0) - y_0\| \le \varepsilon. \tag{2.105}$$

Definition 2.48 A *mild solution* to the Cauchy problem (2.96) is a function $y \in C([0, T]; X)$ with the property that for each $\varepsilon > 0$ there is an ε-approximate solution z of by (2.96) on $[0, T]$ such that $\|y(t) - z(t)\| \le \varepsilon$ for all $t \in [0, T]$ and $y(0) = y_0$.

Let us note that every strong solution $y \in C([0, T]; X) \cap W^{1,1}((0, T]; X)$ to (2.96) is a mild solution. Indeed, for any $\varepsilon > 0$ let $0 = t_0 \le t_1 \le \cdots \le t_N$ be an ε-discretization of $[0, T]$ such that y is differentiable in all t_i and

$$\frac{dy}{dt}(t_i) + Ay(t_i) \ni f(t_i), \quad i = 1, ..., N,$$
$$\int_{t_{i-1}}^{t_i} \|f(t) - f(t_i)\| dt \le \varepsilon(t_i - t_{i-1}), \quad i = 1, ..., N,$$
$$\|y(t) - y(s)\| \le \varepsilon, \quad \forall t, s \in [t_{i-1}, t_i], \ i = 1, ..., N.$$

We note that, since $y : [0, T] \to X$ is a.e. differentiable on $(0, T)$ and continuous on $[0, T]$, while $f \in L^1(0, T; X)$, for each ε there is such a set $\{t_i\}_{i=1}^N$. Moreover, we have $\|y(t) - y(s)\| \le \varepsilon$ for all $t, s \in [0, T]$. Then, the step function $z : [0, T] \to X$ defined by $z = y(t_i)$ on $(t_{i-1}, t_i]$ is a solution to the ε-discretization D_A^ε $(0 = t_0, t_1, ..., t_n; f_1, ..., f_n)$, and we have that $\|y(t) - z(t)\| \le \varepsilon$ for all $t \in [0, T]$, as claimed.

Theorem 2.49, essentially due to M.G. Crandall, is the main result of the existence theory for problem (2.96).

Theorem 2.49 *Let A be ω-accretive, $y_0 \in \overline{D(A)}$, and $f \in L^1(0, T; X)$. For each $\varepsilon > 0$, let problem* (2.96) *have an ε-approximate solution. Then, the Cauchy problem* (2.96) *has a unique mild solution y. Moreover, there is a continuous function $\delta = \delta(\varepsilon)$ such that $\delta(0) = 0$ and if z is an ε-approximate solution of* (2.96), *then*

$$\|y(t) - z(t)\| \le \delta(\varepsilon) \quad \text{for } \ t \in [0, T - \varepsilon]. \tag{2.106}$$

Let $f, g \in L^1(0, T; X)$ and $y, \bar{y}$ be mild solutions to (2.96) *corresponding to f and g, respectively. Then,*

$$\begin{aligned} \|y(t) - \bar{y}(t)\| &\le e^{\omega(t-s)} \|y(s) - \bar{y}(s)\| \\ &+ \int_s^t e^{\omega(t-\tau)} [y(\tau) - \bar{y}(\tau), f(\tau) - g(\tau)]_s d\tau \quad \text{for } 0 \le s < t \le T. \end{aligned} \tag{2.107}$$

Here and everywhere in the following, $\overline{D(A)}$ is the closure of the domain $D(A)$ of A in the space X. Theorem 2.49 should be viewed as a conditional existence result, namely, the existence of a unique mild solution for (2.96) is the consequence of two assumptions on A: the quasi-accretivity and the existence of an ε-approximate solution. The latter is implied by the quasi-m-accretivity or, eventually, by a weaker condition of this type. We postpone the proof of Theorem 2.49 and pause briefly to point out a few consequences of this fundamental result.

Theorem 2.50 *Let K be a closed convex cone of X and let A be ω-accretive in $X \times X$ such that*

$$D(A) \subset K \subset \bigcap_{0<\lambda<\lambda_0} R(I + \lambda A) \quad \textit{for some } \lambda_0 > 0. \tag{2.108}$$

Let $y_0 \in \overline{D(A)}$ and $f \in L^1(0,T;X)$ be such that $f(t) \in K$, a.e. $t \in (0,T)$. Then, problem (2.96) *has a unique mild solution y. If y and $\bar{y}$ are two mild solutions to* (2.96) *corresponding to f and g, respectively, then*

$$\begin{aligned} \|y(t) - \bar{y}(t)\| &\le e^{\omega(t-s)}\|y(s) - \bar{y}(s)\| \\ &+ \int_s^t e^{\omega(t-\tau)}[y(\tau) - \bar{y}(\tau), f(\tau) - g(\tau)]_s d\tau \quad \textit{for } 0 \le s < t \le T. \end{aligned} \tag{2.109}$$

Proof. Let $f \in L^1(0,T;X)$ and let f_i be the nodal approximation of f, that is,

$$f_i = \frac{1}{t_i - t_{i-1}} \int_{t_{i-1}}^{t_i} f(s)ds, \quad i = 1, 2, ..., N,$$

where $\{t_i\}_{i=1}^N$, $t_0 = 0$, is a partition of the interval $[0, t_N]$ such that $t_i - t_{i-1} < \varepsilon$, $t - \varepsilon < t_N < T$. By assumption (2.108), it follows that, for ε small enough, the function $z = z_i$ on $(t_{i-1}, t_i]$, $z_0 = y_0$, is well defined by (2.104) and it is an ε-approximate solution to (2.96). (It is readily seen by assumption (2.97) and the ω-accretivity of A that equation (2.104) has a unique solution $\{z_i\}_{i=0}^N$.) Thus, Theorem 2.49 is applicable and so problem (2.96) has a unique solution satisfying (2.109). □

Corollary 2.51 *Let A be quasi-m-accretive. Then, for each $y_0 \in \overline{D(A)}$ and $f \in L^1(0,T;X)$ there is a unique mild solution y to* (2.96).

In the sequel, we frequently refer to the map $(y_0, f) \to y$ from $\overline{D(A)} \times L^1(0,T;X)$ to $C([0,T];X)$ as the *nonlinear evolution associated with* A. It should be noted that, in particular, the range condition (2.108) holds if $K = X$ and A is quasi-m-accretive in $X \times X$.

In the special case when $f \equiv 0$, if A is ω-accretive and the following range condition holds

$$R(I + \lambda A) \supset \overline{D(A)} \quad \text{for all small } \lambda > 0, \tag{2.110}$$

then condition (2.108) holds and so we have, by Theorem 2.49:

Theorem 2.52 (Crandall and Liggett [60]) *Let A be quasi-accretive, satisfying the range condition* (2.110) *and $y_0 \in \overline{D(A)}$. Then, the Cauchy problem*

$$\begin{aligned} &\frac{dy}{dt} + Ay \ni 0, \quad t > 0, \\ &y(0) = y_0, \end{aligned} \tag{2.111}$$

has a unique mild solution y. Moreover, y is given by the exponential formula

$$y(t) = \lim_{n \to \infty} \left(I + \frac{t}{n} A \right)^{-n} y_0, \quad \forall t > 0, \tag{2.112}$$

uniformly in t on compact intervals.

Indeed, in this case, if $t_0 = 0$, $t_i = i\varepsilon$, $i = 1, \dots, N$, then the solution z_ε to the ε-discretization $D_A^\varepsilon(0 = t_0, t_1, \dots, t_N)$ is given by the iterative scheme

$$z_\varepsilon(t) = (I + \varepsilon A)^{-i} y_0 \quad \text{for } t \in ((i-1)\varepsilon, i\varepsilon].$$

Hence, by (2.106), we have

$$\|y(t) - (I + \varepsilon A)^{-i} y_0\| \le \delta(\varepsilon) \quad \text{for } (i-1)\varepsilon < t \le i\varepsilon, \ i = 1, \dots, N,$$

which implies the exponential formula (2.112) with uniform convergence on compact intervals. We note that, in particular, the range conditions (2.108) and (2.110) are automatically satisfied if A is quasi-m-accretive, that is, if $\omega I + A$ is m-accretive for some real ω. The solution y to (2.111) given by exponential formula (2.112) is also denoted by $e^{-At} y_0$.

Corollary 2.53 *Let A be quasi-m-accretive and let $y_0 \in \overline{D(A)}$. Then the Cauchy problem* (2.111) *has a unique mild solution y given by the exponential formula* (2.112).

In particular, Corollary 2.53 implies existence and uniqueness of a mild solution y to (2.111) if A is continuous and quasi-accretive.

Another immediate consequence of Theorem 2.52 is

Corollary 2.54 *Let A be an ω-accretive operator satisfying condition* (2.110) *and let K be a closed subset of X such that $(I + \lambda A)^{-1} K \subset K$, $\forall \lambda > 0$. Then, for each $y_0 \in K$, the mild solution y to* (2.111) *satisfies $y(t) \in K$, $\forall t \in [0, T]$.*

We now apply Theorem 2.50 to the mild solutions $y = y(t)$ and $\bar{y} = x$ to the equations

$$\begin{aligned} &y' + Ay \ni f \text{ in } (0, T), \\ &y' + Ay \ni v \text{ in } (0, T), \ v \in Ax. \end{aligned}$$

We have, by (2.109),

$$\begin{aligned} \|y(t) - x\| \le e^{\omega(t-s)} \|y(s) - x\| + \int_s^t [y(\tau) - x, f(\tau) - v]_s e^{\omega(t-\tau)} d\tau, \\ \forall\, 0 \le s < t \le T, \ [x, v] \in A. \end{aligned} \tag{2.113}$$

Such a function $y \in C([0,T];X)$ is called an *integral solution* to equation (2.96).

The convergence theorem can be made more precise for the autonomous equation (2.111), that is, for $f \equiv 0$. (See, e.g., [16], p. 139.)

Corollary 2.55 *Let A be quasi-accretive and satisfy condition* (2.110), *and let $y_0 \in \overline{D(A)}$. Let y be the mild solution to problem* (2.111) *and let y_ε be an ε-approximate solution to* (2.111) *with $y_\varepsilon(0) = y_0$. Then,*

$$\|y_\varepsilon(t) - y(t)\| \le C_T(\|y_0 - x\| + |Ax|\,(\varepsilon + t^{\frac{1}{2}}\varepsilon^{\frac{1}{2}})), \quad \forall t \in [0,T], \tag{2.114}$$

for all $x \in D(A)$. In particular, we have

$$\left\| y(t) - \left(I + \frac{t}{n}A\right)^{-n} y_0 \right\| \le C_T(\|y_0 - x\| + tn^{\frac{1}{2}}|Ax|) \tag{2.114$'$}$$

for all $t \in [0,T]$ and $x \in D(A)$. Here, C_T is a positive constant independent of x and y_0 and $|Ax| = \inf\{\|z\|;\ z \in Ax\}$.

In particular, Theorems 2.49, 2.52 and their corollaries are true for continuous and quasi-accretive operators $A : X \to X$. (As is well known, in an infinite dimensional Banach space the Cauchy problem (2.96) is not well posed for continuous operators A.)

Proof of Theorem 2.49. Here, we shall point out only the main steps of the proof which are marked by the lemmas which follow. (We refer to [15] for the complete proof.)

Let z be a solution to an ε-discretization $D_A^\varepsilon(0 = t_1, t_1, ..., t_N; f_1, ..., f_N)$ and let w be a solution to $D_A^\varepsilon(0 = s_0, s_1, ..., s_M; g_1, ..., g_M)$ with the nodal values z_i and w_j, respectively. We set $a_{ij} = \|z_i - w_j\|$, $\delta_i = (t_i - t_{i-1})$, $\gamma_j = (s_j - s_{j-1})$.

Lemma 2.56 *For all $1 \le i \le N$, $1 \le j \le M$, we have*

$$\begin{aligned} a_{ij} \le \left(1 - \omega\,\tfrac{\delta_i\gamma_j}{\delta_i+\gamma_j}\right)^{-1} &\left(\tfrac{\gamma_j}{\delta_i+\gamma_j}\,a_{i-1,j} + \tfrac{\delta_i}{\delta_i+\gamma_j}\,a_{i,j-1}\right. \\ &\left. + \frac{\delta_i\gamma_j}{\delta_i+\gamma_j}\,[z_i - w_j, f_i - g_j]_s\right). \end{aligned} \tag{2.115}$$

Moreover, for all $[x,v] \in A$ we have

$$a_{i,0} \le \alpha_{i,1}\|z_0 - x\| + \|w_0 - x\| + \sum_{k=1}^{i} \alpha_{i,k}\delta_k(\|f_k\| + \|v\|), \quad 0 \le i \le N, \tag{2.116}$$

and

$$a_{0,j} \le \beta_{j,1}\|w_0 - x\| + \|z_0 - x\| + \sum_{k=1}^{j} \beta_{j,k}\gamma_k(\|g_k + \|v\|), \quad 0 \le j \le M, \tag{2.117}$$

where

$$\alpha_{i,k} = \prod_{m=k}^{i}(1 - \omega\delta_m)^{-1}, \quad \beta_{j,k} = \prod_{m=k}^{j}(1 - \omega\gamma_m)^{-1}. \tag{2.118}$$

Proof. We have

$$f_i + \delta_i^{-1}(z_{i-1} - z_i) \in Az_i, \quad g_j + \gamma_j^{-1}(w_{j-1} - w_j) \in Aw_j, \tag{2.119}$$

and, because A is ω-accretive, this yields (see (2.85))

$$[z_i - w_j, f_i + \delta_i^{-1}(z_{i-1} - z_i) - g_j - \gamma_j^{-1}(w_{j-1} - w_j)]_s \geq -\omega\|z_i - w_j\|.$$

Hence,

$$\begin{aligned}
-\omega\|z_i - w_j\| &\leq [z_i - w_j, f_i - g_j]_s + \delta_i^{-1}[z_i - w_j, z_{i-1} - z_i]_s \\
&\quad + \gamma_j^{-1}[z_i - w_j, w_j - w_{j-1}]_s \\
&\leq [z_i - w_j, f_i - g_j]_s - \delta_i^{-1}(\|z_i - w_j\| - \|z_{i-1} - w_j\|) \\
&\quad - \gamma_j^{-1}(\|z_i - w_j\| - \|z_i - w_{j-1}\|),
\end{aligned}$$

and rearrranging we obtain (2.115).

To get estimates (2.116), (2.117), we note that, inasmuch as A is ω-accretive, we have (see (2.69))

$$\|z_i - x\| \leq (1 - \delta_i\omega)^{-1}\|z_i - x + \delta_i(f_i + \delta_i^{-1}(z_{i-1} - z_i) - v)\|,$$

respectively,

$$\|w_j - x\| \leq (1 - \gamma_j\omega)^{-1}\|w_j - x + \gamma_j(g_j + \gamma_j^{-1}(w_{j-1} - w_j) - v)\|,$$

for all $[x, v] \in A$. Hence,

$$\begin{aligned}
\|z_i - x\| &\leq (1 - \delta_i\omega)^{-1}\|z_{i-1} - x\| + (1 - \delta_i\omega)^{-1}\delta_i(\|f_i\| + \|v\|) \\
\|w_j - x\| &\leq (1 - \gamma_j\omega)^{-1}\|w_{j-1} - x\| + (1 - \gamma_j\omega)^{-1}\gamma_j(\|g_i\| + \|v\|)
\end{aligned}$$

and (2.116), (2.117) follow by a simple calculation. □

In order to get, by (2.115), explicit estimates for a_{ij}, we invoke a technique frequently used in stability analysis of finite difference numerical schemes.

Namely, consider the solution ψ to the linear first order hyperbolic equation

$$\frac{\partial\psi}{\partial t}(t, s) + \frac{\partial\psi}{\partial s}(t, s) - \omega\psi(t, s) = \varphi(t, s) \quad \text{for} \quad 0 \leq t \leq T, \ 0 \leq s \leq T, \tag{2.120}$$

with the boundary conditions

$$\psi(t, s) = b(t - s) \quad \text{for} \quad t = 0 \quad \text{or} \quad s = 0, \tag{2.121}$$

where $b \in C([-T, T])$ and φ will be defined later on.

We approximate (2.120) by the difference equations

$$\frac{\psi_{i,j} - \psi_{i-1,j}}{\delta_i} + \frac{\psi_{i,j} - \psi_{i,j-1}}{\gamma_j} - \omega\psi_{ij} = \varphi_{i,j}, \ \text{for } i = 1, ..., N, \ j = 1, ..., M, \tag{2.122}$$

where $\delta_i = t_i - t_{i-1}$, $\gamma_j = s_j - s_{j-1}$, and $\varphi_{i,j}$ is a piecewise constant approximation of φ defined below.

In the following, we take

$$\varphi(t, s) = \|f(t) - g(s)\|, \quad \varphi_{i,j} = \|f_i - g_j\|, \quad i = 1, ..., N, \ j = 1, ..., M,$$

where f_i and g_j are the nodal approximations of $f, g \in L^1(0,T;X)$, respectively. We get, via the characteristics method,

$$\psi(t,s) = G(b,\varphi)(t,s)$$
$$= \begin{cases} e^{\omega s}b(t-s) + \int_0^s e^{\omega(s-\tau)}\varphi(t-s+\tau,\tau)d\tau & \text{if } 0 \le s < t \le T, \\ e^{\omega t}b(t-s) + \int_0^t e^{\omega(t-\tau)}\varphi(\tau,s-t+\tau)d\tau & \text{if } 0 \le t < s \le T. \end{cases} \tag{2.123}$$

We set $\Omega = (0,T) \times (0,T)$, and let

$$\|\varphi\|_\Omega = \inf\{\|f\|_{L^1(0,T)} + \|g\|_{L^1(0,T)};\ |\varphi(t,s)| \le |f(t)| + |g(s)|, \text{ a.e. } (t,s) \in \Omega\}. \tag{2.124}$$

Let $\Omega(\Delta) = [0,t_N] \times [0,s_M]$ and let $B : [-s_M, t_N] \to \mathbb{R}$, $\phi : \Omega(\Delta) \to \mathbb{R}$ be the piecewise constant functions defined by

$$B(r+s) = b_{ij} \quad \text{for } t_{i-1} < r \le r_i,\ -s_j \le s < -s_{j-1},$$
$$\phi(t,s) = \phi_{i,j} \quad \text{for } (t,s) \in (t_{i-1},t_i] \times (s_{j-1},s_j],$$

where $b_{i,j}, \phi_{i,j} \in \mathbb{R}$, $b(0) = B(0)$. (We note that $B = B_\varepsilon$ and $\Phi = \Phi_\varepsilon$ depend on ε, but in the following we drop ε.)

Denote by $\Psi = H_\Delta(B,\phi)$ the piecewise constant function on Ω defined by

$$\Psi = \psi_{i,j} \quad \text{on } (t_{i-1},t_i] \times (s_{j-1},s_j], \tag{2.125}$$

that is, the solution to (2.122), (2.123).

Lemma 2.57 below, which provides the convergence of the finite difference scheme (2.120) as $m(\Delta) \to 0$, is the main step of the proof. Since its proof is quite technically though elementary, we omit it and refer to [15].

Lemma 2.57 *Let $b \in C([-T,T])$ and $\varphi \in L^1(\Omega)$ be given. Then,*

$$\|G(b,\varphi) - H_\Delta(B,\phi)\|_{L^\infty(\Omega(\Delta))} \to 0 \tag{2.126}$$

as $m(\Delta) + \|b - B\|_{L^\infty(-s_M,t_N)} + \|\varphi - \phi\|_{\Omega(\Delta)} \to 0$.

Proof of Theorem 2.49. We apply Lemma 2.57, where $\varphi(t,s) = \|f(t) - g(s)\|$, $\phi = \{\phi_{i,j}\}$, $\phi_{i,j} = \|f_i - g_j\|$, $1 \le j \le M$, $1 \le i \le N$, f_i and g_j are the nodal values of f and g, respectively, and

$$B(t) = b_{i,0} \quad \text{for } t_{i-1} < t \le t_i,\ i = 1,\dots,N,$$
$$B(s) = b_{0,j} \quad \text{for } -s_j < s \le -s_{j-1},\ j = 1,\dots,M,$$

$$B(t) \to b(t) = e^{\omega t}\|z_0 - x\| + \|w_0 - x\| + \int_0^t e^{\omega(t-\tau)}(\|f(\tau)\| + \|v\|)d\tau,\ \forall t \in [0,T].$$

Then, by Lemma 2.57, we see that, for every $\eta > 0$, we have

$$\|z(t) - w(s)\| \le G(b,\varphi)(t,s) + \eta,\quad \forall s,t \in [0,T], \tag{2.127}$$

as soon as $0 < \varepsilon < \nu(\eta)$.

If $f \equiv g$ and $z_0 = w_0$, then $G(b, \varphi)(t, t) = e^{\omega t} b(0) = 2e^{\omega t} \|z_0 - x\|$ and so,

$$\|z(t) - w(t)\| \leq \eta + 2e^{\omega t} \|z - x\|, \quad \forall x \in D(A),\ t \in [0, T],$$

for all $0 < \varepsilon \leq \nu(\eta)$. Because $\|z_0 - s_0\| \leq \varepsilon$, $y_0 \in \overline{D(A)}$, and x is arbitrary in $D(A)$, it follows that the sequence z_ε of ε-approximate solutions satisfies the Cauchy criterion and so $y(t) = \lim_{\varepsilon \to 0} z_\varepsilon(t)$ exists uniformly on $[0, T]$. Now, we take the limit as $\varepsilon \to 0$ with $s = t + h$, $g \equiv f$, and $z_0 = w_0 = y_0$. We get

$$\begin{aligned}
&\|y(t+h) - y(t)\| \leq G(b, \varphi)(t + h, t) = e^{\omega t}(e^{\omega h} + 1)\|y_0 - x\| \\
&\quad + \int_0^h e^{\omega(h-\tau)}(\|f(\tau)\| + \|v\|)d\tau + \int_0^t e^{\omega(t-\tau)}\|f(\tau + h) - f(\tau)\|d\tau, \ \forall [x, v] \in A,
\end{aligned}$$

and therefore y is continuous on $[0, T]$. Now, by (2.127) we have, for $f \equiv g$, $t = s$,

$$\|z(t) - y(t)\| \leq \delta(\varepsilon), \quad \forall t \in [0, T],$$

where z is any ε-approximate solution and $\delta(\varepsilon) \to 0$ as $\varepsilon \to 0$. Finally, we take $t = s$ in (2.127) and let ε tend to zero. Then, by (2.123), we get the inequality

$$\|y(t) - \bar{y}(t)\| \leq e^{\omega t}\|y(0) - \bar{y}(0)\| + \int_0^t e^{\omega(t-\tau)}\|f(\tau) - g(\tau)\|d\tau.$$

To obtain (2.107), we apply (2.127), where $\varphi(t, s) = [y(t) - \bar{y}(t), f(t) - g(s)]_s$ and $t = s$. Then, by (2.123), we see that

$$G(h, \varphi)(t, t) = e^{\omega t}\|y(0) - \bar{y}(0)\| + \int_0^t e^{\omega(t-s)}[y(s) - \bar{y}(s), f(s) - g(s)]_s ds,$$

and so (2.107) follows for $s = 0$ and, consequently, for all $s \in (0, t)$. □

Remark 2.58 An important consequence of Theorem 2.49 is that for the Cauchy problem (2.96) with quasi-m-accretive operators A or, more generally, for quasi-accretive operators satisfying the range condition (2.110), the mild solution y is obtained as limit of the finite difference scheme (see (2.104))

$$\begin{aligned}
&y(t) = \lim_{h \to 0} y_h(t) \ \text{ in } X,\ \forall t \in [0, T], \\
&y_h(t) = y_h^i, \quad \forall t \in [ih, (i+1)h),\ i = 0, 1, ..., N = \left[\tfrac{T}{h}\right], \\
&y_h^{i+1} + hAy_h^{i+1} \ni y_h^i + f_i^h; \quad f_i^h = \textstyle\int_{ih}^{(i+1)h} f(t)dt, \quad i = 0, 1, ..., N, \\
&y_h^0 = y_0.
\end{aligned} \tag{2.128}$$

This means that Theorem 2.49 is not a simple abstract existence result, but provides also a constructive way to obtain the solution to the Cauchy problem (2.96) by a convergent finite difference scheme. This fact is important not only for the numerical approximation of the solution, but it is also useful to deriving estimates for the mild solution y in the case where it is not differentiable, and so one cannot work directly on equation (2.96).

In particular, Theorem 2.49 applies to the finite dimensional Cauchy problem

$$\frac{dy}{dt} + a(y) + b(y) \ni f(t), \quad t \in (0,T),$$
$$y(0) = y_0,$$

where $a : \mathbb{R}^N \to \mathbb{R}^N$ is a maximal monotone (equivalently, m-accretive) graph in the Euclidean space $\mathbb{R}^N$, while $b : \mathbb{R}^N \to \mathbb{R}^N$ is Lipschitzian and $f \in L^1(0,T;\mathbb{R}^N)$. It is applicable in particular to discontinuous and monotone functions $a : \mathbb{R}^N \to \mathbb{R}^N$ which become maximal monotone by filling the jumps in a discontinuous point.

The simplest example is the Cauchy problem

$$\frac{dy}{dt} + \text{sign}_0(y) = 0 \quad t \geq 0,$$
$$y(0) = y_0,$$

where $\text{sign}_0(y) = 1$ for $y > 0$, $\text{sign}_0(y) = -1$ for $y < 0$, $\text{sign}_0(0) = 0$ which is not well posed, but if we replace the function sign_0 by $\text{sign}(y) = \text{sign}_0(y)$ for $y \neq 0$, $\text{sign}(0) = [-1,1]$, it has a unique solution y by virtue of Theorem 2.49.

Taking into account that any continuous and accretive operator A in a Banach space is m-accretive, Theorem 2.49 applies to this case Too. However, in an infinite dimensional space, the continuity assumption is excessively restrictive because it excludes from applications the important class of partial differential equations with boundary value conditions (see examples in Chapters 3 and 4).

Regularity of mild solutions

A question of great interest is that of circumstances under which the mild solutions constructed above are strong solutions. One may construct simple examples which show that in a general Banach space this might be false. However, if the space is reflexive, then under natural assumptions on f and y_0, the answer is positive. Namely, we have

Theorem 2.59 *Let X be reflexive, K be a closed cone of X and let A be closed and ω-accretive operator which satisfies condition* (2.108). *Let $y_0 \in D(A)$ and $f \in W^{1,1}([0,T];X)$ be such that $f(t) \in K$, $\forall t \in [0,T]$. Then, problem* (2.96) *has a unique mild strong solution y which is strong solution and $y \in W^{1,\infty}([0,T];X)$. Moreover, y satisfies the estimate*

$$\left\|\frac{dy}{dt}(t)\right\| \leq e^{\omega t}|f(0)-Ay_0| + \int_0^t e^{\omega(t-s)} \left\|\frac{df}{ds}(s)\right\| ds, \quad a.e.\ t \in (0,T), \tag{2.129}$$

where $|f(0) - Ay_0| = \inf\{\|w\|;\ w \in f(0) - Ay_0\}$.

In particular, we have the following theorem.

Theorem 2.60 *Let X be a reflexive Banach space and let A be an ω-m-accretive operator. Then, for each $y_0 \in D(A)$ and $f \in W^{1,1}([0,T];X)$, problem* (2.96) *has a unique strong solution $y \in W^{1,\infty}([0,T];X)$ that satisfies estimate* (2.129).

Proof of Theorem 2.59. Let y be the mild solution to problem (2.96) provided by Theorem 2.50. We apply estimate (2.109), where $y(t) := y(t+h)$ and $g(t) := f(t+h)$. We get

$$\begin{aligned}\|y(t+h) - y(t)\| &\le \|y(h) - y(0)\|e^{\omega t} + \int_0^t \|f(s+h) - f(s)\|e^{\omega(t-s)}ds \\ &\le Ch + \|y(h) - y(0)\|e^{\omega t},\end{aligned}$$

because $f \in W^{1,1}([0,T];X)$ (see Theorem 1.34). Now, applying the same estimate (2.109) to y and y_0, we get

$$\|y(h) - y_0\| \le \int_0^h \|f(s) - \xi\|e^{\omega(h-s)}ds \le \int_0^h |Ay_0 - f(s)|ds,\ \forall \xi \in Ay_0,\ h \in [0,T].$$

We may conclude, therefore, that the mild solution y is Lipschitz on $[0,T]$. Then, by Theorem 1.33, it is, a.e., differentiable and belongs to $W^{1,\infty}([0,T];X)$. Moreover, we have

$$\left\|\frac{dy}{dt}(t)\right\| = \lim_{h\to 0} \frac{\|y(t+h) - y(t)\|}{h} \le e^{\omega t}|Ay_0 - f(0)| + \int_0^t \left\|\frac{df}{ds}(s)\right\| e^{\omega(t-s)}ds,$$
$$\text{a.e. } t \in (0,T).$$

Now, let $t \in [0,T]$ be such that

$$\frac{dy}{dt}(t) = \lim_{h\to 0} \frac{1}{h}(y(t+h) - y(t))$$

exists. By inequality (2.113), we have

$$\|y(t+h) - x\| \le e^{\omega h}\|y(t) - x\| + \int_t^{t+h} e^{\omega(t+h-s)}[y(\tau) - x, f(\tau) - w]_s d\tau,\ \forall [x,w] \in A.$$

Noting that

$$[v - x, u - v]_s \le \|u - x\| - \|v - x\|,\quad \forall u, v, x \in X,$$

we get

$$\begin{aligned}[y(t) - x, y(t+h) - y(t)]_s &\le (e^{\omega h} - 1)\|y(t) - x\| \\ &\quad + \int_t^{t+h} e^{\omega(t+h-\tau)}[y(\tau) - x, f(\tau) - w]_s d\tau.\end{aligned}$$

Because the bracket $[u,v]_s$ is upper semicontinuous in (u,v), and positively homogeneous and continuous in v (see Proposition 2.38), this yields

$$\left[y(t) - x, \frac{dy}{dt}(t)\right]_s - \omega\|y(t) - x\| \le [y(t) - x, f(t) - w]_s,\quad \forall [x,w] \in A.$$

Taking into account part (v) of Proposition 2.38, this implies that there is $\xi \in J(y(t) - x)$ such that (J is the duality mapping)

$$\left(\frac{dy}{dt}(t) - \omega(y(t) - x) - f(t) - w, \xi\right) \le 0. \tag{2.130}$$

Inasmuch as the function y is differentiable in t, we have

$$y(t-h) = y(t) - h\,\frac{d}{dt}\,y(t) + hg(h), \tag{2.131}$$

where $g(h) \to 0$ for $h \to 0$. On the other hand, by condition (2.108), for every h sufficiently small and positive, there are $[x_h, w_h] \in A$ such that

$$y(t-h) + hf(t) = x_h + hw_h.$$

Substituting successively in (2.130) and in (2.131) we get

$$(1-\omega h)\|y(t) - x_h\| \leq h\|g(h)\|, \quad \forall h \in (0, \lambda_0).$$

Hence, $x_h \to y(t)$ and $w_h \to f(t) - \frac{dy(t)}{dt}$ as $h \to 0$. Because, as seen earlier, A is closed, we conclude that

$$\frac{dy}{dt}\,(t) + Ay(t) \ni f(t),$$

as claimed. □

In particular, Theorems 2.49–2.60 remain true for equations of the form

$$\begin{cases} \dfrac{dy}{dt}\,(t) + Ay(t) + Fy(t) \ni f(t), \quad t \in [0,T], \\ y(0) = y_0, \end{cases} \tag{2.132}$$

where A is m-accretive in $X \times X$ and $F : X \to X$ is Lipschitzian. Indeed, in this case, as easily seen, the operator $A+F$ is quasi-m-accretive, that is, $A + F + \omega I$ is m-accretive for $\omega = \|F\|_{\mathrm{Lip}}$.

More can be said about the regularity of a strong solution to problem (2.96) if the space X is uniformly convex. Namely, we have (see, e.g., [15], p. 143).

Theorem 2.61 *Let A be ω-m-accretive, $f \in W^{1,1}([0,T]; X)$, $y_0 \in D(A)$ and let X be uniformly convex along with the dual X^*. Then, the strong solution to problem (2.96) is everywhere differentiable from the right, $\left(\frac{d^+}{dt}\right) y$ is right continuous, and*

$$\frac{d^+}{dt}\,y(t) + (Ay(t) - f(t))^0 = 0, \ \forall t \in [0,T), \tag{2.133}$$

$$\left\|\frac{d^+}{dt}\,y(t)\right\| \leq e^{\omega t}\|(Ay_0 - f(0))^0\| + \int_0^t e^{\omega(t-s)}\left\|\frac{df}{ds}\,(s)\right\| ds, \ \forall t \in [0,T). \tag{2.134}$$

Here, $(Ay - f)^0$ is the element of minimum norm in the set $Ay - f$.

We are given a Hilbert space H and a reflexive Banach space V such that $V \subset H$ continuously and densely. Denote by V' the dual space. Then, identifying H with its own dual, we may write $V \subset H \subset V'$ algebraically and topologically.

The norms of V and H are denoted $\|\cdot\|$ and $|\cdot|$, respectively. We denote by (v_1, v_2) the pairing between $v_1 \in V'$ and $v_2 \in V$; if $v_1, v_2 \in H$, this is the ordinary inner product in H. Finally, we denote by $\|\cdot\|_*$ the norm of V' (which is the dual norm). In addition to these spaces, we are given a single-valued, quasi-monotone operator $A : V \to V'$. This means that

$$(Ay - Az, y - z) \geq -\alpha|y - z|_H^2, \quad \forall y, z \in V, \tag{2.135}$$

for some $\alpha \in \mathbb{R}$. We assume also that A is demicontinuous and coercive from V to V'. We note the following simple application of Theorem 2.61.

Theorem 2.62 *Let $f \in W^{1,1}([0,T];H)$ and $y_0 \in V$ be such that $Ay_0 \in H$. Then, there exists one and only one function $y : [0,T] \to V$ that satisfies*

$$y \in W^{1,\infty}([0,T];H), \quad Ay \in L^\infty(0,T;H), \tag{2.136}$$

$$\begin{cases} \dfrac{dy}{dt}(t) + Ay(t) = f(t), & a.e.\ t \in (0,T), \\ y(0) = y_0. \end{cases} \tag{2.137}$$

Proof. Define the operator $A_H : H \to H$,

$$A_H u = Au, \quad \forall u \in D(A_H) = \{u \in V;\ Au \in H\},$$

which is the realization of A in $H \times H$. By hypothesis, the operator $u \to u + Au$ is monotone, demicontinuous, and coercive from V to V'. Hence, it is surjective (see, e.g., Corollary 2.2) and so, A_H is m-accretive (maximal monotone) in $H \times H$. Then, we may apply Theorem 2.61 to conclude the proof. □

If the space X^* is uniformly convex, A is quasi-m-accretive, $f \in W^{1,1}([0,T];X)$, and $y_0 \in D(A)$, then the strong solution $y \in W^{1,\infty}([0,T];X)$ to problem (2.96) (see Theorem 2.59) can be obtained by a simpler proof as

$$y(t) = \lim_{\lambda \to 0} y_\lambda(t) \quad \text{in } X, \text{ uniformly on } [0,T], \tag{2.138}$$

where $y_\lambda \in C^1([0,T];X)$ are the solutions to the Yosida approximating equation

$$\begin{cases} \dfrac{dy_\lambda}{dt}(t) + A_\lambda y_\lambda(t) = f(t), & t \in [0,T], \\ y_\lambda(0) = y_0, \end{cases} \tag{2.139}$$

where $A_\lambda = \lambda^{-1}(I - (I + \lambda A)^{-1})$ for $0 < \lambda < \lambda_0$. Indeed, by Lemma 2.56, we have

$$\frac{1}{2}\frac{d}{dt}\|y_\lambda(t) - y_\mu(t)\|^2 + (A_\lambda y_\lambda(t) - A_\mu y_\mu(t), J(y_\lambda(t) - y_\mu(t))) = 0,$$
$$\text{a.e. } t \in (0,T), \quad \text{for all } \lambda, \mu \in (0, \lambda_0).$$

Since A is ω-accretive and $A_\lambda y \in A(I + \lambda A)^{-1} y$, we have

$$\begin{aligned} &\frac{1}{2}\frac{d}{dt}\|y_\lambda(t) - y_\mu(t)\|^2 + (A_\lambda y_\lambda(t) - A_\mu y_\mu(t), J(y_\lambda(t) - y_\mu(t))) \\ &\quad - J((I + \lambda A)^{-1} y_\lambda(t) - (1 + \mu A)^{-1} y_\mu(t))) \\ &\le \omega\|(1 + \lambda A)^{-1} y_\lambda(t) - (1 + \mu A)^{-1} y_\mu(t)\|^2, \quad \text{a.e. } t \in (0,T). \end{aligned} \tag{2.140}$$

On the other hand, multiplying the equation

$$\frac{d^2 y_\lambda}{dt^2} + \frac{d}{dt} A_\lambda y_\lambda(t) = \frac{df}{dt}, \quad \text{a.e. } t \in (0,T),$$

by $J\left(\frac{dy_\lambda}{dt}\right)$, it yields

$$\frac{1}{2}\frac{d}{dt}\left\|\frac{dy_\lambda}{dt}(t)\right\|^2 \le \left\|\frac{df}{dt}(t)\right\| \left\|\frac{dy_\lambda}{dt}(t)\right\| + \omega\left\|\frac{dy_\lambda}{dt}(t)\right\|, \quad \text{a.e. } t \in (0,T),$$

because A_λ is ω-accretive. This implies that

$$\left\|\frac{dy_\lambda}{dt}(t)\right\| \le e^{\omega t}\left\|\frac{dy_\lambda}{dt}(0)\right\| + \int_0^t e^{\omega(t-s)}\left\|\frac{df}{ds}(s)\right\| ds$$
$$\le e^{\omega t}|Ay_0 - f(0)| + \int_0^t e^{\omega(t-s)}\left\|\frac{df}{ds}(s)\right\| ds. \tag{2.141}$$

Hence, $\|A_\lambda y_\lambda(t)\| \le C$, $\forall \lambda \in (0,\lambda_0)$, and $\|y_\lambda(t) - (1+\lambda A)^{-1}y_\lambda(t)\| \le C\lambda$. Since J is uniformly continuous on bounded sets, it follows by (2.140) that

$$\frac{1}{2}\frac{d}{dt}\|y_\lambda(t) - y_\mu(t)\|^2 \le \omega\|(I+\lambda A)^{-1}y_\lambda(t) - (I+\mu A)^{-1}y_\mu(t)\|^2$$
$$+(\|A_\lambda y_\lambda(t)\|+\|A_\mu y_\mu(t)\|)\|J(y_\lambda(t)-y_\mu(t))-J((I+\lambda A)^{-1}y_\lambda(t)-(I+\mu A)^{-1}y_\mu(t))\|$$
$$\le \omega\|y_\lambda(t) - y_\mu(t)\|^2 + C(\lambda+\mu)$$
$$+\|J(y_\lambda(t) - y_\mu(t)) - J((I+\lambda A)^{-1}y_\lambda(t) - (1+\mu A)^{-1}y_\mu(t))\|,$$

because $\|(I+\lambda A)^{-1}y_\lambda - y_\lambda\| = \lambda\|A_\lambda y_\lambda\| \le C\lambda$. Then, taking into account that J is uniformly continuous and that $\{\|A_\lambda y_\lambda\|\}$ is bounded, the latter implies, via Gronwall's lemma, that $\{y_\lambda\}$ is a Cauchy sequence in the space $C([0,T];X)$ and $y(t) = \lim_{\lambda\to 0} y_\lambda(t)$ exists in X uniformly on $[0,T]$. Let $[x,w]$ be arbitrary in A and let $x_\lambda = x + \lambda w$. Multiplying equation (2.92) by $J(y_\lambda(t) - x_\lambda)$ and integrating on $[s,t]$, we get

$$\frac{1}{2}\|y_\lambda(t) - x_\lambda\|^2 \le \frac{1}{2}\|y_\lambda(s) - x_\lambda\|^2 e^{\omega(t-s)} + \int_s^t e^{\omega(t-\tau)}(f(\tau) - w, J(y_\lambda(\tau) - x_\lambda))d\tau,$$

and, letting $\lambda \to 0$,

$$\frac{1}{2}\|y(t) - x\|^2 \le \frac{1}{2}\|y(s) - x\|^2 e^{\omega(t-s)} + \int_s^t e^{\omega(t-\tau)}(f(\tau) - w, J(y_\lambda(\tau) - x))d\tau,$$

because J is continuous. This yields

$$\left(\frac{y(t)-y(s)}{t-s}, J(y(s)-x)\right) \le \frac{1}{2}\|y(s) - x\|^2(e^{\omega(t-s)} - 1)(t-s)^{-1}$$
$$+\frac{1}{t-s}\int_s^t e^{\omega(t-\tau)}(f(\tau) - w, J(y_\lambda(\tau) - x))d\tau, \tag{2.142}$$

because, as seen earlier,

$$\frac{1}{2}\|y(t) - x\|^2 - \frac{1}{2}\|y(s) - x\|^2 \ge (y(t) - x, J(y(s) - x)).$$

By (2.141), we see that y is absolutely continuous on $[0,T]$ and $\frac{dy}{dt} \in L^\infty(0,T;X)$. Hence, y is, a.e., differentiable on $(0,T)$. If $s = t_0$ is a point where y is differentiable, by (2.142) we see that

$$\left(f(t_0) - \frac{dy}{dt}(t_0) - w + \omega(y(t_0) - x), J(y(t_0) - x)\right) \ge 0, \qquad \forall [x,w] \in A.$$

On the other hand, since the operator $A+\omega I$ is m-accretive, it is maximal accretive in $X \times X$ and so the latter implies that

$$f(t_0) - \frac{dy}{dt}(t_0) \in Ay(t_0).$$

Hence, y is the strong solution to problem (2.96). □

Continuous semigroups of contractions

Definition 2.63 Let K be a closed subset of a Banach space X. A *continuous semigroup of contractions on* K is a family of mappings $\{S(t);\ t \geq 0\}$ that maps K into itself and has the following properties:

(i) $S(t+s)x = S(t)S(s)x,\ \forall x \in K,\ t,s \geq 0.$

(ii) $S(0)x = x,\ \forall x \in K.$

(iii) For every $x \in K$, the function $t \to S(t)x$ is continuous on $[0,\infty)$.

(iv) $\|S(t)x - S(t)y\| \leq \|x-y\|,\ \forall t \geq 0,\ x,y \in K.$

More generally, if instead of (iv) we have

(v) $\|S(t)x - S(t)y\| \leq e^{\omega t}\|x-y\|,\ \forall t \geq 0,\ x,y \in K,$

we say that $S(t)$ is a *continuous* ω*-quasi-contractive semigroup* on K.

In the special case where X is a space of Lebesgue integrable functions or Sobolev spaces on some set $\Omega \subset \mathbb{R}^N$, we shall also refer to the semigroup $S(t)$ as a *semiflow.*

The operator $A_0 : D(A_0) \subset K \to X$, defined by

$$A_0 x = \lim_{t \downarrow 0} \frac{S(t)x - x}{t}, \quad x \in D(A_0), \tag{2.143}$$

where $D(A_0)$ is the set of all $x \in K$ for which the limit (2.143) exists, is called the *infinitesimal generator* of the semigroup $S(t)$.

There is a close relationship between the continuous semigroups of contractions and accretive operators. Indeed, it is easily seen that in this case the operator $-A_0$ is accretive in $X \times X$. More generally, if $S(t)$ is quasi-contractive, then $-A_0$ is ω-accretive. Keeping in mind the theory of C_0-semigroups of linear contractions, one might suspect that there is a one-to-one correspondence between the class of continuous semigroups of contractions and that of m-accretive operators. Indeed, as seen in Theorem 2.52, if X is a Banach space and A is an ω-accretive mapping satisfying the range condition (2.110) (in particular, if A is ω-m-accretive), then, for every $y_0 \in \overline{D(A)}$, the Cauchy problem (2.111) has a unique mild solution $y(t) = S_A(t)y_0 = e^{-At}y_0$ given by the exponential formula (2.112), that is,

$$S_A(t)y_0 = \lim_{n\to\infty} \left(I + \frac{t}{n}A\right)^{-n} y_0. \tag{2.144}$$

For this reason, $S_A(t)$ also denoted by e^{-At} is called the *continuous semigroup generated by* A. We have the following.

Proposition 2.64 *$S_A(t)$ is a continuous ω-quasi-contractive semigroup on $K = \overline{D(A)}$.*

Proof. It is obvious that conditions (ii)–(iv) are satisfied as a consequence of Theorem 2.52. To prove (i), we note that, for a fixed $s > 0$, $y_1(t) = S_A(t+s)x$ and $y_2(t) = S_A(t)S_A(s)x$ are both mild solutions to the problem

$$\begin{cases} \dfrac{dy}{dt} + Ay = 0, \quad t \geq 0, \\ y(0) = S_A(s)x, \end{cases} \tag{2.145}$$

and so, by uniqueness of the solution we have $y_1 \equiv y_2$.

Let us assume now that X, X^* are uniformly convex Banach spaces and that A is an ω-accretive set, that is, closed and satisfies the condition

$$\overline{\text{conv}\, D(A)} \subset \bigcap_{0<\lambda<\lambda_0} R(I + \lambda A) \quad \text{for some } \lambda_0 > 0. \tag{2.146}$$

Then, for every $x \in D(A)$, $S_A(t)x$ is differentiable from the right on $[0, +\infty)$ and

$$-A^0x = \lim_{t \downarrow 0} \frac{S_A(t)x - x}{t}, \quad \forall x \in D(A).$$

Hence, $-A^0 \subset A_0$, where A_0 is the infinitesimal generator of $S_A(t)$. $\square$

As a matter of fact, we may prove in this case the following partial extension of Hille–Philips theorem in continuous semigroups of contractions.

Proposition 2.65 *Let X and X^* be uniformly convex and let A be an ω-accretive and closed set of $X \times X$ satisfying condition* (2.146). *Then, there is a continuous ω-quasi-contractive semigroup $S(t)$ on $\overline{D(A)}$, whose generator A_0 coincides with $-A^0$.*

Proof. For simplicity, we assume that $\omega = 0$. We have already seen that A^0 (the minimal section of A) is single-valued, everywhere defined on $D(A)$, and $-A_0x = A^0x$, $\forall x \in D(A)$. Here, A_0 is the infinitesimal generator of the semigroup $S_A(t)$ defined on $\overline{D(A)}$ by the exponential formula (2.112). We prove that $D(A_0) = D(A)$. Let $x \in D(A_0)$. Then

$$\limsup_{h \downarrow 0} \frac{\|S_A(t+h)x - S_A(t)x\|}{h} < \infty, \quad \forall t \geq 0,$$

and, by the semigroup property (i), it follows that $t \to S_A(t)x$ is Lipschitz continuous on every compact interval $[0, T]$. Hence, $t \to S_A(t)x$ is a.e. differentiable on $(0, \infty)$ and

$$\frac{d}{dt}\, S_A(t)x = A_0 S_A(t)x, \quad \text{a.e. } t > 0.$$

Because $y(t) = S_A(t)x$ is a mild solution to (2.111), that is, a.e. differentiable and $\left(\frac{d}{dt}\right) y(0) = A_0x$, it follows by Theorem 2.60 that $S_A(t)x$ is a strong solution to (2.111), that is,

$$\frac{d}{dt}\, S_A(t)x + A^0 S_A(t)x = 0, \quad \text{a.e. } t > 0.$$

Now,

$$-A_0x = \lim_{h\downarrow 0} \frac{1}{h} \int_0^h A^0 S_A(t)x\,dt,$$

and this implies as in the proof of Theorem 2.61 that $x \in D(A)$ and $-A_0x \in Ax$. This completes the proof. □

A problem of interest is if any semigroup of contractions $S(t)$ on a closed convex set $K \subset X$, or in particular on X, is generated by an m-accretive operator A (as is the case for linear C_0-semigroups of contractions). The answer is positive if X is a Hilbert space and was proved by Y. Kōmura [82] (see also [14]).

If X is reflexive, then as seen earlier $S(t)(D(A)) \subset D(A)$, $\forall t \geq 0$, but in general Banach spaces X this is not true because the function $t \to S(t)x$ is not differentiable for $x \in D(A)$. We have, however, the following weaker result.

Proposition 2.66 *Let A be ω-accretive such that $\overline{D(A)} \subset R(I+\lambda A)$, $\forall \lambda \in \left(0, \frac{1}{\omega}\right)$. If $S(t) = e^{-tA}$ is the semigroup generated by A, we have*

$$\|S(t)x - S(s)x\| \leq \exp(\omega(t-s))|t-s|\,|Ax|, \ \forall x \in D(A), \tag{2.147}$$

for $0 \leq s \leq t$.

Proof. By (2.144), we have

$$\begin{aligned}
\|S(t)x - x\| &= \lim_{n\to\infty} \|J^n_{\frac{t}{n}}x - x\| \leq \limsup_{n\to\infty} \sum_{j=1}^n \|J^j_{\frac{t}{n}}x - J^{j-1}_{\frac{t}{n}}x\| \\
&\leq \limsup_{n\to\infty} \sum_{j=1}^n \left(1 - \frac{t}{n}\omega\right)^{-j+1} \|J_{\frac{t}{n}}x - x\| \\
&\leq t\|A_{\frac{t}{n}}x\| \lim_{n\to\infty} \sum_{j=1}^n \left(1 - \frac{t}{n}\omega\right)^{-j+1} \\
&\leq t|Ax| \frac{e^{\omega t} - 1}{t\omega} \leq t|Ax|, \ \forall x \in D(A).
\end{aligned}$$

Taking into account that

$$\|S(t)x - S(s)x\| \leq \exp(\omega(t-s))\|S(t-s)x - x\| \leq \exp(\omega(t-s))|t-s|\,|Ax|,$$

we get (2.147), as claimed. It turns out that (2.147) extends to all $x \in \widehat{D}(A)$, the *generalized domain* of A defined by

$$\widehat{D}(A) = \left\{ x \in \mathcal{D} = \bigcup_{\nu>0} \bigcap_{0<\lambda<\nu} R(I+\lambda A); |Ax| < \infty \right\},$$

where (see Proposition 2.33) $|Ax| = \sup_{\lambda>0} \|A_\lambda x\| = \lim_{\lambda\downarrow 0} \|A_\lambda x\|$.

Clearly, $D(A) \subset \widehat{D}(A) \subset \overline{D(A)}$. □

Proposition 2.67 *Under hypotheses of Proposition* 2.66, *inequality* (2.147) *holds for all $x \in \widehat{D}(A)$. Moreover, $S(t)(\widehat{D}(A)) \subset \widehat{D}(A)$, $\forall t \geq 0$ and*

$$|AS(t)x| \leq \exp(\omega t)|Ax|, \quad \forall x \in \widehat{D}(A), \ t > 0. \tag{2.148}$$

Proof. Since, as easily seen, the function $x \to |Ax|$ is lower semicontinuous and $D(A)$ is dense in $\widehat{D}(A)$, clearly (2.147) extends to all of $\widehat{D}(A)$. We also have

$$\mu(x) = \liminf_{h \downarrow 0} \frac{1}{h} \|S(h)x - x\| \geq |Ax|, \quad \forall x \in \widehat{D}(A), \tag{2.149}$$

which, by (2.147) implies equality in (2.149) and so, we get (2.148) by substituting x with $S(t)x$.

To prove (2.149), we note that by the accretivity of A (for simplicity, take $\omega = 0$) we have

$$\limsup_{h \downarrow 0} (h^{-1}(S(h)x - x), z) \leq (u - x, x), \ \forall (u, v) \in A,$$

where $z \in J(u - x)$. This follows by proving it first for $S_\lambda(t) = e^{-tA_\lambda}$ and letting afterwards $\lambda \to 0$, but we omit the details. Then, we get

$$\mu(x)\|x - u\| \geq -{}_X(u - x, v)_{X^*}, \ \forall (u, v) \in A,$$

which, for $u = J_\lambda x, \ v = A_\lambda x$, yields

$$\|A_\lambda x\| \leq \mu(x), \quad \forall \lambda \in (0, \infty),$$

and, for $\lambda \to 0$, one obtains (2.149), as claimed. □

In nonreflexive Banach spaces X it is hard to identify in specific examples the generalized domain $\widehat{D}(A)$ but, by virtue of Proposition 2.67 and more precisely of (2.149), one can view it as the set of all $x \in \overline{D(A)}$, where $t \to S(t)x$ is locally Lipschitz on $[0, \infty)$.

Remark 2.68 If A is m-accretive, $f \equiv 0$, and y_e is a stationary (equilibrium) solution to (2.96) (i.e., $0 \in Ay_e$), then we see by estimate (2.109) that the solution $y = y(t)$ to (2.96) is bounded on $[0, \infty)$. More precisely, we have

$$\|y(t) - y_e\| \leq \|y(0) - y_e\|, \quad \forall t \geq 0.$$

Moreover, if A is strongly accretive (i.e., $A - \gamma I$ is accretive for some $\gamma > 0$), then

$$\|y(t) - y_e\| \leq e^{-\gamma t}\|y(0) - y_0\|, \quad \forall t \geq 0,$$

which amounts to saying that the trajectory $\{y(t), \ t \geq 0\}$ approaches as $t \to \infty$ the equilibrium solution y_e of the system. This means that the dynamic system associated with (2.96) is *dissipative* and, in this sense, sometimes we refer to equations of the form (2.96) as *dissipative systems.*

Gradient systems

Here, we consider problem (2.96) in the case where A is the subdifferential $\partial\varphi$ of a lower semicontinuous convex function φ from a Hilbert space H to $\overline{\mathbb{R}} = (-\infty, +\infty]$. In other words, consider the problem

$$\begin{cases} \dfrac{dy}{dt}(t) + \partial\varphi(y(t)) \ni f(t), & \text{in } (0, T), \\ y(0) = y_0, \end{cases} \tag{2.150}$$

in a real Hilbert space H with the scalar product $(\cdot,\cdot)$ and norm $|\cdot|$. It turns out that the nonlinear evolution generated by $A = \partial\varphi$ on $\overline{D(A)}$ has, for $y_0 \in \overline{D(A)}$, regularity properties that in the linear case are characteristic to analytic semigroups.

We recall that, if $\varphi : H \to \overline{\mathbb{R}}$ is a lower semicontinuous, convex function, then its subdifferential $A = \partial\varphi$ is maximal monotone (equivalently, m-accretive) in $H \times H$ and $\overline{D(A)} = \overline{D(\varphi)}$ (see Theorem 2.15). Then, by Theorem 2.50, for every $y_0 \in \overline{D(A)}$ and $f \in L^1(0,T;H)$ the Cauchy problem (2.150) has a unique mild solution $y \in C([0,T];H)$, which is a strong solution if $y_0 \in D(A)$ and $f \in W^{1,1}([0,T];H)$ (Theorem 2.59).

Theorem 2.69 below, essentially due to H. Brezis [33], amounts to saying that y remains a strong solution to (2.150) on every interval $[\delta,T]$ even if $y_0 \notin D(A)$ and f is not absolutely continuous. In other words, the evolution e^{-tA} generated by $A = \partial\varphi$ has a smoothing effect on initial data and on the right-hand side f of (2.150). (Everywhere in the following, the Hilbert space H is identified with its own dual.)

Theorem 2.69 *Let $f \in L^2(0,T;H)$ and $y_0 \in \overline{D(A)}$. Then the mild solution y to problem (2.96) belongs to $W^{1,2}([\delta,T];H)$ for every $0 < \delta < T$, and*

$$y(t) \in D(A), \quad a.e.\ t \in (0,T), \tag{2.151}$$

$$t^{\frac{1}{2}}\frac{dy}{dt} \in L^2(0,T;H) \quad \varphi(u) \in L^1(0,T), \tag{2.152}$$

$$\frac{dy}{dt}(t) + \partial\varphi(y(t)) \ni f(t), \quad a.e.\ t \in (0,T). \tag{2.153}$$

Moreover, if $y_0 \in D(\varphi)$, then

$$\frac{dy}{dt} \in L^2(0,T;H), \quad \varphi(y) \in W^{1,1}([0,T]). \tag{2.154}$$

The proof relies on the chain differentiation formula

$$\frac{d}{dt}\varphi(y(t)) = \left(\xi(t), \frac{dy}{dt}(t)\right), \quad \text{a.e. } t \in (0,T) \tag{2.155}$$

for all $y \in W^{1,2}([0,T];H)$ and $\xi(t) \in L^2(0,T;H)$, $\xi(t) \in \partial\varphi(y(t))$, a.e. $t \in (0,T)$. Then, by (2.153) it follows that

$$t\left|\frac{dy}{dt}(t)\right|^2 + t\frac{d}{dt}\varphi(y(t)) = t\left(f(t), \frac{dy}{dt}(t)\right), \quad \text{a.e. } t \in (0,T), \tag{2.156}$$

from which (2.152)–(2.154) follow. We also have

Theorem 2.70 *Let $S(t) = e^{-At}$ be the continuous semigroup of contractions generated by $A = \partial\varphi$ on $\overline{D(A)}$. Then, $S(t)\overline{D(A)} \subset D(A)$ for all $t > 0$, and*

$$\left|\frac{d^+}{dt}S(t)y_0\right| = |A^0 S(t)y_0| \le |A^0 x| + \frac{1}{t}|x - y_0|, \quad \forall t > 0, \tag{2.157}$$

for all $y_0 \in \overline{D(A)}$ and $x \in D(A)$.

Proof. By (2.156), we see that

$$\int_0^t s\left|\frac{dy}{ds}(s)\right|^2 ds + t\varphi(y(t)) \leq \int_0^T \varphi(y(s))ds, \quad \forall t > 0.$$

Next, we multiply (2.150) by $y(t) - x$ and integrate on $(0, t)$ to get

$$\frac{1}{2}|y(t) - x|^2 + \int_0^t \varphi(y(s))ds \leq \frac{1}{2}|y(0) - x|^2 + t\varphi(x).$$

This yields

$$\begin{aligned}\int_0^t s\left|\frac{dy}{ds}(s)\right|^2 &\leq \frac{1}{2}(|y(0) - x|^2 - |y(t) - x|^2 + t(\varphi(x) - \varphi(y(t))) \\ &\leq \frac{1}{2}(|y(0) - x|^2 - |y(t) - x|^2 + t(A^0 x, x - y(t)) \\ &\leq \frac{1}{2}|y(0) - x|^2 + \frac{t^2|A^0 x|^2}{2}, \quad \forall t > 0.\end{aligned}$$

Because the function $t \to |\frac{d}{dt}\, y(t)|$ (and, consequently, $t \to |\frac{d^+}{dt}\, y(t)|$) is monotonically decreasing, this implies the desired inequality. □

Equations of the form (2.150) are called in literature *conservative systems* because, as seen by (2.155), the energy of the system remains constant, that is,

$$\varphi(y(t)) + \int_0^t |Ay(s)|^2 ds \equiv \text{constant}, \quad \forall t \geq 0.$$

The function φ is called the *potential* of the system.

A nice feature of nonlinear semigroups generated by $A = \partial\varphi$ is that the orbit $\gamma(y_0) = \{e^{-tA}, t \geq 0\}$ is weakly convergent to a stationary (equilibrium) solution $\xi \in A^{-1}0$. Namely, one has the following result.

Theorem 2.71 *Let $A = \partial\varphi$, where $\varphi : H \to (-\infty, +\infty]$ is a convex l.s.c. function such that $(\partial\varphi)^{-1}(0) \neq \emptyset$. Then, for each $y_0 \in \overline{D(A)}$ there is $\xi \in (\partial\varphi)^{-1}(0)$ such that*

$$\xi = w\text{-}\lim_{t\to\infty} e^{-At} y_0. \tag{2.158}$$

Proof. The proof is based on the formula (2.155) and the property of the potential $\varphi : H \to [0, \infty)$ to be a Lyapunov function for $S_A(t)$. Namely, if we multiply, scalarly in H, the equation

$$\frac{d}{dt}\, y(t) + Ay(t) \ni 0, \quad \text{a.e. } t > 0,$$

by $y(t) - x$, where $x \in (\partial\varphi)^{-1}(0)$, we obtain by (2.155) that

$$\frac{1}{2}\frac{d}{dt}|y(t) - x|^2 \leq 0, \quad \text{a.e. } t > 0,$$

because, by the monotonicity of A, $(Ay(t), y(t) - x) \geq 0$, $\forall t \geq 0$. This implies that the orbit $\gamma(y_0)$ is bounded. Denote by K the weak *omega*-limit set associated to $\gamma(y_0)$, that is, $K = \left\{ w\text{-}\lim_{t_n\to\infty} y(t_n) \right\}$.

Let us notice that $K \subset (\partial\varphi)^{-1}(0)$. Indeed, if $y(t_n) \rightharpoonup \xi$, for some $\{t_n\} \to \infty$, then we see by (2.157) that

$$\lim_{n\to\infty} \frac{dy}{dt}(t_n) = 0$$

and since A is demiclosed, this implies that $0 \in A\xi$ (i.e., $\xi \in A^{-1}(0) = (\partial\varphi)^{-1}(0)$). On the other hand, $t \to |y(t) - x|^2$ is decreasing for each $x \in (\partial\varphi)^{-1}(0)$ and, in particular, for each $x \in K$.

Let ξ_1, ξ_2 be two arbitrary elements of K given by

$$\xi_1 = w\text{-}\lim_{n'\to\infty} y(t_{n'}), \quad \xi_2 = w\text{-}\lim_{n''\to\infty} y(t_{n''}),$$

where $t_{n'} \to \infty$ and $t_{n''} \to \infty$ as $n' \to \infty$ and $n'' \to \infty$, respectively.

Because $\lim_{t\to\infty} |y(t) - x|^2$ exists for each $x \in K \subset (\partial\varphi)^{-1}(0)$, we have

$$\lim_{n'\to\infty} |y(t_{n'}) - \xi_1|^2 = \lim_{n''\to\infty} |y(t_{n''}) - \xi_1|^2,$$

$$\lim_{n''\to\infty} |y(t_{n''}) - \xi_2|^2 = \lim_{n'\to\infty} |y(t_{n'}) - \xi_2|^2.$$

The latter implies by an elementary calculation that $|\xi_1 - \xi_2|^2 = 0$. Hence, K consists of a single point and this completes the proof of (2.158)). □

There is a discrete version of Theorem 2.71 which asserts that the sequence $\{y_n\}$ defined by

$$y_{n+1} = y_n - h\partial\varphi(y_{n+1}), \quad n = 0, 1, ..., \quad h > 0,$$

is weakly convergent in H to an element $\xi \in (\partial\varphi)^{-1}(0)$, that is, to a minimum point for φ on H. This result is known in convex optimization as the steepest descent algorithm (see R.T. Rockafellar [95]).

If, under assumptions of Theorem 2.71, the trajectory $\{y(t)\}_{t\geq 0}$ is relatively compact in H (this happens for instance if each level set $\{x;\ \varphi(x) \leq \lambda\}$ is compact), then (2.158) is strengthening to

$$y(t) = e^{-At}y_0 \to \xi \quad \text{strongly in } H \text{ as } t \to \infty.$$

Asymptotic behavior of semigroups

In most physical problems, equation (2.145) describes the evolution (or transition) of an open system to an equilibrium state $y_\infty \in A^{-1}(0)$ and, in general, such a behavior cannot be rigorously derived from a numerical simulation.

In fact, the *longtime behavior of trajectories* $\{y(t); t > 0\}$ to the nonlinear semigroup $S(t) = e^{-tA}$ and their convergence for $t \to \infty$ to an equilibrium solution $\xi \in A^{-1}(0)$ was largely studied in the literature by different methods including the Lyapunov function method (A. Pazy [93]) and the dynamic topology.

The Lyapunov function for the semigroup $S(t) : [0, \infty) \times K \to K$ is a continuous function $V : K \to [0, \infty)$ which is nonincreasing on every trajectory of $S(t)$, that is,

$$V(S(t)y_0) \leq V(S(s)y_0), \quad 0 \leq t \leq s < \infty, \ y_0 \in K. \tag{2.159}$$

Formally, if $-A$ is the generator of $S(t)$, then the latter holds if V is Gateaux differentiable and

$$(Ay, \partial V(y)) \geq 0, \quad \forall y \in D(A).$$

However, it is hard to check this condition if the semigroup $S(t)$ is not differentiable in t, as is the case in nonreflexive Banach spaces X. (In such a situation, one could however use the approximating formula (2.128).)

An alternative approach is the LaSalle invariance principle developed for nonlinear semigroups in Banach spaces by C. Dafermos and M. Slemrod [64] and we shall briefly present it in the following.

Let A be accretive in the Banach space X such that, for some $\lambda_0 > 0$,

$$\overline{D(A)} \subset \bigcap_{0<\lambda<\lambda_0} R(\lambda I + A). \tag{2.160}$$

Then, as seen earlier (Theorem 2.52), there is a continuous semigroup of contractions $S(t) = e^{-tA}$ on $K = \overline{D(A)}$ given by (2.112) and so leaves invariant the set K. For each $y_0 \in K$, $\{S(t)y_0, t \geq 0\} = \gamma(y_0)$ is the orbit throughout y_0 and $\omega(y_0) = \{\xi \in K;$ $\xi = \lim_{t_n \to \infty} S(t_n)y_0\}$ is the *omega-limit set* of y_0. This concept was introduced by Birkhoff and used later on by LaSalle to sharpen the Lyapunov stability theory for finite differential systems. We summarize in Theorem 2.72 below the main properties of $\omega(y_0)$.

Theorem 2.72 *If $\omega(y_0) \neq \emptyset$, then $\omega(y_0)$ is invariant under the semigroup $S(t)$ and $S(t)$ is an homeomorphism of $\omega(y_0)$ onto itself. Moreover, for each $t \geq 0$, $S(t)$ is an isometry on $\omega(y_0)$ and, if $\xi \in K$ is a fixed point of S (i.e., $S(t)\xi = \xi$, $\forall t \geq 0$), then $\omega(y_0) \subset \{y \in X; \|y - \xi\| = r\}$, where $0 \leq r \leq \|\xi - y_0\|$.*

Proof. The proof of the invariance of $\omega(y_0)$ as well as that $S(t)\omega(y_0) = \omega(y_0)$, $\forall t \geq 0$, is standard in the theory of dynamical systems and follows by the semigroup property $S(t+s) = S(t)S(s)$ and by the continuity of $t \to S(t)x$, $\forall x \in K$.

As regards the isometry property

$$\|S(t)x - S(t)y\| = \|x - y\|, \quad \forall t \geq 0, \ x, y \in \omega(x_0),$$

it is immediate by the contraction property of $S(t)$ and the definition of $\omega(x_0)$. We omit the details. It should be said, however, that the existence of a fixed point for $S(t)$ is, in general, a delicate problem because in a nonreflexive Banach space the set $\{\xi;\ S(t)\} = \xi$, $\forall t \geq 0\}$ does not coincide with $A^{-1}(0)$. □

The convergence of $S(t)y_0$ for $t \to \infty$ is a specific property of open systems far from thermodynamic equilibrium and implies in particular the existence of a bounded attractor associated with the semigroup $S(t)$. It should be noted also that the elements of $\omega(y_0)$ are reversible states of the system.

The condition $\omega(y_0) \neq \emptyset$ in Theorem 2.72 is, in particular, satisfied if the orbit $\gamma(y_0)$ is relatively compact in X. Quite surprising, this property is implied by the compactness of the resolvent operator $(I + \lambda A)^{-1}$. Namely, we have

Theorem 2.73 *Assume that A is accretive $0 \in A_0$ and $\overline{D(A)} \subset R(I+\lambda A)$, $\forall \lambda > 0$. If $J_\lambda = (I+\lambda A)^{-1}$ is compact in X for some $\lambda > 0$, then $\gamma(y_0)$ is precompact in X for any $y_0 \in K = \overline{D(A)}$.*

Proof. Taking into account the resolvent equation

$$J_\mu(y_0) = J_\lambda\left(\frac{\lambda}{\mu}y_0 + \left(1-\frac{\lambda}{\mu}\right)J_\mu(y_0)\right),$$

it suffices to prove that J_λ is compact for some $\lambda > 0$. On the other hand, we have

$$\|S(t)y_0\| \le \|S(t)y_0 - S(t)\xi\| + \|S(t)\xi\| \le \|\xi\| + \|y_0 - \xi\|,\ \forall \xi \in A^{-1}0,$$

and so $\gamma(u_0)$ is a bounded set of X.

Let $y_0 \in K = \overline{D(A)}$. There is $\{y_l\} \subset D(A)$, $y_l \to y_0$ as $l \to \infty$. For any sequence $\{t_n\} \subset R^+$, we can find by the diagonal process a subsequence, denoted again by $\{t_n\}$, such that $\{J_{\frac{1}{k}}S(t_n)y_l\}$ is Cauchy for all positive integers k, l. Then, by

$$\begin{aligned}\|S(t_n)x - S(t_m)x\| &\le \|S(t_n)x - S(t_n)y_l\| + \left\|\left(\tfrac{1}{k}\right)A_{\frac{1}{k}}S(t_n)y_l\right\| \\ &\quad + \|J_{\frac{1}{k}}S(t_n)y_l - J_{\frac{1}{k}}S(t_m)y_l\| \\ &\quad + \left\|\left(\tfrac{1}{k}\right)A_{\frac{1}{k}}S(t_m)y_l\right\| + \|S(t_m)y_l - S(t_m)x\| \\ &\le 2\|x - y_l\| + \left(\tfrac{2}{k}\right)|Ay_l| + \|J_{\frac{1}{k}}S(t_n)y_l - J_{\frac{1}{k}}S(t_m)y_l\|\end{aligned}$$

because $y_l \in \widehat{D}(A)$ and so, by Proposition 2.67,

$$\|A_{\frac{1}{k}}S(t)y_l\| \le |AS(t)y_l| \le |Ay_l|,\ \forall t \ge 0.$$

This implies that $\lim_{n\to\infty} S(t_n)y_0$ exists and so the orbit $\gamma(y_0)$ is compact. □

In some situations, the semigroup $S(t) = e^{-tA}$ has the *finite extinction property.* Namely, we have

Theorem 2.74 *Let $A : X \to X$ be quasi-m-accretive in a Banach space X and $x \in \overline{D(A)}$. Assume that there are $\gamma \ge 0$ and $\rho > \gamma\|x\|$ such that*

$$(\xi, J(u)) \ge \rho\|u\| - \gamma\|u\|^2,\quad \forall u \in D(A),\ \xi \in Au. \tag{2.161}$$

Then, we have

$$S(t)x = 0,\quad \forall t \ge T(x) = \frac{1}{\gamma}\log\left(\frac{\rho}{\rho - \gamma\|x\|}\right). \tag{2.162}$$

Proof. We recall that $u(t) = S(t)x$ is given by: $u(t) = \lim_{h\to 0} u_h(t)$, $\forall t \ge 0$, where

$$u_h(t) = u_h^i,\quad \forall t \in [ih, (i+1)h),\ i = 0, 1,$$

$$u_h^{i+1} + hAu_h^{i+1} \ni u_h^i,\quad i = 0, 1, ...,;\ u_h^0 = x.$$

By (2.161), we get

$$\frac{1}{2}\exp(-2\gamma(i+1)h)\|u_h^{i+1}\|^2 + h\rho\exp(-2\gamma(i+1)h)\|u_h^{i+1}\| \le \frac{1}{2}\exp(-2\gamma ih)\|u_h^i\|^2,\quad \forall i = 0, 1,$$

and, therefore,

$$\frac{1}{2}\exp(-2\gamma jh)\|u_h(jh)\|^2 + h\rho\sum_{i=1}^{j}\exp(-2\gamma(j+1)h)\|u_h^{i+1}\| \le \frac{1}{2}\|x\|^2.$$

Then, for $h \to 0$, we get

$$\exp(-2\gamma t)\|u(t)\|^2 + 2\rho\int_0^t \exp(-2\gamma s)\|u(s)\|ds \le \|x\|^2, \ \forall t \ge 0.$$

Solving this integral inequality yields

$$\|u(t)\| + \frac{\rho}{\gamma}(1-\exp(-\gamma t))\exp(\gamma t) \le \exp(\gamma t)\|x\|, \ \forall t \ge 0,$$

and, therefore,

$$\|u(t)\| = 0 \ \text{ for } t \ge \frac{1}{\gamma}\log\left(\rho(\rho-\gamma\|x\|)^{-1}\right),$$

as claimed. □

A standard example is the operator

$$Au = \rho \text{ sign } u + A_0u, \ u \in D(A_0) \subset X,$$

where sign $u = u\|u\|^{-1}$ if $u \ne 0$; sign $0 = \{u; \ \|u\| \le 1\}$ and A_0 is a quasi-m-accretive operator in X such that $(A_0u, J(u)) \ge -\gamma\|u\|^2$, $\forall u \in D(A_0)$. (Here, J is the duality mapping.) By Proposition 2.43, we see that A is quasi-m-accretive and also that condition (2.161) holds.

We note that condition (2.161) is implied (and in Hilbert spaces equivalent) with

$$0 \in \text{int } A(0). \tag{2.163}$$

Indeed, by (2.163), we have

$$(Ay - z, J(y)) \ge -\gamma\|y\|^2, \ \ \forall y \in D(A).$$

Then, for $z = \rho Ay\|Ay\|^{-1}$ we get (2.161).

Approximation of nonlinear semigroups

One might expect the solution to Cauchy problem (2.96) to be continuous with respect to the operator A, that is, with respect to small structural variations of the problem. We show below that this indeed happens in a certain precise sense and for a certain notion of convergence defined in the space of quasi-m-accretive operators.

Consider in a general Banach space X a sequence A_n of subsets of $X \times X$. The subset of $X \times X$, $\liminf A_n$ is defined as the set of all $[x, y] \in X \times X$ such that there are sequences x_n, y_n, $y_n \in A_nx_n$, $x_n \to x$ and $y_n \to y$ as $n \to \infty$. If A_n are quasi-m-accretive, there is a simple resolvent characterization of $\liminf A_n$.

Proposition 2.75 *Let $A_n + \omega I$ be m-accretive for $n = 1, 2...$ and some $\omega \ge 0$. Then $A \subset \liminf A_n$ if and only if*

$$\lim_{n\to\infty}(I+\lambda A_n)^{-1}x = (I+\lambda A)^{-1}x, \ \ \forall x \in X, \ \textit{for } 0 < \lambda < \omega^{-1}. \tag{2.164}$$

Proof. Assume that (2.164) holds and let $[x, y] \in A$ be arbitrary but fixed. Then,

$$(I + \lambda A)^{-1}(x + \lambda y) = x, \quad \forall \lambda \in (0, \omega^{-1})$$

and, by (2.164), we have

$$(I + \lambda A_n)^{-1}(x + \lambda y) \to (I + \lambda A)^{-1}(x + \lambda y) = x.$$

In other words, $x_n = (I + \lambda A_n)^{-1}(x + \lambda y) \to x$ as $n \to \infty$ and $x_n + \lambda y_n = x + \lambda y$, $y_n \in A x_n$. Hence, $y_n \to y$ as $n \to \infty$, and so $[x, y] \in \liminf A_n$.

Conversely, let us assume now that $A \subset \liminf A_n$. Let x be arbitrary in X and let $x_0 = (I + \lambda A)^{-1}x$, that is, $x_0 + \lambda y_0 = x$, where $y_0 \in A x_0$. Then, there are $[x_n, y_n] \in A_n$ such that $x_n \to x_0$ and $y_n \to y_0$ as $n \to \infty$. We have

$$x_n + \lambda y_n = z_n \to x_0 + \lambda y_0 = x \quad \text{as } n \to \infty.$$

Hence, $(I + \lambda A_n)^{-1}x \to x_0 = (I + \lambda A)^{-1}y_0$ for $0 < \lambda < \omega^{-1}$, as claimed. □

In the literature, such a convergence is called *convergence in the sense of graphs.*

Theorem 2.76 below, due to H. Brezis and A. Pazy [47], is the nonlinear version of the Trotter–Kato theorem from the theory of C_0-semigroups and, roughly speaking, it amounts to saying that if A_n is convergent to A in the sense of graphs, then the dynamic (evolution) generated by A_n is uniformly convergent to that generated by A.

Theorem 2.76 *Let A_n be ω-m-accretive in $X \times X$, $f^n \in L^1(0, T; X)$ for $n = 1, 2, \ldots$ and let y_n be mild solution to the problem*

$$\frac{dy_n}{dt}(t) + A_n y_n(t) \ni f^n(t) \quad \text{in } [0, T], \quad y_n(0) = y_0^n. \tag{2.165}$$

Let $A \subset \liminf A_n$ and assume that

$$\lim_{n \to \infty} \left(\int_0^T \|f^n(t) - f(t)\| dt + \|y_0^n - y_0\| \right) = 0. \tag{2.166}$$

Then, $y_n(t) \to y(t)$ uniformly on $[0, T]$, where y is the mild solution to (2.96).

We omit the proof which is based on some sharp estimates provided by Theorem 2.49 and refer to [16].

In applications, (2.165) is a smooth approximation of equation (2.96) or has a simpler structure and so, Theorem 2.76 can be used either for the numerical treatment or to get sharp estimates for $y(t) \equiv S(t)y_0$ which cannot be obtained directly from the original system. A few examples will be given later on for some specific systems.

Time-dependent Cauchy problems

Here, we shall briefly treat the time-varying Cauchy problem

$$\begin{cases} \dfrac{dy}{dt}(t) + A(t)y(t) = f(t), & t \in [0, T], \\ y(0) = y_0, \end{cases} \tag{2.167}$$

where $\{A(t)\}_{t \in [0,T]}$ is a family of quasi-m-accretive operators in $X \times X$.

The existence problem for the time-dependent Cauchy problem (2.167) is a difficult one and not completely solved even for linear m-accretive operators $A(t)$. In general, one cannot expect a positive and convenient answer to the existence problem for (2.167) if one takes into account that the standard existence results for (2.167) require the regularity of the function $t \to (I+\lambda A(t))^{-1}$ (see, e.g., [61]), while in most applications to partial differential equations the domain $D(A(t))$ might not be independent of time or the resolvent $(I+\lambda A(t))^{-1}$ is not smooth as a function of t. As a matter of fact, the situation is worse in the nonlinear case if $t \to (I+\lambda A(t))^{-1}$ is not regular. This is illustrated by the following simple example

$$\frac{dy}{dt} + H(y - \alpha(t)) \ni 0, \quad t \geq 0; \; y(0) = y_0,$$

where H is the Heviside multivalued function $H(r) = 1$ for $r \geq 0$, $H(0) = [0,1]$, $H(r) = 0$ for $r < 0$ and $\alpha : [0, \infty) \to \mathbb{R}$ is a continuous function. (In this case, $A(t)y = H(y - \alpha(t))$, $\forall\, y \in \mathbb{R}$.)

The above Cauchy problem can be rewritten as

$$\frac{dz}{dt} + H(z) \ni -\alpha', \; t > 0; \quad z(0) = y_0 - \alpha(0),$$

which clearly has a mild solution z if and only if α' is a bounded measure (for instance, if α is with bounded variation).

However, we can identify a general class of time-dependent problems of the form (2.167) which are well posed.

Nonlinear demicontinuous evolutions in duality pair of spaces

Let V be a reflexive Banach space and H be a real Hilbert space identified with its own dual such that $V \subset H \subset V'$ algebraically and topologically. Denote by $\|\cdot\|$ the norm of V and by $|\cdot|_H$ the norm of H.

Theorem 2.77 *Let $\{A(t);\; t \in [0,T]\}$ be a family of nonlinear and demicontinuous operators from V to V' satisfying the following assumptions:*

(i) *The function $t \to A(t)u(t)$ is measurable from $[0,T]$ to V' for every measurable function $u : [0,T] \to V$.*

(ii) $(A(t)u, u) \geq \omega\|u\|^p - C_1|u|_H^2 + C_2$, $\forall u \in V$, $t \in [0,T]$.

(iii) $\|A(t)u\|_{V'} \leq C_3(|u|_H + \|u\|^{p-1} + 1)$, $\forall u \in V$, $t \in [0,T]$, *where* $\omega > 0$, $p > 1$.

(iv) $(A(t)u - A(t)\bar{u}, u - \bar{u}) \geq -C_4|u - \bar{u}|_H^2$, $\forall\, u, \bar{u} \in V$.

Then, for every $y_0 \in H$ and $f \in L^q(0,T;V')$, $\frac{1}{p} + \frac{1}{q} = 1$, there is a unique absolutely continuous function $y \in W^{1,q}([0,T];V')$ that satisfies

$$\begin{gathered} y \in C([0,T];H) \cap L^p(0,T;V), \\ \frac{dy}{dt}(t) + A(t)y(t) = f(t), \quad a.e.\ t \in (0,T), \\ y(0) = y_0. \end{gathered} \tag{2.168}$$

Proof. For the sake of simplicity, we assume that $p \geq 2$. We also set

$$\widetilde{A}(t) = A(t)y + \lambda y, \quad \forall\, y \in V, \ t \in [0, T],$$

where $\lambda > 0$ is such that $\lambda \geq \max(C_1, C_4)$. Then, by (ii)–(iv) it follows that, for each t, $\widetilde{A}(t) : V \to V'$ is monotone and

$$(\widetilde{A}(t)u, u) \geq \omega\|u\|^p + C_2, \ \forall\, u \in V, \tag{2.169}$$

$$\|\widetilde{A}(t)u\|_{V'} \leq C_3(\|u\|^{p-1} + 1) + \lambda|u|_H^2, \ \forall, u \in V. \tag{2.170}$$

Moreover, equation (2.168) can be rewritten as

$$\frac{dy}{dt} + \widetilde{A}(t)y - \lambda y = f, \quad t \in [0, T].$$

Equivalently,

$$\frac{dz(t)}{dt} + e^{-\lambda t}\widetilde{A}(t)(e^{\lambda t}z(t)) = e^{-\lambda t}f(t), \ t \in [0, T].$$

Let $A^*(t) : V \to V'$ be defined by

$$A^*(t)u \equiv e^{-\lambda t}\widetilde{A}(t)(e^{\lambda t}u), \quad \forall\, u \in V, \ t \in [0, T].$$

Then, equation (2.168) reduces via the substitution $u(t) = e^{-\lambda t}y(t)$ to

$$\begin{aligned} &\frac{du}{dt} + A^*(t)u = f^*, \quad t \in [0, T], \\ &u(0) = u_0, \end{aligned} \tag{2.171}$$

where $\widetilde{f}(t) \equiv e^{-\lambda t}f(t)$. We also note that $A^*(t) : V \to V'$ is monotone and satisfies (2.169)–(2.170). Hence, to prove the theorem it suffices to show that the Cauchy problem (2.171) has a unique absolutely continuous solution $u \in W^{1,q}([0, T]; V') \cap C([0, T]; H) \cap L^p(0, T; V)$. To this purpose, consider the spaces $\mathcal{V} = L^p(0, T; V)$, $\mathcal{H} = L^2(0, T; H)$, $\mathcal{V}' = L^q(0, T; V')$. Clearly, we have

$$\mathcal{V} \subset \mathcal{H} \subset \mathcal{V}'$$

algebraically and topologically.

Let $y_0 \in H$ be arbitrary and fixed and let $B : \mathcal{V} \to \mathcal{V}'$ be the operator

$$Bv = \frac{dv}{dt}, \quad v \in D(B) = \left\{v \in \mathcal{V};\ \frac{dv}{dt} \in \mathcal{V}', \ v(0) = y_0\right\},$$

where $\frac{d}{dt}$ is considered in the sense of vectorial distributions on $(0, T)$. We note that $D(B) \subset W^{1,q}(0, T; V') \cap L^q(0, T; V) \subset C([0, T]; H)$, so that $v(0) = y_0$ makes sense. The idea is to write (2.171) as an equation of the form $\mathcal{A}y = f$, where $\mathcal{A} : \mathcal{V} \to \mathcal{V}'$ is a nonlinear monotone, demicontinuous and coercive operator and derive existence by Theorem 2.8 and Corollary 2.9.

Let us check that B is maximal monotone in $\mathcal{V} \times \mathcal{V}'$. Because B is clearly monotone, it suffices to show that $R(B + \Phi_p) = \mathcal{V}'$, where $\Phi_p(v(t)) = F(v(t))\|v(t)\|^{p-2}$, $v \in \mathcal{V}$, and $F : V \to V'$ is the duality mapping of V. Indeed, for every $g \in \mathcal{V}'$ the equation

$$Bv + \Phi_p(v) = g,$$

equivalently,

$$\frac{dv}{dt} + F(v)\|v\|^{p-2} = g \text{ in } [0,T], \quad v(0) = y_0,$$

has, by Theorem 2.62, a unique solution

$$v \in C([0,T];H) \cap L^p(0,T;V), \quad \frac{dv}{dt} \in L^q(0,T;V').$$

(Renorming the spaces V and V', via Asplund's theorem, we may assume that V and V' are strictly convex and F is demicontinuous and that so is the operator $v \to F(v)\|v\|^{p-2}$.) Hence, B is maximal monotone in $\mathcal{V} \times \mathcal{V}'$.

Define the operator $A_0 : \mathcal{V} \to \mathcal{V}'$ (the realization of A^* in the pair $(\mathcal{V}, \mathcal{V}')$) by

$$(A_0 v)(t) = A^*(t)v(t), \quad \text{a.e. } t \in (0,T).$$

Clearly, A_0 is monotone, demicontinuous, and coercive from $\mathcal{V}$ to $\mathcal{V}'$ because so is $A^*(t) : V \to V'$. Then, by Corollaries 2.9 and 2.14, $\mathcal{A} = A_0 + B$ is maximal monotone and surjective. Hence, $R(A_0 + B) = \mathcal{V}'$, which completes the proof.

The proof in the case $1 < p < 2$ is completely similar if we take $\mathcal{V} = L^p(0,T;V) \cap L^2(0,T;H)$ and replace as above $A(t)$ by $A(t) + \lambda I$ for some $\lambda > 0$. The details are left to the reader. □

It should be said that Theorem 2.77 applies neatly to the nonlinear parabolic boundary value problem

$$\begin{aligned} \frac{\partial y}{\partial t}(x,t) - \sum_{|\alpha| \le m} D^\alpha(A_\alpha(t,x,y,...,D^m y)) &= f(x,t), \ (x,t) \in \Omega \times (0,T) \\ y(x,0) &= y_0(x), \ x \in \Omega \\ D^\beta y &= 0 \text{ on } \partial\Omega \text{ for } |\beta| < m, \end{aligned} \tag{2.172}$$

where $A_\alpha : [0,T] \times \Omega \times \mathbb{R}^{mN} \to \mathbb{R}^{mN}$ are measurable in (t,x), continuous in other variables and satisfy the conditions

$$|A_\alpha(t,x,\xi)| \le C(|\xi|^{p-1} + g(x)),$$

$$\sum_{|\alpha| \le m} (A_\alpha(t,x,\xi) - A_\alpha(t,x,\bar{\xi}))(\xi_\alpha - \bar{\xi}_\alpha) \ge 0,$$

$$\sum_{|\alpha| \le m} A_\alpha(t,x,\xi)\xi_\alpha \ge C|\xi|^p,$$

for all $(t,x,\xi) \in (0,T) \times \Omega \times \mathbb{R}^{mN}$, where $\{\xi_j\}_{j=1}^{mN}$, $g \in L^p(Q)$, $\frac{1}{p} + \frac{1}{q} = 1$. Here, one applies Theorem 2.77 for $V = W_0^{m,p}(\Omega), V' = W^{-m,q}(\Omega)$ and $A(t) : V \to V'$ defined by

$$(A(t)y,z) = \sum_{|\alpha| \le m} \int_\Omega A_\alpha(t,x,y(x),...,D^m y(x)) \cdot D^\alpha y(x)dx, \ \forall y,z \in W_0^{m,p}(\Omega).$$

We also note (see [15], p. 183) that Theorem 2.77 can be used to treat the existence for the Itô stochastic differential equation

$$dX + AX\,dt = dW, \ t \in (0,T), \quad X(t) = X_0, \tag{2.173}$$

where $A : D(A) \subset H \to H$ is m-accretive and W is a Wiener process in the space H. Namely, by the substitution $y(t) = X(t) - W(t)$, equation (2.173) reduces to the random differential equation

$$\begin{aligned} &\frac{dy}{dt} + A(y + W(t)) = 0, \quad t \in (0,T), \\ &y(0) = X_0. \end{aligned} \tag{2.174}$$

Since the function $t \to W(t)$ is not with bounded variation, the operator $A(t) \equiv A(y + W(t))$ and, implicitly, its resolvent $(I + \lambda A(t))^{-1}$ is not smooth as function of t. However, if A is the restriction to H of a demicontinuous monotone operator from V to V' satisfying (i)–(iii), then Theorem 2.77 is applicable to equation (2.174) for a suitable V-valued Wiener process $W(t)$.

Comments. The main results of the theory of nonlinear maximal monotone operators in Banach spaces are essentially due to Minty [85, 86] and Browder [50]). Other important contributions are due to Brezis [42], Lions [78] and Rockafellar [94, 96, 97], Moreau [88], mainly in connection with the theory of subdifferential type operators.

The general theory of nonlinear m-accretive operators in Banach spaces has been developed in the works of Kato [79] and Crandall and Pazy [62,63] in connection with the theory of semigroups of nonlinear contractions and nonlinear Cauchy problem in Banach spaces, which is presented later on.

The existence theory for the Cauchy problem associated with nonlinear m-accretive operators in Banach spaces begins with the influential pioneering papers of Kōmura [82] and Kato [80] in Hilbert spaces. The theory was subsequently extended in a more general setting by several authors mentioned below. The main result of Section 2.3 is due to Crandall and Evans [59] (see also Crandall [57]), and Theorem 2.52 has been previously proved by Crandall and Liggett [60]. The existence and uniqueness of integral solutions for problem (2.96) is due to Bénilan [30]. Theorems 2.60 and 2.61 were established in a particular case in Banach space by Kōmura [82] (see also Kato [80]) and later extended in Banach spaces with uniformly convex duals by Crandall and Pazy [61, 62].

Chapter 3

Nonlinear Elliptic Boundary Value Problems

Mathematicians are like Frenchmen; if one speaks to them, they translate into their own language and then it will be very soon something entirely different.

J.W. von Goethe

We shall apply here the general theory presented in Chapter 2 pertaining to the existence theory for nonlinear infinite dimensional equations with maximal monotone and m-accretive operators to some classes of nonlinear boundary value problems on open subsets $\Omega \subset \mathbb{R}^N$. In particular, the theory covers semilinear elliptic problems, nonlinear diffusion equations, nonlinear stationary Fokker–Planck equations and the conservation law equation. Most of these problems are not treated in their most general framework because, as pointed out earlier, the emphasis is not put on the generality but on the method. As a matter of fact, more general existence results can be obtained by combining the above maximal monotonicity and m-accretiveness theory with specific compactness arguments based on the Schauder theorem or the degree theory in Banach spaces, but we shall not develop this approach here.

3.1 Nonlinear Elliptic Problems of Divergence Type

We consider the Dirichlet boundary value problem

$$\lambda y - \operatorname{div} \beta(\nabla y) \ni f, \quad x \in \Omega, \tag{3.1}$$

$$y = 0 \quad \text{on } \partial\Omega, \tag{3.2}$$

where Ω is a bounded and open domain of $\mathbb{R}^N$ with smooth boundary $\partial\Omega$. (Of class C^1, for instance.) Here, $\beta : \mathbb{R}^N \to 2^{\mathbb{R}^N}$ is a maximal monotone graph in $\mathbb{R}^N \times \mathbb{R}^N$ such that $0 \in \beta(0)$ and λ is a nonnegative constant. Here and everywhere in the following, we use the standard notations

$$\nabla \cdot z(x) = \operatorname{div} z(x) = \sum_{i=1}^{N} \frac{\partial z_i}{\partial x_i}(x), \quad \nabla y(x) = \left\{ \frac{\partial}{\partial x_i} y(x) \right\}_{i=1}^{N}, \ x \in \Omega,$$

for $y, z_i \in W^{1,1}(\Omega)$, $i = 1, ..., N$.

In particular, problem (3.1)–(3.2) describes the equilibrium state of diffusion processes where the diffusion flux $\vec{q}$ is a nonlinear function of the gradient ∇y of local density y. In the special case, where β is a potential function (i.e., $\beta = \nabla j$, $j : \mathbb{R}^N \to \mathbb{R}^N$), then the functional $\phi(y) = \int_\Omega j(\nabla y)dx + \frac{\lambda}{2}\int_\Omega y^2 dx$ can be viewed as the energy of the system and equation (3.1) describes the critical points of ϕ. The elliptic character of equation (3.1) is given by monotonicity assumption on β.

It should be said that equation (3.1) with boundary condition (3.2) might be highly nonlinear and so the best one can expect from the existence point of view is a weak solution.

Definition 3.1 Let $f \in W^{1,q}(\Omega)$. The function $y \in L^1(\Omega)$ is said to be a weak solution to the Dirichlet problem (3.1)–(3.2) if $y \in W_0^{1,1}(\Omega)$ and there is $\eta \in (L^q(\Omega))^N$, $1 < p < \infty$, such that $y\psi \in L^1(\Omega)$, $\forall \psi \in W_0^{1,p}(\Omega)$ and

$$\eta(x) \in \beta(\nabla y(x)), \quad \text{a.e. } x \in \Omega, \tag{3.3}$$

$$\lambda \int_\Omega y\psi\, dx + \int_\Omega \eta(x) \cdot \nabla\psi(x)dx = f(\psi),\ \forall \psi \in W_0^{1,p}(\Omega),\ \tfrac{1}{p} + \tfrac{1}{q} = 1. \tag{3.4}$$

Similarly, the function y is said to be a weak solution to equation (3.1) with the Neumann boundary value condition

$$\beta(\nabla y(x)) \cdot n(x) = 0 \quad \text{on } \partial\Omega, \tag{3.5}$$

if $y \in W^{1,1}(\Omega)$, $y\psi \in L^1(\Omega)$, $\forall \psi \in W^{1,1}(\Omega)$, and there is $\eta \in (L^p(\Omega))^N$ which satisfies (3.3), and (3.4) holds for all $\psi \in W^{1,q}(\Omega)$. (Here n is, as usually, the outward normal to $\partial\Omega$.)

The first existence result for problem (3.1)–(3.2) concerns the case where β is single-valued.

Theorem 3.2 *Assume that $\beta : \mathbb{R}^N \to \mathbb{R}^N$ is continuous, monotonically increasing, and*

$$|\beta(r)| \le C_1(1 + |r|^{p-1}),\ \forall r \in \mathbb{R}^N, \tag{3.6}$$

$$\beta(r) \cdot r \ge \omega|r|^p - C_2, \quad \forall r \in \mathbb{R}^N, \tag{3.7}$$

where $\omega > 0$, $p > 1$, $\frac{2N}{N+2} \le p$. Then, for each $f \in W^{-1,q}(\Omega)$ and $\lambda > 0$, there is a unique weak solution $y \in W_0^{1,p}(\Omega)$ to problem (3.1)–(3.2). Moreover, if $f \ge 0$, a.e. in Ω, then $y \ge 0$, a.e. in Ω and, if $f \in L^\infty(\Omega)$, then $y \in L^\infty(\Omega)$.

Proof. We apply Corollary 2.9 to the operator $T : X \to X^*$, where $X = W_0^{1,1}(\Omega)$, $X^* = W^{-1,q}(\Omega)$, defined by

$$(v, Tu) = \int_\Omega \beta(\nabla u(x)) \cdot \nabla v(x)dx + \lambda \int_\Omega u(x)v(x)\, dx,\ \forall u, v \in X = W_0^{1,p}(\Omega). \tag{3.8}$$

Here, $(\cdot,\cdot)$ is the duality pairing on $W_0^{1,p}(\Omega) \times W^{-1,q}(\Omega)$, that is,

$$(v, w) = w(v), \quad \forall v \in W_0^{1,p}(\Omega),\ w \in W^{-1,q}(\Omega).$$

It is easily seen that T is monotone and demicontinuous. Indeed, if $u_j \to u$ strongly in $X = W_0^{1,p}(\Omega)$, then, for $j \to \infty$, $\nabla u_j \to \nabla u$ strongly in $L^p(\Omega)$ and, by continuity of β, we have on a subsequence $\beta(\nabla u_j) \to \beta(\nabla u)$, a.e. on Ω. On the other hand, by (3.6) we have that $\{\beta(\nabla u_j)\}$ is bounded in $L^q(\Omega)$ and therefore it is weakly sequentially compact in $L^q(\Omega)$. Hence, we also have (eventually, on a subsequence)

$$\beta(\nabla u_j) \rightharpoonup \beta(\nabla u) \quad \text{in} \quad (L^q(\Omega))^N.$$

Then, we infer that

$$\lim_{j\to\infty} \int_\Omega \beta(\nabla u_j) \cdot \nabla v \, dx = \int_\Omega \beta(\nabla u) \cdot \nabla v \, dx, \quad \forall v \in X$$

and also

$$\lim_{j\to\infty} \int_\Omega u_j v \, dx = \int_\Omega uv \, dx,$$

because $W^{1,p}(\Omega) \subset L^2(\Omega)$ by Theorem 1.12. Hence,

$$Tu_j \rightharpoonup Tu \quad \text{in} \quad X^* = W^{-1,q}(\Omega).$$

It is also clear by (3.7) that T is coercive, that is,

$$(u, Tu) \geq \omega \int_\Omega |\nabla u|^p dx - C_2, \quad \forall u \in X.$$

Hence, $R(T) = W^{-1,q}(\Omega)$, as claimed.

Assume that $f \geq 0$, a.e. in Ω, then for $\psi = y^- \subset W_0^{1,p}(\Omega)$ it follows by (3.3), (3.4) that $y^- = 0$, a.e. in Ω, that is, $y \geq 0$, a.e. in Ω. If $f \in L^\infty(\Omega)$, then $-|f|_\infty \leq f \leq |f|_\infty$, a.e. in Ω and, therefore,

$$\lambda \left(y - \frac{1}{\lambda} |f|_\infty \right) - \operatorname{div}(\beta(\nabla y) - \beta(\nabla |f|_\infty) \leq 0 \ \text{ in } \mathcal{D}'(\Omega),$$

and taking in (3.4) $\psi = \left(y - \frac{1}{\lambda} |f|_\infty\right)$ yields

$$\left(y - \frac{1}{\lambda} |f|_\infty \right)^- = 0, \quad \text{a.e. in } \Omega.$$

(Here, $|\cdot|_9$ is the $L^\infty(\Omega)$-norm.) Hence, $y \leq \frac{1}{\lambda} |f|_\infty$, a.e. in Ω. Similarly, it follows $y \geq -\frac{1}{\lambda} |f|_\infty$, a.e. in Ω.

This completes the proof. □

In particular, it follows that, for each $f \in L^2(\Omega)$, problem (3.1)–(3.2) has a unique solution. If $\lambda = 0$, we still have a solution $y \in W_0^{1,p}(\Omega)$, but in general it is not unique. A similar existence result follows for problem (3.1)–(3.5), namely, the following.

Theorem 3.3 *Under the assumptions of Theorem 3.2, for each $f \in (W^{1,p}(\Omega))^*$ and $\lambda > 0$ there is a unique weak solution $y \in W^{1,p}(\Omega)$ to problem (3.1)–(3.5).*

Proof. One applies again Corollary 2.9 to the operator $T : W^{1,p}(\Omega) \to (W^{1,p}(\Omega))^*$ defined by (3.8) for all $v \in W^{1,p}(\Omega)$. (Here, $(W^{1,p}(\Omega))^*$ is the dual space of $W^{1,p}(\Omega)$.) It follows as in the previous case that T is monotone and demicontinuous. As regards the coercivity, we note that by (3.7)–(3.8) we have

$$(u, Tu) \geq \omega \int_\Omega |\nabla u|^p dx + \lambda \int_\Omega u^2 dx. \tag{3.9}$$

(This time $(\cdot,\cdot)$ is the duality pairing on $W^{1,p}(\Omega) \times (W^{1,p}(\Omega))^*$.)

Recalling that (see Theorem 1.12 and (1.19))

$$\|u\|_{W^{1,p}(\Omega)} \leq C(\|\nabla u\|_{L^p(\Omega)} + \|u\|_{L^q(\Omega)}), \quad \forall u \in W^{1,p}(\Omega),$$

for $1 \leq q \leq \frac{Np}{N-p}$, $N > p$ and $q \geq 1$ for $N \geq p$, we see, by (3.9), that

$$(u, Tu) \geq \omega \|u\|^\alpha_{W^{1,p}(\Omega)}, \quad \forall u \in W^{1,p}(\Omega),$$

where $\alpha = \max\{p, 2\}$ and therefore T is coercive, as desired. □

The above existence results extend to general maximal monotone (multivalued) graphs $\beta \subset \mathbb{R}^N \times \mathbb{R}^N$ satisfying assumptions of the form (3.6) and (3.7), that is,

$$\sup\{|w|;\ w \in \beta(r)\} \leq C_1(1 + |r|^{p-1}),\ \forall r \in \mathbb{R}^N, \tag{3.10}$$

$$w \cdot r \geq \omega |r|^p - C_2,\ \forall (w, r) \in \beta. \tag{3.11}$$

(Here, and everywhere in the following, we denote by $|r|$ the Euclidean norm of $r \in \mathbb{R}^N$.) For simplicity, we shall assume that $f \in L^2(\Omega)$.

Theorem 3.4 *Let β be a maximal monotone graph in $\mathbb{R}^N \times \mathbb{R}^N$ satisfying conditions* (3.10) *and* (3.11) *for $\omega > 0$, and $p > 1$. Then, for each $f \in L^2(\Omega)$ and $\lambda > 0$ there is a unique weak solution $y \in W_0^{1,p}(\Omega)$ to problem* (3.1)–(3.2) (*respectively, a unique weak solution $y \in W^{1,p}(\Omega)$ to problem* (3.1)–(3.5)) *in the following sense*

$$\lambda \int_\Omega y\psi + \int_\Omega \eta \cdot \nabla\psi\, dx = \int_\Omega f\psi\, dx,\ \forall \psi \in W_0^{1,p}(\Omega) \cap L^2(\Omega)$$
$$(\textit{respectively},\ \forall \psi \in W^{1,p}(\Omega) \cap L^2(\Omega)), \tag{3.12}$$

where $\eta \in \beta(\nabla y)$, a.e. in Ω.

Of course, if p is such that $W^{1,p}(\Omega) \subset L^2(\Omega)$. (By the Sobolev embedding Theorem 1.1, this happens, for instance, if $p \geq \frac{2N}{N-2}$, then (3.12) coincides with (3.4).)

Proof. We prove the existence theorem in the case of problem (3.1)–(3.2) only, the other case (i.e., the Neumann boundary condition (3.5)) being completely similar. We first assume that $f \in W^{-1,q}(\Omega) \cap L^2(\Omega)$.

We consider the Yosida approximation of β_ε, that is,

$$\beta_\varepsilon(r) = \frac{1}{\varepsilon}(r - ((1 + \varepsilon\beta)^{-1}r) \in \beta((1 + \varepsilon\beta)^{-1})),\ \ \forall r \in \mathbb{R}^N,\ \varepsilon > 0, \tag{3.13}$$

and the corresponding approximating problem

$$\begin{cases} \lambda y_\varepsilon - \operatorname{div}(\beta_\varepsilon(\nabla y_\varepsilon) + \varepsilon \nabla y_\varepsilon) = f & \text{in } \Omega, \\ \beta_\varepsilon(\nabla y_\varepsilon) = 0 & \text{on } \partial\Omega, \end{cases} \tag{3.14}$$

which, by Theorem 3.2 has a unique solution $y_\varepsilon \in H_0^1(\Omega)$. Indeed, β_ε is Lipschitz and it is readily seen that conditions (3.10) and (3.11) hold with $p = 2$ (with constants C independent of ε). On the other hand, by (3.10) and (3.11), we see that

$$\begin{aligned} |\beta_\varepsilon(r)| &< \sup\{|w|;\ w \in \beta((1+\varepsilon\beta)^{-1}r)\} \le C_1(|(1+\varepsilon\beta)^{-1}r|^{p-1} + 1) \\ &\le C_3(|r|^{p-1} + 1), \quad \forall r \in \mathbb{R}^N,\ \forall \varepsilon > 0, \end{aligned} \tag{3.15}$$

and

$$\begin{aligned} \beta_\varepsilon(r) \cdot r &= \beta_\varepsilon(r) \cdot (1+\varepsilon\beta)^{-1}r + \varepsilon|\beta_\varepsilon(r)|^2 \\ &\ge \varepsilon|(1+\varepsilon\beta)^{-1}r|^p + C_4(\varepsilon|r|^{p-1} + 1) \\ &\ge \omega|r|^p + C_5\varepsilon|r|^p + C_6, \quad \forall r \in \mathbb{R}^N,\ \varepsilon > 0. \end{aligned} \tag{3.16}$$

(The constants C_i arising in (3.15) and (3.16) are independent of ε.) Therefore, we have

$$\lambda \int_\Omega y_\varepsilon \psi\, dx + \int_\Omega (\beta_\varepsilon(\nabla y_\varepsilon) + \varepsilon \nabla y_\varepsilon) \cdot \nabla\psi\, dx = \int_\Omega f\psi\, dx, \qquad \forall \psi \in H_0^1(\Omega), \tag{3.17}$$

and so, for $\psi = y_\varepsilon$, we obtain that

$$\begin{aligned} \lambda \int_\Omega y_\varepsilon^2 dx + \varepsilon \int_\Omega |\nabla y_\varepsilon|^2 dx + \int_\Omega \beta_\varepsilon(\nabla y_\varepsilon) \cdot (1+\varepsilon\beta)^{-1} \nabla y_\varepsilon dx \\ + \varepsilon \int_\Omega |\beta_\varepsilon(\nabla y_\varepsilon)|^2 dx = \int_\Omega f y_\varepsilon dx. \end{aligned} \tag{3.18}$$

Taking into account that $\beta_\varepsilon(\nabla y_\varepsilon) \in \beta((1+\varepsilon\beta)^{-1}\nabla y_\varepsilon)$, it follows by (3.16) and (3.18) that

$$\begin{aligned} \lambda \int_\Omega y_\varepsilon^2 dx + \varepsilon \int_\Omega |\nabla y_\varepsilon|^2 dx + \omega \int_\Omega |(1+\varepsilon\beta)^{-1} \nabla y_\varepsilon|^p dx \\ + \varepsilon \int_\Omega |\beta_\varepsilon(\nabla y_\varepsilon)|^2 dx \le C \int_\Omega |f|^2 dx, \quad \forall \varepsilon > 0. \end{aligned} \tag{3.19}$$

(Here and everywhere in the sequel, C is a positive constant independent of ε.) In particular, it follows by (3.19) that

$$\int_\Omega |(1+\varepsilon\beta)^{-1}\nabla y_\varepsilon - \nabla y_\varepsilon|^2 dx \to 0 \quad \text{as } \varepsilon \to 0, \tag{3.20}$$

because $\varepsilon^2|\beta_\varepsilon(r)|^2 = |(1+\varepsilon\beta)^{-1}r - r|^2$, $\forall r \in \mathbb{R}^N$. Moreover, by (3.15) we see that

$$\|\beta_\varepsilon(\nabla y_\varepsilon)\|_{L^q(\Omega)} \le C(\|(1+\varepsilon\beta)^{-1}\nabla y_\varepsilon\|_{L^p(\Omega)}^p + 1).$$

Then, on a subsequence again denoted ε, we have by (3.19) and (3.20),

$$y_\varepsilon \to y \qquad \text{weakly in } L^2(\Omega) \cap W^{1,p}(\Omega), \tag{3.21}$$

$$(1+\varepsilon\beta)^{-1}\nabla y_\varepsilon \to \nabla y \qquad \text{weakly in } (L^p(\Omega))^N, \tag{3.22}$$

$$\beta_\varepsilon(\nabla y_\varepsilon) \to \eta \qquad \text{weakly in } (L^q(\Omega))^N, \tag{3.23}$$

as $\varepsilon \to 0$. Taking into account (3.17) and (3.22), (3.23), we obtain by the weak semicontinuity of the L^p-norm that

$$\begin{aligned}&\lambda \int_\Omega |\nabla y|^2 dx + \int_\Omega |\nabla y|^p dx \le C \int_\Omega f^2 dx, \quad \text{and}\\ &\lambda \int_\Omega y\psi\, dx + \int_\Omega \eta \cdot \nabla\psi\, dx = \int_\Omega f\psi\, dx, \quad \forall \psi \in W_0^{1,p}(\Omega) \cap L^2(\Omega).\end{aligned} \tag{3.24}$$

Because $f \in W^{-1,q}(\Omega) \cap L^2(\Omega)$, the latter extends to all of $\psi \in W_0^{1,p}(\Omega)$. To complete the proof, it suffices to show that

$$\eta(x) \in \beta(\nabla y(x)), \quad \text{a.e. } x \in \Omega. \tag{3.25}$$

To this end, we start with the obvious inequality

$$\int_\Omega (\beta_\varepsilon(\nabla y_\varepsilon) - \zeta) \cdot ((1 + \varepsilon\beta)^{-1} \nabla y_\varepsilon - u) dx \ge 0, \tag{3.26}$$

for all $u \in L^p(\Omega)$ and $\zeta \in (L^q(\Omega))^N$ such that $\zeta(x) \in \beta(u(x))$, a.e. $x \in \Omega$. (This is an immediate consequence of monotonicity of β because, by (3.13), $\beta_\varepsilon(y) \in \beta((1+\varepsilon\beta)^{-1}y)$, $\forall y \in \mathbb{R}^N$, $\forall \varepsilon > 0$.)

Letting ε tend to zero in (3.26), by (3.22)–(3.23) we obtain that

$$\int_\Omega (\eta - \zeta) \cdot (y - u) dx \ge 0.$$

Now, choosing $u = (1+\beta)^{-1}(\eta + y)$ and $\zeta = \eta + y - u \in \beta(u)$, we obtain that

$$\int_\Omega (y - u)^2 dx = 0.$$

Hence, $y = u$ and $\eta = \zeta \in \beta(u)$, a.e. in Ω. This completes the proof of existence for $f \in W^{-1,q}(\Omega) \cap L^2(\Omega)$.

If $f \in L^2(\Omega)$, consider a sequence $\{f_n\} \subset W^{-1,q}(\Omega) \cap L^2(\Omega)$ strongly convergent to f in $L^2(\Omega)$. If y_n are corresponding solutions to problem (3.12), we obtain, by the monotonicity of β,

$$\lambda \int_\Omega |y_n - y_m|^2 dx \le \|f_n - f_m\|_{W^{-1,q}(\Omega)} \|y_n - y_m\|_{W_0^{1,p}(\Omega)},$$

whereas, by estimate (3.24), we see that $\{y_n\}$ is bounded in $W_0^{1,p}(\Omega)$. Hence, on a subsequence, again denoted $\{n\}$, we have

$$\begin{aligned} y_n &\to y \quad \text{strongly in } L^2(\Omega) \text{ and weakly in } W_0^{1,p}(\Omega)\\ \eta_n \in \beta(\nabla y_n) &\to \eta \quad \text{weakly in } (L^q(\Omega))^N. \end{aligned}$$

Clearly, (y, η) verify (3.12) and arguing as above it follows also $\eta \in \beta(\nabla y)$, a.e. in Ω. This completes the proof of existence. The uniqueness is immediate by the monotonicity of β. □

We have chosen β multivalued not only for the sake of generality, but because this case arises naturally in specific problems. For instance, if β is the subdifferential

∂j of a lower-semicontinuous convex function, that is, not differentiable, then β is necessarily multivalued and this situation occurs, for instance, in the description of stationary (equilibrium) states of systems with nondifferentiable energy.

Define the operator $A : D(A) \subset L^2(\Omega) \to L^2(\Omega)$,

$$\begin{cases} D(A) = \{y \in W_0^{1,p}(\Omega);\ \exists \eta \in (L^q(\Omega))^N;\ \eta(x) \in \beta(\nabla y(x)), \\ \qquad\qquad\qquad\qquad\qquad\qquad \text{a.e. } x \in \Omega,\ \operatorname{div} \eta \in L^2(\Omega)\}, \\ Ay = \{-\operatorname{div} \eta\},\quad \forall y \in D(A). \end{cases} \tag{3.27}$$

If β is single-valued, then A can be simply represented as

$$\begin{cases} Ay = -\operatorname{div} \beta(\nabla y),\quad \forall y \in D(A) \\ D(A) = \{y \in W_0^{1,p}(\Omega);\ \operatorname{div} \beta(\nabla y) \in L^2(\Omega)\}. \end{cases} \tag{3.28}$$

We have the following theorem.

Theorem 3.5 *The operator A is maximal monotone (equivaqlently, m-accretive) in $L^2(\Omega) \times L^2(\Omega)$. Moreover, if $\beta = \partial j$, where $j : \mathbb{R}^N \to \mathbb{R}$ is a continuous convex function, then $A = \partial \varphi$, where $\varphi : L^2(\Omega) \to \overline{\mathbb{R}}$ (the energy function) is given by*

$$\varphi(y) = \begin{cases} \displaystyle\int_\Omega j(\nabla y)dx & \text{if } y \in W_0^{1,p}(\Omega) \cap L^2(\Omega),\ j(\nabla y) \in L^1(\Omega) \\ +\infty & \text{otherwise.} \end{cases} \tag{3.29}$$

Proof. Arguing as in Proposition 2.23, it follows that φ is convex and lower-semicontinuous. On the other hand, because (3.28) is taken in the variational sense (3.12), that is, in the sense of distributions on Ω, we have

$$(Ay, \psi) = \int_\Omega \beta(\nabla y) \cdot \nabla \psi\, dx,\quad \forall \psi \in L^2(\Omega) \cap W_0^{1,p}(\Omega). \tag{3.30}$$

(Here $(\cdot,\cdot)$ is the above duality defined by the scalar product of $L^2(\Omega)$ and also the duality pairing on $W_0^{1,p}(\Omega) \times W^{1-q}(\Omega)$, $\frac{1}{p} + \frac{1}{q} = 1$.) This yields, of course,

$$(Ay - Az, y - z) \geq 0,\quad \forall y, z \in W_0^{1,p}(\Omega) \cap L^2(\Omega)$$

and so it remains true for all $y, z \in D(A)$. Hence A is monotone in $L^2(\Omega) \times L^2(\Omega)$.

To prove the maximal monotonicity, consider the equation

$$\lambda y + Ay \ni f, \tag{3.31}$$

where $\lambda > 0$ and $f \in L^2(\Omega)$. Taking into account (3.30), we rewrite (3.31) as

$$\lambda \int_\Omega y\psi + \int_\Omega \eta \cdot \nabla \psi\, dx = \int_\Omega f\psi\, dx,\quad \forall \psi \in W_0^{1,p}(\Omega) \cap L^2(\Omega), \tag{3.32}$$

where $\eta \in (L^q(\Omega))^N$, $\eta(x) \in \beta(\nabla y(x))$, a.e. $x \in \Omega$.

On the other hand, by Theorem 3.4, there is a solution y to (3.12) and therefore to (3.31), because by (3.30) it also follows that

$$\operatorname{div} \eta(\psi) = -\int_\Omega f\psi + \lambda \int_\Omega fy \leq C\|\psi\|_{L^2(\Omega)},\qquad \forall \psi \in W_0^{1,p}(\Omega) \cap L^2(\Omega)$$

and, therefore, $\operatorname{div} \eta \in L^2(\Omega)$. Hence A is maximal monotone, as claimed.

Now, if β is a subgradient maximal monotone graph of the form ∂j, it is easily seen that $A \subset \partial\varphi$, that is,

$$\varphi(y) - \varphi(z) \leq \int_\Omega \eta(y-z)dx, \qquad \forall \eta \in Ay,\ y, z \in L^2(\Omega).$$

Because A is maximal in the class of monotone operators, we have therefore $A = \partial\varphi$, as claimed. □

Remark 3.6 If instead of φ we consider the function $\phi : W_0^{1,p}(\Omega) \to \overline{\mathbb{R}}$ defined by

$$\phi(y) = \begin{cases} \displaystyle\int_\Omega j(\nabla y)dx & \text{if } y \in W_0^{1,p}(\Omega), \\ +\infty, & \text{otherwise,} \end{cases}$$

we see that ϕ is convex and lower-semicontinuous on $W_0^{1,p}(\Omega)$ (see Proposition 1.9). Moreover, its subdifferential $\partial\phi : W_0^{1,p}(\Omega) \to W^{-1,q}$, $\frac{1}{p} + \frac{1}{q} = 1$, is given by $\partial\phi = G$, where

$$G(y) = \{-\text{div } \eta\},\ \eta \in (L^q(\Omega))^N,\ \eta(x) \in \beta(\nabla y(x)),\ \text{a.e. in } \Omega.$$

Here is the argument. It is easily seen by the Gauss–Ostrogodski formula that $G \subset \partial\phi$. On the other hand, arguing as in the proof of Theorem 3.4, one can show that, for each $f \in W^{-1,q}(\Omega)$, the equation $J(y) + G(y) \ni f$ has a solution y, where $J : W_0^{1,p}(\Omega) \to W^{-1,q}(\Omega)$ is the duality mapping of the space $W_0^{1,p}(\Omega)$. (See (1.5), (1.6).) This means that G is maximal monotone in $W_0^{1,p}(\Omega) \to W^{-1,q}(\Omega)$ and so $G = \partial\phi$, as claimed.

It turns out that in the special case, where $\beta = \partial j$, assumptions (3.10) and (3.11) can be weakened to

$$\lim_{|r|\to\infty} \frac{j(r)}{|r|} = \lim_{|p|\to\infty} \frac{j^*(p)}{|p|} = +\infty, \tag{3.33}$$

$$\inf j = j(0) = 0, \qquad \lim_{|r|\to\infty} \frac{j(-r)}{j(r)} < \infty. \tag{3.34}$$

Here j^* is the conjugate of j, that is, $j^*(p) = \sup\{(p \cdot u) - j(u);\ u \in \mathbb{R}^N\}$. By $|\cdot|$ we denote, as usual, the Euclidean norm in $\mathbb{R}^N$.

An example of such a function j is

$$j(r) \equiv (1 + |r|)\log(1 + |r|) - |r|$$

with the conjugate

$$j^*(p) \equiv \exp(|p|) - |p| - 1.$$

We have

Theorem 3.7 *Under assumptions* (3.33) *and* (3.34), *problem* (3.1)–(3.5) *have, for each* $\lambda > 0$ *and* $f \in L^2(\Omega)$, *a unique weak solution* $y^* \in W^{1,1}(\Omega)$ *in the following sense*

$$\begin{cases} \int_\Omega (\lambda y v + \eta \cdot \nabla v) dx = \int_\Omega f v \, dx, & \forall v \in C^1(\overline{\Omega}) \\ \eta \in (L^1(\Omega))^N, \ \eta(x) \in \beta(\nabla y(x)), & a.e.\ x \in \Omega \\ j^*(\eta) \in L^1(\Omega), \ j(\nabla y) \in L^1(\Omega). \end{cases} \tag{3.35}$$

Moreover, y^* *is the unique minimizer of problem*

$$\min \left\{ \frac{\lambda}{2} \int_\Omega |y(x) - \frac{1}{\lambda} f(x)|^2 dx + \int_\Omega j(\nabla y(x)) dx; \ y \in W^{1,1}(\Omega) \right\}. \tag{3.36}$$

Proof. We assume for simplicity $\lambda = 1$. The existence of a unique minimizer u^* for problem (3.36) is an immediate consequence of Proposition 1.6 and of the fact that, under the first of conditions (3.33), the function is strictly convex

$$\varphi : L^2(\Omega) \to \overline{\mathbb{R}} = (-\infty, +\infty], \ \varphi(u) = \int_\Omega j(\nabla u(x)) dx + \frac{1}{2} \int_\Omega (u - f)^2 dx$$

and weakly lower-semicontinuous in the space $L^2(\Omega)$. Here is the argument. It follows by (3.33) that for each $\mu > 0$ the set $\mathcal{M} = \{y \in W^{1,1}(\Omega); \ \varphi(y) \le \mu\}$ is bounded in $W^{1,1}(\Omega)$, that is,

$$\|\nabla y\|_{(L^1(\Omega))^N} \le C, \quad \forall y \in \mathcal{M}$$

and

$$\left\{ \int_E |\nabla y(x)| dx; \ E \subset \Omega, \ u \in \mathcal{M} \right\}$$

is uniformly absolutely continuous. Indeed, by the first relation in (3.33), for any Lebesgue measurable set $E \subset \Omega$ and $n \in \mathbb{N}$, we have

$$\begin{aligned} \int_E |\nabla y(x)| dx &= \int_{E \cap \{x; |\nabla y(x)| \ge n\}} |\nabla y(x)| dx + \int_{E \cap \{x; |\nabla y(x)| < n\}} |\nabla y(x)| dx \\ &\le (\delta(n))^{-1} \int_{E \cap \{x; |\nabla y(x)| \ge n\}} j(\nabla y(x)) dx + n m(E) \\ &\le (\delta(n))^{-1} \mu + n m(E), \quad \forall n \in \mathbb{N}, \end{aligned} \tag{3.37}$$

where $\delta(n) \to \infty$ as $n \to \infty$ and m is the Lebesgue measure. This implies, as claimed, that for $\int_E |\nabla y(x)| dx \le \varepsilon$, $m(E) \le \eta(\varepsilon)$, where $\eta(\varepsilon) \to 0$ for $\varepsilon \to 0$.

Then, by the Dunford–Pettis theorem (Theorem 1.30), the set $\mathcal{M}$ is weakly compact in $W^{1,1}(\Omega)$. Hence, if $\{y_n\} \subset \mathcal{M}$ is weakly convergent to y in $L^2(\Omega)$, it follows that $\nabla y_n \to \nabla y$ weakly in $(L^1(\Omega))^N$ and, because the convex integrand $v \to \int_\Omega j(v)$ is weakly lower-semicontinuous in $(L^1(\Omega))^N$, we infer that

$$\liminf_{n \to \infty} \varphi(y_n) \ge \varphi(y).$$

Hence φ is weakly lower-semicontinuous (equivalently, lower-semicontinuous), and $\varphi(y) \to +\infty$ as $\|y\|_{L^2(\Omega)} \to \infty$. Then, as mentioned above, there is a unique solution $y \in W^{1,1}(\Omega) \cap L^2(\Omega)$ to the variational problem (3.36).

In order to prove that y^* is a solution to (3.35), we start with the approximating equation

$$\min\left\{\int_\Omega \left(j_\varepsilon(\nabla y) + \frac{\varepsilon}{2}\,|\nabla y(x)|^2 + \frac{1}{2}\,|y-f|^2\right)dx;\ y \in H^1(\Omega)\right\}, \tag{3.38}$$

where $j_\varepsilon \in C^1(\mathbb{R}^N)$ is the function (see (2.27)),

$$j_\varepsilon(p) = \inf\left\{\frac{1}{2\varepsilon}\,|v-p|^2 + j(v);\ v \in \mathbb{R}^N\right\}.$$

Problem (3.38) has a unique solution $y_\varepsilon \in H^1(\Omega)$ which, as easily seen, satisfies the elliptic boundary value problem

$$\begin{aligned} y_\varepsilon - \varepsilon\Delta y_\varepsilon - \operatorname{div}_x(\partial j_\varepsilon(\nabla y_\varepsilon)) = f \quad &\text{in } \Omega,\\ (\varepsilon\nabla y_\varepsilon + \partial j_\varepsilon(\nabla y_\varepsilon))\cdot n = 0 \quad &\text{on } \partial\Omega, \end{aligned} \tag{3.39}$$

where $n = \vec{n}$ is the outward normal to $\partial\Omega$. Equivalently,

$$\int_\Omega ((\varepsilon\nabla y_\varepsilon + \partial j_\varepsilon(\nabla y_\varepsilon))\cdot\nabla v + y_\varepsilon v)dx = \int_\Omega fv\,dx, \quad \forall v \in H^1(\Omega). \tag{3.40}$$

(The Gâteaux differential of the function arising in (3.38) is just the operator from the left-hand side of (3.39) or (3.40).) We recall that (see Theorem 2.16)

$$\begin{aligned} \partial j_\varepsilon(p) &= \frac{1}{\varepsilon}\,(p - (1+\varepsilon\beta)^{-1}p) \in \beta((1+\varepsilon\beta)^{-1}p), \quad \forall p \in \mathbb{R}^N,\\ j_\varepsilon(p) &= \frac{1}{2\varepsilon}\,|p - (1+\varepsilon\beta)^{-1}p|^2 + j((1+\varepsilon\beta)^{-1}p). \end{aligned}$$

Then, it is readily seen by (3.38) that on a subsequence, again denoted $\{\varepsilon\} \to 0$, we have

$$\begin{aligned} y_\varepsilon &\to y^* && \text{weakly in } L^2(\Omega),\\ ((1+\varepsilon\beta)^{-1}\nabla y_\varepsilon - \nabla y_\varepsilon) &\to 0 && \text{strongly in } L^2(\Omega;\mathbb{R}^N),\\ (1+\varepsilon\beta)^{-1}\nabla y_\varepsilon &\to \nabla y^* && \text{weakly in } L^1(\Omega;\mathbb{R}^N),\\ \varepsilon\nabla y_\varepsilon &\to 0 && \text{strongly in } L^2(\Omega;\mathbb{R}^N). \end{aligned} \tag{3.41}$$

The latter follows by the obvious inequality

$$\begin{aligned} &\int_\Omega\left(j((1+\varepsilon\beta)^{-1}\nabla y_\varepsilon) + \frac{1}{2\varepsilon}\,|\nabla y_\varepsilon - (1+\varepsilon\beta)^{-1}\nabla y_\varepsilon|^2 + \frac{\varepsilon}{2}\,|\nabla y_\varepsilon|^2 + \frac{1}{2}\,|y_\varepsilon - f|^2\right)dx\\ &\quad \le \int_\Omega\left(j(\nabla y_\varepsilon) + \frac{\varepsilon}{2}\,|\nabla y_\varepsilon|^2 + \frac{1}{2}\,|y_\varepsilon - f|^2\right)dx \qquad (3.42)\\ &\quad \le \int_\Omega\left(j_\varepsilon(\nabla v) + \frac{\varepsilon}{2}\,|\nabla v|^2 + \frac{1}{2}\,|v - f|^2\right)dx, \quad \forall v \in H^1(\Omega). \end{aligned}$$

On the other hand, by (3.42) and the first condition in (3.33), it follows as above via the Dunford–Pettis theorem (Theorem 1.30) that $\{(1+\varepsilon\beta)^{-1}\nabla y_\varepsilon\}$ is weakly compact in $L^1(\Omega;\mathbb{R}^N)=(L^1(\Omega))^N$ and so (3.41) follows. Then, taking into account the weak lower-semicontinuity of functional φ in $L^1(\Omega;\mathbb{R}^N)$, we see that

$$\int_\Omega\left(j(\nabla y^*)+\frac{1}{2}\,|y^*-f|^2\right)dx\le\int_\Omega\left(j(\nabla v)+\frac{1}{2}\,|v-f|^2\right)dx,\quad\forall v\in W^{1,1}(\Omega),$$

that is, y^* is optimal in problem (3.36).

Now, we recall the inequality (see Proposition 1.7)

$$j(v)+j^*(p)\ge v\cdot p,\quad\forall v,p\in\mathbb{R}^N \tag{3.43}$$

with equality if and only if $p\in\beta(v)=\partial j(v)$. Taking into account (3.40), where $v=y_\varepsilon$, this yields

$$\begin{aligned}\int_\Omega(j((1+\varepsilon\beta)^{-1}\nabla y_\varepsilon)+j^*(\partial j(\nabla y_\varepsilon)))dx&=\int_\Omega(1+\varepsilon\beta)^{-1}\nabla y_\varepsilon\cdot\partial j(\nabla y_\varepsilon)dx\\&=\int_\Omega\nabla y_\varepsilon\cdot\partial j(\nabla y_\varepsilon)dx-\frac{1}{\varepsilon}\int_\Omega|\partial j_\varepsilon(\nabla y_\varepsilon)|^2dx\le\int_\Omega|f|^2dx.\end{aligned}$$

(Here, $\partial j(\nabla y_\varepsilon)$ is any section of $\partial j(\nabla y_\varepsilon)$.)

Then, by inequality (3.42), we see that $\{\int_\Omega j^*(\partial j(\nabla y_\varepsilon))dx\}$ is bounded and so, by the second condition in (3.33) and, again by the Dunford–Pettis theorem, we infer as above that $\left\{\int_E\partial j(y_\varepsilon);\ E\subset\Omega\right\}$ is uniformly absolutely continuous and therefore $\{\partial j(\nabla y_\varepsilon)\}$ is weakly compact in $(L^1(\Omega))^N$. Here, one uses the same argument as above (see (3.37)) starting from the inequality

$$\begin{aligned}\int_E|\partial j(\nabla y_\varepsilon)|dx&\le\int_{E\cap[|\partial j(\nabla y_\varepsilon)|\ge n]}|\partial j(\nabla y_\varepsilon)|dx+\int_{E\cap[|\partial j(\nabla y_\varepsilon)\le n]}|\partial j(\nabla y_\varepsilon)|dx,\\&\le C\delta(n)+nm(E),\quad\forall n\in\mathbb{N},\ E\subset\Omega.\end{aligned}$$

Hence, we may assume that for $\varepsilon\to0$,

$$\partial j(\nabla y_\varepsilon)\to\eta\quad\text{weakly in }(L^1(\Omega))^N,$$

and, letting $\varepsilon\to0$ in (3.40), it follows that

$$\int_\Omega(y^*v+\nabla v\cdot\eta)dx=\int_\Omega fv\,dx,\quad\forall v\in C^1(\overline{\Omega}). \tag{3.44}$$

To conclude the proof, it remains to be shown that

$$\eta(x)\in\beta(\nabla y^*(x)),\quad\text{a.e. }x\in\Omega. \tag{3.45}$$

To this aim, we notice that, by virtue of (3.41) and the conjugacy equality, it follows by the weak lower-semicontinuity of the convex integrand in $L^1(\Omega)$ that

$$\begin{aligned}\int_\Omega(j(\nabla y^*)+j(\eta))dx&\le\liminf_{\varepsilon\to0}\int_\Omega(1+\varepsilon\beta)^{-1}\nabla y_\varepsilon\cdot\partial j(\nabla y_\varepsilon)dx\\&\le\liminf_{\varepsilon\to0}\int_\Omega\nabla y_\varepsilon\cdot\partial j(\nabla y_\varepsilon)dx.\end{aligned} \tag{3.46}$$

On the other hand, by (3.40) and (3.42), we see that

$$\lim_{\varepsilon\to 0}\int_\Omega \nabla y_\varepsilon\cdot\partial j(\nabla y_\varepsilon)dx=-\int_\Omega (y^*-f)y^*dx. \tag{3.47}$$

We also have by (3.43) that

$$\begin{aligned}\nabla y^*\cdot\eta&\le j(\nabla y^*)+j^*(\eta), &&\text{a.e. in }\Omega\\ -\nabla y^*\cdot\eta&\le j(-\nabla y^*)+j^*(\eta)\le Cj(\nabla y^*)+j(\eta), &&\text{a.e. in }\Omega.\end{aligned}$$

(The second inequality follows by the convexity of j^*.) Hence, $\nabla y^*\cdot\eta\in L^1(\Omega)$ and so, by (3.44), (3.46) and (3.47), we see that

$$\int_\Omega (j(\nabla y^*)+j^*(\eta)-\nabla y^*\cdot\eta)dx\le 0,$$

because (3.44) extends by density to all $v\in W^{1,1}(\Omega)$ such that $\nabla v\cdot\eta\in L^1(\Omega)$. Recalling that $j^*(\nabla y^*)+j(\eta)-\nabla y^*\cdot\eta\ge 0$, a.e. in Ω, we infer that

$$j(\nabla y^*(x))+j^*(\eta(x))=\nabla y^*(x)\cdot\eta(x),\quad \text{a.e. } x\in\Omega,$$

which implies (3.45), as claimed. Hence, y^* is a weak solution in sense of (3.35).

Conversely, any weak solution y^* to (3.35) minimizes the functional φ. Indeed, we have

$$\begin{aligned}\varphi(y^*)-\varphi(v)&\le\int_\Omega\left(j(\nabla y^*)-j(v)+\frac{1}{2}\left(|y^*-f|^2-|v-f|^2\right)\right)dx\\ &\le\int_\Omega(\eta\cdot(\nabla y^*-\nabla v)+(y^*-f)(u^*-v))dx=0,\quad \forall v\in C^1(\overline{\Omega}).\end{aligned}$$

The latter inequality extends to all $v\in D(\varphi)\in\{z\in L^2(\Omega);\ \varphi(z)<\infty\}$. □

Remark 3.8 In particular, it follows by Theorem 3.7 that the operator A, defined by (3.27) in sense of definition (3.35), is maximal monotone in $L^2(\Omega)\times L^2(\Omega)$.

Remark 3.9 Theorem 3.2 extends to nonlinear m-order elliptic boundary value problems of the form

$$\begin{aligned}\sum_{|\alpha|\le m} D^\alpha A_\alpha(x,y,D^\gamma y)&=f(x), && x\in\Omega,\ |\gamma|\le m,\\ D^\alpha y&=0 && \text{on } \partial\Omega,\ |\alpha|<m,\end{aligned} \tag{3.48}$$

where $A_\alpha:\Omega\times\mathbb{R}^{mN}\to\mathbb{R}^{mN}$ are measurable functions in x continuous in other variables and satisfy the following conditions.

(i) $\sum_{|\alpha|\le m}(A_\alpha(x,\xi)-A_\alpha(x,\eta))\cdot(\xi-\eta)\ge 0,\ \forall\xi,\eta\in\mathbb{R}^{mN}$.

(ii) $\sum_{|\alpha|\le m}A_\alpha(x,\xi)\cdot\xi\ge\omega\|\xi\|^p-C,\ \forall\xi\in\mathbb{R}^{mN}$, where $\omega>0,\ p>1$.

(iii) $\|A_\alpha(x,\xi)\|\le C_1\|\xi\|^{p-1}+C_2,\ \forall\xi\in\mathbb{R}^{mN},\ x\in\Omega$.

Here D^γ is the multi-index $\{D_{x_j}^{\gamma_j},\ j=1,...,N,\ \gamma_j \le m\}$, and $\|\cdot\|$ is the Euclidean norm in $\mathbb{R}^{mN}$.

Indeed, as easily seen, under these assumptions, the operator $A : X \to X^*$, $X = W_0^{m,p}(\Omega)$, $X^* = W^{-m,q}(\Omega)$, defined by

$$(Ay, z) = \sum_{|\alpha| \le m} \int_\Omega A_\alpha(x, y(x), D^\gamma y(x)) \cdot D^\alpha z(x) dx, \quad \forall y, z \in W_0^{m,p}(\Omega)$$

is monotone and coercive.

Moreover, it is demicontinuous, because if $y_n \to y$ in $W_0^{m,p}(\Omega)$, then by Theorem 1.12 it follows that $\{D^\gamma y_n\}$ is compact in $L^{\frac{Np}{N-p}}(\Omega)$ for all $|\gamma| \le m$ and so, on a subsequence, $D^\gamma y_n \to D^\gamma y$ in $L^{\frac{Np}{N-p}}(\Omega)$. This implies that $A_\alpha(x, y, D^\gamma y_n) \to A_\alpha(x, y, D^\gamma y)$ weakly in $L^q(\Omega)$, $\frac{1}{q} = 1 - \frac{1}{p}$, and by the continuity of $v \to A_\alpha(\cdot, v)$ it also follows that $A_\alpha(x, y(x), D^\gamma y_n(x)) \to A_\alpha(x, y(x), D^\gamma y(x))$, a.e. $x \in \Omega_\delta$. Hence, $(Ay_n, z) \to (Ay, z)$, $\forall z \in W_0^{m,p}(\Omega)$, and so A is demicontinuous. Then, the existence of a generalized solution $y \in W_0^{m,p}(\Omega)$ to problem (3.48) for $f \in L^2(\Omega)$ (that is, $Ay = f$) follows as in the previous case by Corollary 2.9. The details are left to the reader.

Consider now the case where

$$j(r) \equiv g(|r|), \quad r \in \mathbb{R}^N,$$

where $g : [0, \infty) \to \mathbb{R}$ is a convex continuous function which satisfies

$$\begin{aligned} &\lim_{r \to \infty} \frac{g(r)}{r} = \lim_{|p| \to \infty} \frac{g^*(p)}{|p|} = +\infty, \\ &\inf_{r \ge 0} g(r) = g(0) = 0. \end{aligned} \tag{3.49}$$

Then, problem (3.1)–(3.2) where $\beta = \partial j$ reduces to

$$\begin{aligned} &\lambda y - \operatorname{div}(\partial g(\eta)) \ni f && \text{in } \Omega, \\ &\eta(x) \in \operatorname{sign}(\nabla y(x)), && \text{a.e. } x \in \Omega, \\ &y = 0 && \text{on } \partial\Omega, \end{aligned} \tag{3.50}$$

where $\operatorname{sign} : \mathbb{R}^N \to \mathbb{R}^N$ is the multivalued maximal monotone mapping

$$\begin{aligned} &\operatorname{sign} r = \tfrac{r}{|r|} \quad \text{for } r \ne 0; \\ &\operatorname{sign} 0 = \{\xi \in \mathbb{R}^N;\ |\xi| \le 1\}. \end{aligned} \tag{3.51}$$

We note, however, that the special case $g(r) \equiv |r|$, that is,

$$\begin{aligned} &\lambda y - \operatorname{div}(\operatorname{sign}(\nabla y)) \ni f && \text{in } \Omega, \\ &y = 0 && \text{on } \partial\Omega, \end{aligned} \tag{3.52}$$

is ruled out from Theorem 3.7. This singular case will be treated later on in the space of functions with bounded variations on Ω.

Variational approach to image denoising

The nonlinear diffusion techniques and PDE-based variational models are very popular in image denoising and restoring (see, e.g., Rudin, Osher and Fatemi [99]). A gray value image is defined by a function f from a given domain Ω of $\mathbb{R}^N$, to $\mathbb{R}$. In each point $x \in \Omega$, $f(x)$ is the light intensity of the corrupted image located in x. Then, a restored (denoised) image $u : \Omega \to \mathbb{R}$ is computed from the minimization problem (3.38), that is,

$$\text{Minimize}\left\{\frac{\lambda}{2}\int_\Omega (u(x)-f(x))^2dx + \int_\Omega j(\nabla u(x))dx,\ u \in X(\Omega)\right\}, \tag{3.53}$$

where $\lambda > 0$, $j : \mathbb{R}^N \to \mathbb{R}$ is a given function and $X(\Omega)$ is a Sobolev space of functions on Ω. The term $j(\nabla u)$ arising here is taken in order to smooth (mollify) the observation u. In order for the minimization problem to be well posed, one must assume that j is convex and lower-semicontinuous and $X(\Omega)$ must be taken, in general, as a distribution space on Ω, for instance, the Sobolev space $W^{1,p}(\Omega)$, where $p \geq 1$. In this case, as seen earlier, problem (3.53) is equivalent to the nonlinear diffusion equation

$$\begin{cases} -\text{div}(\beta(\nabla y)) + \lambda(y-f) = 0 & \text{in } \Omega, \\ \beta(\nabla y(x)) \cdot n = 0 & \text{on } \partial\Omega, \end{cases} \tag{3.54}$$

where $\beta : \mathbb{R}^N \to \mathbb{R}^N$ is the subdifferential of j and $n = n(x)$ is the normal to $\partial\Omega$ at x. (If $X(\Omega) = W_0^{1,p}(\Omega)$, then in the above problem the boundary condition is: $y = 0$ on $\partial\Omega$.) It turns out that equation (3.54) can be used as a successful filtering process to restoring the original corrupted image f.

In the first (naive) image processing models, j was taken quadratic (linear Gaussian filters) and most of the subsequent models have considered in (3.53) functions j of the form

$$j(\nabla y) \equiv |\nabla y|^p,\quad p \geq 1,$$

and $X(\Omega) = W^{1,p}(\Omega)$. As mentioned above, $p > 1$, then $y \in W^{1,p}(\Omega)$ and so, the term $j(\nabla y)$ in the above minimization problem has a smoothing effect in restoring the degraded image f while preserving edges. For the second objective, equation (3.52), that is $p = 1$, $j(\nabla y) \equiv |\nabla y|$ and $X(\Omega) = W^{1,1}(\Omega))$ might be apparently more appropriate. However, in this case, the functional arising in (3.53), that is $y \to \int_\Omega |\nabla y| dx$ is not lower-semicontinuous in $L^2(\Omega)$, and so problems (3.53), (3.54) are not well posed. Thus, for existence in problems (3.53)–(3.54), the Sobolev space $W^{1,1}(\Omega)$ must be replaced by a large space where this happens and the natural choice is the space $BV(\Omega)$ of functions u with bounded variations, and instead of the Sobolev norm $\int_\Omega |\nabla y| dx$ to take the total variation functional of y. (This functional framework is briefly discussed below.) The case treated in Theorem 3.7 is an intermediate one between $L^p(\Omega)$ with $p > 1$ and $BV(\Omega)$ and so, for p sufficiently close to 1, leads to acceptable results. (On this case, see also the work [29].)

The BV approach to the nonlinear elliptic equations with singular diffusivity.

As mentioned earlier, the existence theory for equation (3.1) developed above fails for $p = 1$, the best example being, perhaps, in the case where $\beta = \partial j$, $j(u) = |u|$ mentioned above, that is, of the singular diffusion equation (see (3.52))

$$y - \operatorname{div}(\operatorname{sign}(\nabla y)) \ni f \quad \text{in } \Omega, \tag{3.55}$$

with the Dirichlet boundary value condition

$$y = 0 \quad \text{on } \partial\Omega, \tag{3.56}$$

or the Neumann condition

$$\nabla y \cdot n = 0 \quad \text{on } \partial\Omega. \tag{3.57}$$

This equation comes formally from variational problems with nondifferentiable energy and it is our aim here to give a rigorous meaning to it. As noticed earlier, this equation is relevant in image restoration as well as in mathematical modeling of faceted crystal growth. Formally, (3.55) is equivalent with the minimization problem (for Dirichlet null boundary condition)

$$\min\left\{\frac{1}{2}\int_\Omega |y-f|^2 dx + \int_\Omega |\nabla y| dx;\ y \in W_0^{1,1}(\Omega)\right\} \tag{3.58}$$

or

$$\min\left\{\frac{1}{2}\int_\Omega |y-f|^2 dx + \int_\Omega |\nabla y| dx;\ y \in W^{1,1}(\Omega)\right\} \tag{3.59}$$

in the case of Neumann boundary conditions.

However, as mentioned earlier, problems (3.58) or (3.59) are not well posed in the $W^{1,1}(\Omega)$-setting, the main reason being that the energy functionals

$$\varphi(y) = \begin{cases} \displaystyle\int_\Omega |\nabla y(x)| dx & \text{if } y \in W_0^{1,1}(\Omega), \\ +\infty & \text{otherwise} \end{cases}$$

$$\psi(y) = \begin{cases} \displaystyle\int_\Omega |\nabla y(x)| dx & \text{if } y \in W^{1,1}(\Omega), \\ +\infty & \text{otherwise} \end{cases}$$

are not lower-semicontinuous on $L^2(\Omega)$. As mentioned earlier, this fact suggests replacing the space $W^{1,1}(\Omega)$ by a larger space and more precisely by the space $BV(\Omega)$ defined in Section 1.3. Namely, consider the function $\varphi : L^2(\Omega) \to (-\infty, +\infty]$, defined by

$$\Phi(y) = \begin{cases} |Dy|(\Omega) & \text{if } y \in BV(\Omega) \\ +\infty & \text{otherwise,} \end{cases} \tag{3.60}$$

respectively,

$$\Psi(y) = \begin{cases} |Dy|(\Omega) & \text{if } y \in BV(\Omega) \\ +\infty & \text{otherwise.} \end{cases} \tag{3.61}$$

Here H^{N-1} is the Hausdorff $(N-1)$ dimensional measure in R^N and $\gamma(u)$ is the trace of u on $\partial\Omega$. It follows that (see Theorem 13.4.2 in [4] the functions Φ and Ψ are the *lower-semicontinuous envelopes in* $L^1(\Omega)$ of φ and ψ, respectively, that is, for all $y \in L^1(\Omega)$,

$$\liminf_{n\to\infty} \varphi(y_n) \geq \Phi(y) \text{ if } y_n \to y \text{ in } L^1(\Omega),$$

and there exists $\{y_n\} \to y$ in $L^1(\Omega)$ such that

$$\limsup_{n\to\infty} \varphi(y_n) \leq \Phi(y).$$

Similarly, for the pair (ψ, Ψ). This means that the minimization problems

$$\min \frac{1}{2}\int_\Omega |y-f|^2 dx + |Dy|(\Omega); \qquad y \in BV(\Omega), \tag{3.62}$$

$$\min \frac{1}{2}\int_\Omega |y-f|^2 dx + |Dy|(\Omega); \qquad y \in BV(\Omega), \tag{3.63}$$

which replace (3.58) and (3.59), respectively, have unique solutions $y \in BV(\Omega)$. If we denote by $\partial\Phi, \partial\Psi : L^2(\Omega) \to L^2(\Omega)$ the subdifferentials of functions Φ and Ψ, that is,

$$\partial\Phi(y) = \left\{\eta \in L^2(\Omega);\ \Phi(y) - \Phi(z) \leq \int_\Omega \eta(y-z)dx,\ \forall y, z \in BV(\Omega)\right\}, \tag{3.64}$$

$$\partial\Psi(y) = \left\{\xi \in L^2(\Omega);\ \Psi(y) - \Psi(z) \leq \int_\Omega \xi(y-z)dx,\ \forall y, z \in BV(\Omega)\right\}, \tag{3.65}$$

we may write equivalently (3.62) and (3.63) as

$$y + \partial\Phi(y) \ni f \tag{3.66}$$

respectively,

$$v + \partial\Psi(v) \ni f. \tag{3.67}$$

The solutions y and v to equations (3.66) (respectively, (3.67)) are to be viewed as variational (generalized) solutions to (3.56) and (3.57) and, respectively, (3.66) and (3.67).

Taking into account that for $y \in W^{1,1}(\Omega) \subset BV(\Omega)$ we have $\|Dy\| = |\nabla y|_{L^1(\Omega)}$, it follows that, if $y \in W_0^{1,1}(\Omega)$ and $\eta = -\text{div}\left(\frac{\nabla y}{|\nabla y|}\right) \in L^2(\Omega)$, then $\eta \in \partial\Phi(y)$. Similarly, if $y \in W^{1,1}(\Omega)$, $\nabla y \cdot n = 0$ on $\partial\Omega$ and $\xi = -\text{div}\left(\frac{\nabla y}{|\nabla y|}\right) \in L^2(\Omega)$, then $\xi \in \partial\Psi(y)$. Of course, in general, one might not expect that $y \in W^{1,1}(\Omega)$ and so, the above calculation remains formal.

However, one can give a meaning to equation (3.66), that is, to (3.55)–(3.56), by approximating it by the elliptic boundary value problem

$$\begin{aligned} &y - \varepsilon\Delta y - \text{div}(\text{sign}(\nabla y)) \ni f && \text{in } \Omega, \\ &y = 0 && \text{on } \Omega, \end{aligned} \tag{3.68}$$

with the obvious modification in the case of the boundary conditions (3.57), that is, for equation (3.67).

As seen earlier, by Theorem 3.5, where $j(r) \equiv \frac{\varepsilon}{2}|r|^2 + |r|$, it follows that for each $f \in L^2(\Omega)$ equation (3.68) has, for each $\varepsilon > 0$, a unique solution $y_\varepsilon \in H_0^1(\Omega)$. It turns out that, for $\varepsilon \to 0$, y_ε approximates the solution $y \in BV(\Omega)$ to equation (3.66). Namely, we have

Theorem 3.10 *Let Ω be a bounded, convex domain of $\mathbb{R}^N$ with C^2-boundary. Then equation (3.88) has a unique solution $y_\varepsilon = y_\varepsilon(f) \in H_0^1(\Omega) \cap H^2(\Omega)$ which satisfies*

$$\int_\Omega \left(\frac{1}{2}|y_\varepsilon|^2 + \varepsilon|\nabla y_\varepsilon|^2 + |\nabla y_\varepsilon|\right) dx \le \frac{1}{2}\int_\Omega |f|dx, \ \forall \varepsilon > 0, \tag{3.69}$$

$$\int_\Omega (|\nabla y_\varepsilon|^2 + \varepsilon|\Delta y_\varepsilon|^2)dx \le \frac{1}{2\varepsilon}\int_\Omega |f|^2 dx, \tag{3.70}$$

$$\|y_\varepsilon(f) - y_\varepsilon(\bar{f})\|_{L^2(\Omega)} \le \|f - \bar{f}\|_{L^2(\Omega)}, \ \forall f, \bar{f} \in L^2(\Omega). \tag{3.71}$$

Moreover, for $\varepsilon \to 0$,

$$y_\varepsilon \to y \quad in\ L^2(\Omega), \tag{3.72}$$

where $y \in BV(\Omega)$ is the solution to (3.66).

The proof is based on the following result due to H. Brezis (personal communication).

Proposition 3.11 *Let Ω be a bounded, convex domain of $\mathbb{R}^N$, $N \ge 1$, with smooth boundary (of class C^2). Let $J_\varepsilon = (I + \varepsilon A)^{-1}$. where $A = -\Delta$, $D(A) = H_0^1(\Omega) \cap H^2(\Omega)$. Then, for all $\varepsilon > 0$,*

$$\int_\Omega |\nabla J_\varepsilon(y)|d\xi \le \int_\Omega |\nabla y|d\xi, \quad \forall y \in W_0^{1,1}(\Omega) \cap L^2(\Omega). \tag{3.73}$$

Proof. Proposition 3.11 amounts to saying that the operator A is accretive in the space $W_0^{1,1}(\Omega)$. If we denote by e^{-tA} the C_0-semigroup generated on $L^2(\Omega)$ by $-A$, inequality (3.73) is equivalent to

$$\|e^{-tA}y_0\|_{W_0^{1,1}(\Omega)} \le \|y_0\|_{W_0^{1,1}(\Omega)}, \ \forall y_0 \in W_0^{1,1}(\Omega) \cap L^2(\Omega).$$

We note that, in the case $\Omega = \mathbb{R}^N$, (3.73) follows immediately taking into account that (see Theorem 3.16, below)

$$\|(I - \Delta)^{-1}f\|_{L^1(\mathbb{R}^N)} \le \|f\|_{L^1(\mathbb{R}^N)}, \quad \forall f \in L^1(\mathbb{R}^N).$$

The bounded case considered here is more delicate. Rescaling if necessary, we may assume $\varepsilon = 1$ and so, we reduce (3.73) to

$$\int_\Omega |\nabla u|d\xi \le \int_\Omega |\nabla y|d\xi, \quad \forall y \in W_0^{1,1}(\Omega), \tag{3.74}$$

where

$$u - \Delta u = y \text{ in } \Omega; \quad u = 0\, dS - \text{ almost everywhere on } \partial\Omega,$$

and dS is the surface measure on $\partial\Omega$. Without loss of generality, we may also assume $y \in C_0^\infty(\Omega)$. We set

$$D_i = \frac{\partial}{\partial \xi_i},\quad D_{ij}^2 = \frac{\partial^2}{\partial \xi_i \partial \xi_j},\quad i,j = 1,...,N,$$

$$\varphi(\xi) = |\nabla u(\xi)| = \left(\sum_{i=1}^N |D_i u|^2\right)^{\frac{1}{2}},\quad \varphi_\nu(\xi) = \sqrt{\nu^2 + |\nabla u(\xi)|^2},\ \xi \in \Omega.$$

We shall prove first that

$$\frac{\varphi^2}{\varphi_\nu} - \Delta\varphi_\nu \le |\nabla y| \text{ in } \Omega. \tag{3.75}$$

Indeed, we have

$$\varphi_\nu D_j \varphi_\nu = \sum_{i=1}^N D_i u D_{ij}^2 u,$$

which yields

$$(D_j\varphi_\nu)^2 \le \frac{1}{\varphi_\nu^2} |\nabla u|^2 \sum_{i=1}^N (D_{ij}^2 u)^2 \text{ in } \Omega,$$

and, therefore,

$$|\nabla\varphi_\nu|^2 \le \frac{\varphi^2}{\varphi_\nu^2} \sum_{i,j=1}^N |D_{ij}^2 u|^2 \le \sum_{i=1}^N |D_{ij}^2 u|^2 \text{ in } \Omega.$$

We also have

$$\begin{aligned}\varphi_\nu \Delta\varphi_n u + |\nabla\varphi_\nu|^2 &= \sum_{i,j=1}^N |D_{ij}^2 u|^2 + \sum_{i=1}^N D_i u \Delta D_i u = \sum_{i,j=1}^N |D_{ij}^2 u|^2 + \sum_{i=1}^N D_i u(D_i u - D_i y)\\ &= \sum_{i,j=1}^N |D_{ij}^2 u|^2 + |\nabla u|^2 - \nabla u \cdot \nabla y \le |\nabla\varphi_\nu|^2 + \varphi^2 - \varphi|\nabla y|.\end{aligned}$$

This yields

$$-\varphi_\nu \Delta\varphi_\nu + \varphi^2 \le \varphi|\nabla y| \text{ in } \Omega,$$

which implies (3.74), as claimed.

Now, assume that $0 \in \partial\Omega$ and represent locally the boundary $\partial\Omega$ as $\partial\Omega = \{(\xi', \xi_N); \xi_N = \gamma(\xi')\}$, where γ is a C^2-function in a neighborhood of 0 in $\mathbb{R}^{N-1}$ and $\gamma(0) = 0$, $\nabla\gamma(0) = 0$.

We prove first

$$D_N\varphi_\nu(0) = (D_N u)^2(0)(\nu^2 + (D_N u)^2(0))^{-\frac{1}{2}} \Delta_{\xi'}\gamma(0). \tag{3.76}$$

Indeed, we have

$$\varphi_\nu D_N \varphi_\nu = \sum_{i=1}^N D_i u D_N u.$$

Since $u = 0$ on $\partial\Omega$,

$$u(\xi_1, \xi_2, ..., \xi_{N-1}, \gamma(\xi_1, \xi_2, ..., \xi_{N-1})) = 0$$

and differentiating with respect to ξ_i, $i = 1, ..., N-1$, yields

$$D_i u + D_N u D_i \gamma \equiv 0, \ i = 1, ..., N-1,$$

$$D^2_{ii} u + 2D_{iN} u D_i \gamma + D^2_{NN} u (D_i \gamma)^2 + D_N u D_{ii} \gamma \equiv 0,$$

and so, we get in $\{\xi' = 0, \ \xi_N = 0\}$,

$$D_i u(0) = 0, \quad D_{ii} u(0) + D_N u(0) D_{ii} \gamma(0) = 0.$$

Hence,

$$\Delta u(0) = D_{NN} u(0) - D_N u(0) \Delta_{\xi'} \gamma(0),$$

and since $\Delta u(0) = 0$, this yields

$$D_{NN} u(0) = D_N u(0) \Delta_{\xi'} \gamma(0).$$

Therefore,

$$\sqrt{\nu^2 + (D_N u(0))^2}\, D_N \varphi_\nu(0) = D_N u(0) D_{NN} u(0) = (D_N u)^2(0) \Delta_{\xi'} \gamma(0),$$

as claimed.

Now, let n be the outward normal to $\partial\Omega$. We have

$$\frac{\partial \varphi_\nu}{\partial n}(0) = -D_N \varphi_\nu(0) = \frac{-(D_N u(0))^2}{\sqrt{\nu^2 + (D_N u)^2(0)}} \Delta_{\xi'} \gamma(0).$$

On the other hand, since Ω is convex, we have $\Delta\gamma(0) \geq 0$ and, therefore,

$$\frac{\partial \varphi_\nu}{\partial n}(0) \leq 0.$$

Since the origin 0 can be replaced by an arbitrary point of $\partial\Omega$, we have

$$\frac{\partial \varphi_\nu}{\partial n} \leq 0 \ \text{ on } \partial\Omega.$$

Integrating (3.76) over Ω, by Green's formula we get

$$\int_\Omega \frac{\varphi^2}{\varphi_\nu} d\xi - \int_{\partial\Omega} \frac{\partial \varphi_\nu}{\partial n} dS \leq \int_\Omega |\nabla y| d\xi,$$

and, therefore,

$$\int_\Omega \frac{\varphi^2}{\varphi_\nu} d\xi \leq \int_\Omega |\nabla y| d\xi.$$

Then, letting $\nu \to 0$, we get (3.74), thereby completing the proof of Proposition 3.11. □

Proposition 3.12 *Let $g : [0, \infty) \to [0, \infty)$ be a continuous and convex function of at most quadratic growth such that $g(0) = 0$. Then*

$$\int_\Omega g(|\nabla J_\varepsilon(y)|) d\xi \leq \int_\Omega g(|\nabla y|) d\xi, \quad \forall y \in H^1_0(\Omega), \ \forall \varepsilon > 0.$$

Proof. Since g is of at most quadratic growth, as before we may assume that $y \in C_0^\infty(\Omega)$. Furthermore, without loss of generality, we may assume that $g \in C^2([0,\infty))$. (This can be achieved by regularizing the function g.) As in the previous case, it suffices to assume $\varepsilon = 1$. We set $\phi(\xi) = g(\varphi(\xi))$, $\phi_\nu(\xi) = g(\varphi_\nu(\xi))$, $\xi \in \Omega$, where φ and φ_ν are as above. We have

$$\nabla\phi_\nu = g'(\varphi_\nu)\nabla\varphi_\nu,\ \ \Delta\phi_\nu = g'(\varphi_\nu)\Delta\varphi_\nu + g''(\varphi_\nu)|\nabla\varphi_\nu|^2,\ \ \xi \in \Omega,$$

and so

$$\begin{aligned}\frac{\phi^2}{\phi_\nu} - \Delta\phi_\nu &= g'(\varphi_\nu)\left(\frac{\varphi^2}{\varphi_\nu} - \Delta\varphi_\nu\right) + \frac{g^2(\varphi)}{g(\varphi_\nu)} - g'(\varphi_\nu)\frac{\varphi^2}{\varphi_\nu} - g''(\varphi_\nu)|\nabla\varphi_\nu|^2 \\ &\le g'(\varphi_\nu)|\nabla y| + \frac{g^2(\varphi)}{g(\varphi_\nu)} - g'(\varphi_\nu)\frac{\varphi^2}{\varphi_\nu}.\end{aligned} \tag{3.77}$$

Now, proceeding as in the proof of Proposition 3.11, we take $0 \in \partial\Omega$ and represent locally $\partial\Omega$ as $\{(\xi',\xi_N);\xi_N = \gamma(\xi')\}$, where $\gamma \in C^2$, $\gamma(0) = 0$, $\nabla\gamma(0) = 0$. Since g is increasing, we have

$$D_N\phi_\nu(0) = g'(\varphi_\nu(0))D_N\varphi_\nu(0) = -g'(\varphi_\nu(0))\frac{\partial\varphi_\nu}{\partial n}(0) \ge 0.$$

This yields $\frac{\partial\phi_\nu}{\partial n}(0) = -D_N\varphi_\nu(0) \le 0$ and, therefore, replacing 0 by an arbitrary point of $\partial\Omega$, we obtain that $\frac{\partial\phi_\nu}{\partial n} \le 0$ on $\partial\Omega$. Integrating (3.77) over Ω, we get, therefore,

$$\int_\Omega \frac{\phi^2}{\phi_\nu}dx \le \int_\Omega \left(g'(\varphi_\nu)\left(|\nabla y| - \frac{\varphi^2}{\varphi_\nu}\right) + \frac{g^2(\varphi)}{g(\varphi_\nu)}\right)dx.$$

Letting $\nu \to 0$, we see that

$$\int_\Omega g(\varphi)dx \le \int_\Omega g'(\varphi)(|\nabla y| - \varphi)dx + \int_\Omega g(\varphi)d\xi \le \int_\Omega g(|\nabla y|)dx$$

because $g'(u)(u-v) \ge g(u) - g(v)$, $\forall u,v \in \mathbb{R}^+$.

This completes the proof of Proposition 3.12. □

Let $j_\lambda : \mathbb{R}^N \to \mathbb{R}$, $\lambda > 0$, be the Moreau–Yosida approximation of $j(r) = |r|$ (see (2.27)). Hence,

$$j_\lambda(u) = \begin{cases} \dfrac{1}{2\lambda}\,|u|^2 & \text{for } |u| \le \lambda, \\ |u| - \dfrac{\lambda}{2} & \text{for } |u| > \lambda. \end{cases} \tag{3.78}$$

We have

Corollary 3.13 *For all $\varepsilon > 0$ and $\lambda > 0$, we have*

$$\int_\Omega j_\lambda(\nabla J_\varepsilon(\xi))d\xi \le \int_\Omega j_\lambda(\nabla y(\xi))d\xi,\ \ \forall y \in H_0^1(\Omega) \cap L^2(\Omega).$$

Proof. One applies Proposition 3.12 to the function

$$g(r) = \begin{cases} \frac{1}{2\lambda} r^2 & \text{for } 0 \leq r \leq \lambda, \\ r - \frac{\lambda}{2} & \text{for } r > \lambda, \end{cases}$$

and take into account that $g(|u|) \equiv j_\lambda(u)$. □

Remark 3.14 In Corollary 3.13, the quadratic growth condition on g can be relaxed. If, for example, g growth at most of order $p \in [1, \infty)$, then

$$\int_\Omega g(|\nabla J_\varepsilon(y)|)d\xi \leq \int_\Omega g(|\nabla y|)d\xi, \quad \forall y \in W_0^{1,p}(\Omega) \cap L^2(\Omega).$$

In particular, for $g(u) = |u|^p$, where $1 \leq p < \infty$, we obtain that, for each bounded and convex set $\Omega \subset \mathbb{R}^N$ with C^2-boundary, we have

$$|\nabla J_\varepsilon(y)|_p \leq |\nabla y|_p, \quad \forall y \in W_0^{1,p}(\Omega).$$

The case $p = \infty$ is also true and was earlier proved by Brezis and Stampacchia [48]. More precisely, one has (see Corollary III.6 in [48])

$$(I + \lambda A)^{-1} K \subset K, \quad \forall \lambda > 0, \tag{3.79}$$

where $K = \{u \in W_0^1(\Omega);\ |\nabla u| \leq 1, \text{ a.e. in } \Omega\}$, and this result extends to the nonlinear elliptic operators A of the form

$$Ay = -\sum_{i=1}^N \frac{\partial}{\partial x_i}\left(a_i\left(\frac{\partial y}{\partial x_i}\right)\right),$$

where $a : \mathbb{R}^N \to \mathbb{R}^N$ is continuous and monotone, that is,

$$\sum_{i=1}^N (a_i(\xi_i) - a_i(\bar{\xi}_i))(\xi_i - \bar{\xi}_i) \geq 0, \quad \forall \xi = (\xi_1, ..., \xi_N) \in \mathbb{R}^N.$$

In particular, it follows that the operator A is accretive (and, consequently, m-accretive) in $W_0^{1,p}(\Omega)$ for all $1 \leq p \leq \infty$.

Next, we consider for $\lambda > 0$ the function $\beta_\lambda : \mathbb{R}^N \to \mathbb{R}^N$,

$$\beta_\lambda(u) = \nabla j_\lambda(u) = \begin{cases} \frac{1}{\lambda} u & \text{for } |u| \leq \lambda, \\ \frac{u}{|u|} & \text{for } |u| > \lambda, \end{cases} \tag{3.80}$$

where j_λ is the function (3.78) and $|\cdot|$ is the Euclidean norm of $\mathbb{R}^N$. As easily seen, $\beta_\lambda = (\text{sign})_\lambda$ is just the Yosida approximation of the maximal monotone operator $u \to \operatorname{sign} u$ defined in $\mathbb{R} \times \mathbb{R}$.

Corollary 3.15 *Under assumptions of Proposition* 3.11, *we have for all* $y \in H_0^1(\Omega) \cap H^2(\Omega)$,

$$\int_\Omega \Delta y \operatorname{div}(\beta_\lambda(\nabla y))dx \geq 0, \quad \forall \lambda > 0, \tag{3.81}$$

and

$$\int_\Omega \Delta y \operatorname{div} \eta \, dx \geq 0, \tag{3.82}$$

for all $y \in H_0^1(\Omega) \cap H^2(\Omega)$ *and all* $\eta \in (L^\infty(\Omega))^N$ *such that* $\eta(x) \in \operatorname{sign}(\nabla y(x))$, *a.e.* $x \in \Omega$, $\operatorname{div} \eta \in L^2(\Omega)$.

Proof. We set $A_\varepsilon = \varepsilon^{-1}(I - (I + \varepsilon A)^{-1})$, $\forall \varepsilon > 0$, where $A = -\Delta$, $D(A) = H_0^1(\Omega) \cap H^2(\Omega)$. Since $\beta_\lambda = \nabla j_\lambda$, we have by Corollary 3.13

$$\begin{aligned}\int_\Omega A_\varepsilon(y)\operatorname{div}(\beta_\lambda(\nabla y))dx &= \frac{1}{\varepsilon}\int_\Omega (y - (I + \varepsilon A)^{-1} y) \\ \operatorname{div}(\nabla \beta_\lambda(\nabla y))dx &= \frac{1}{\varepsilon}\int_\Omega (\nabla y - \nabla (I + \varepsilon A)^{-1} y) \\ \beta_\lambda(\nabla y))dx &\geq \frac{1}{\varepsilon}\int_\Omega (j_\lambda(\nabla y) - j_\lambda(\nabla (I + \varepsilon A)^{-1} y))dx \geq 0.\end{aligned}$$

Letting $\varepsilon \to 0$ and taking into account that $\lim_{\varepsilon \to 0} A_\varepsilon(y) = Ay$, $\forall y \in D(A)$, we get (3.81), as claimed.

Now, let η be as in the second part of Corollary 3.15 and let A_ε, $\varepsilon > 0$, be as above. We have

$$\begin{aligned}\int_\Omega A_\varepsilon y \operatorname{div} \eta \, dx &= -\int_\Omega \nabla(A_\varepsilon y) \cdot \eta \, dx = -\frac{1}{\varepsilon}\int_\Omega (\nabla y - \nabla (I + \varepsilon A)^{-1} y) \cdot \eta \, dx \\ &= \frac{1}{\varepsilon}\int_\Omega (\nabla (I + \varepsilon A)^{-1} y \cdot \eta - |\nabla y|)dx.\end{aligned}$$

Taking into account that $z = (I + \varepsilon A)^{-1} y \in H_0^1(\Omega) \cap H^2(\Omega)$ satisfies the equation

$$z - \varepsilon \Delta z = y \quad \text{in } \Omega,$$

we have by (3.73)

$$\int_\Omega |\nabla z| dx \leq \int_\Omega |\nabla y| dx.$$

This yields

$$\int_\Omega (\nabla (I + \varepsilon A)^{-1} y \cdot \eta - |\nabla y|)dx \leq 0$$

because $\|\eta\|_{(L^\infty(\Omega))^N} \leq 1$. Hence

$$\int_\Omega A_\varepsilon y \operatorname{div} \eta \, dx \leq 0, \quad \forall \varepsilon > 0,$$

and recalling that $A_\varepsilon y \to -\Delta y$ in $L^2(\Omega)$ as $\varepsilon \to 0$, we get (3.82), as claimed. □

Proof of Theorem 3.10. As seen by Theorem 3.5, y_ε is the unique solution to the minimization problem

$$y_\varepsilon = \arg\min\left\{\frac{1}{2}\int_\Omega |y(x)-f(x)|^2 dx + \frac{\varepsilon}{2}\int_\Omega |\nabla y(x)|^2 dx + \int_\Omega |\nabla y(x)|dx; y \in H_0^1(\Omega)\right\}$$

that is,

$$y_\varepsilon = \arg\min\left\{\frac{1}{2}\int_\Omega |y(x)-f(x)|^2 dx + \varphi(y); y \in L^2(\Omega)\right\},$$

where φ is given by (3.29) with $j(r) \equiv \frac{\varepsilon}{2}|r|^2 + |r|$. This yields

$$\int_\Omega (|y_\varepsilon|^2 + \varepsilon|\nabla y_\varepsilon|^2 + |\nabla y_\varepsilon|)dx \le C, \quad \forall \varepsilon > 0,$$

where C is independent of ε. Hence, on a subsequence again denoted $\{\varepsilon\} \to 0$, we have

$$\begin{aligned} y_\varepsilon &\to y \ \text{ weakly in } L^2(\Omega), \\ \varepsilon\Delta y_\varepsilon &\to 0 \ \text{ strongly in } H^{-1}(\Omega), \end{aligned} \tag{3.83}$$

and, by Theorem 1.29, it follows also that $y \in BV(\Omega)$, and

$$\Phi(y) = |Dy|(\Omega) \le \liminf_{\varepsilon\to 0} \int_\Omega |\nabla y_\varepsilon| dx \le C. \tag{3.84}$$

Let us proved now that $y_\varepsilon \in H^2(\Omega)$. To this purpose, we approximate equation (3.68) by

$$\begin{aligned} y_{\varepsilon,\lambda} - \varepsilon\Delta y_{\varepsilon,\lambda} - \operatorname{div}\beta_\lambda(\nabla y_{\varepsilon,\lambda}) &= f \ \text{ in } \Omega, \\ y_{\varepsilon,\lambda} &= 0 \ \text{ in } \partial\Omega, \end{aligned} \tag{3.85}$$

where β_λ is given by (3.80). By Theorem 3.5, (3.85) has a unique solution $y_{\varepsilon,\lambda} \in H_0^1(\Omega)$ given by

$$y_{\varepsilon,\lambda} = \arg\min\left\{\frac{1}{2}\int_\Omega |y-f|^2 dx + \frac{\varepsilon}{2}\int_\Omega |\nabla y|^2 dx + \int_\Omega j_\lambda(\nabla y)dx; y \in H_0^1(\Omega)\right\}. \tag{3.86}$$

Moreover, we see by (3.85) and the elliptic regularity (see Theorem 1.25) that $y_{\varepsilon,\lambda} \in H^2(\Omega)$. If we multiply (3.85) by $\Delta y_{\varepsilon,\lambda}$ and integrate on Ω, we get by (3.81) that

$$\int_\Omega (|\nabla y_{\varepsilon,\lambda}|^2 + \frac{\varepsilon}{2}|\Delta y_{\varepsilon,\lambda}|^2)dx \le \frac{1}{2\varepsilon}\int_\Omega |f|^2 dx. \tag{3.87}$$

Hence, for each $\varepsilon > 0$, $\{y_{\varepsilon,\lambda}\}$ is bounded in $H^2(\Omega)$ and so, on a subsequence $\lambda \to 0$, we have

$$y_{\varepsilon,\lambda} \to \widetilde{y}_\varepsilon \text{ weakly in } H^2(\Omega) \text{ and strongly in } H_0^1(\Omega).$$

Moreover, since $\{\beta_\lambda(\nabla y_{\varepsilon,\lambda})\}$ is bounded in $L^2(\Omega)$, we have for $\lambda \to 0$

$$\begin{aligned} \beta_\lambda(\nabla y_{\varepsilon,\lambda}) &\to \eta_\varepsilon \quad \text{weakly in } L^2(\Omega), \\ \operatorname{div}\beta_\lambda(\nabla y_{\varepsilon,\lambda}) &\to \operatorname{div}\eta_\varepsilon \ \text{ weakly in } H^{-1}(\Omega). \end{aligned}$$

Since $\nabla y_{\varepsilon,\lambda} \to \nabla\widetilde{y}_\varepsilon$ strongly in $L^2(\Omega)$ and on a subsequence, a.e. in Ω, we have (see, e.g., Corollary 2.14)

$$\eta_\varepsilon(x) \in \text{sign}(\nabla\widetilde{y}_\varepsilon(x)), \quad \text{a.e. } x \in \Omega.$$

Then, because the operator $r \to \text{sign}\, r$ is maximal monotone in $\mathbb{R}^N \times \mathbb{R}^N$, by letting $\lambda \to 0$ in (3.85), we see that $\widetilde{y}_\varepsilon = y_\varepsilon$ is the solution to (3.68). Moreover, by (3.87) we get, for $\lambda \to 0$, that (3.70) holds.

If $f, \bar{f} \in L^2(\Omega)$ and $y_\varepsilon, \bar{y}_\varepsilon \in H_0^1(\Omega) \cap H^2(\Omega)$ are the corresponding solutions to (3.68), we have

$$y_\varepsilon - \bar{y}_\varepsilon - \varepsilon\Delta(y_\varepsilon - \bar{y}_\varepsilon) - \text{div}(\text{sign}(\nabla(y_\varepsilon - \bar{y}_\varepsilon))) = 0.$$

If we multiply the latter by $y_\varepsilon - \bar{y}$ and integrate on Ω, we get via Green's formula that (3.82) holds.

Now, by (3.70) and (3.82), it follows that, for $\varepsilon \to 0$,

$$\varepsilon\Delta y_\varepsilon \to 0 \text{ in } H^{-1}(\Omega), \quad y_\varepsilon \to y \text{ weakly in } L^2(\Omega),$$

and, therefore,

$$y_\varepsilon - \text{div}\, \eta_\varepsilon \to f \quad \text{strongly in } L^2(\Omega),$$

where $\eta_\varepsilon(x) \in \text{sign}(\nabla y_\varepsilon(x))$, a.e. $x \in \Omega$. Taking into account that

$$-\int_\Omega \text{div}\, \eta_\varepsilon(y_\varepsilon - z)dx \geq \int_\Omega (|\nabla y_\varepsilon| - |\nabla z|)dx, \ \forall z \in H_0^1(\Omega),$$

it follows by (3.84) that

$$|Dy|(\Omega) - |Dz|(\Omega) \leq \limsup_{\varepsilon\to 0} \int_\Omega (f - y_\varepsilon)(y_\varepsilon - z)dx \leq \int_\Omega (f - y)(y - z), \ \forall z \in H_0^1(\Omega).$$

Then, by (3.60), we get

$$\Phi(y) \leq \Phi(z) + \int_\Omega (f - y)(y - z)dx, \ \forall z \in BV(\Omega) \cap L^2(\Omega),$$

and, therefore, $f - y \in \partial\Phi(y)$, that is, y is the solution to (3.66), as claimed.

To prove (3.72), we note that taking into account that $y_\varepsilon \in H_0^1(\Omega) \cap H^2(\Omega)$, by (3.68) we have

$$\int_\Omega (|y_\varepsilon|^2 + \varepsilon(\nabla y_\varepsilon|^2 + |\nabla y_\varepsilon|)dx = \int_\Omega f y_\varepsilon dx.$$

By (3.84), this yields

$$\limsup_{\varepsilon\to 0} \int_\Omega |y_\varepsilon|^2 dx \leq \int_\Omega fy\, dx - \Phi(y),$$

while, by (3.66),

$$\int_\Omega |y|^2 dx + \Phi(y) = \int_\Omega fy\, dx.$$

Hence $\limsup_{\varepsilon\to 0} \|y_\varepsilon\|_{L^2(\Omega)} \leq \|y\|_{L^2(\Omega)}$ and, as $y_\varepsilon \to y$ weakly in $L^2(\Omega)$, this implies (3.72), as claimed. □

3.2 Semilinear Elliptic Operators in $L^p(\Omega)$

In most situations, the m-accretive operators arise as partial differential operators on a domain Ω with appropriate boundary value conditions. These boundary value problems do not have an appropriate formulation in a variational functional setting (as in the case with nonlinear elliptic problems of divergence type treated in Section 3.1) but have, however, an adequate treatment in the framework of m-accretive operator theory. We consider a few significant examples below.

Throughout this section, Ω is a bounded and open subset of $\mathbb{R}^N$ with a smooth boundary (of class C^2, for instance), denoted $\partial\Omega$.

Let β be a maximal monotone graph in $\mathbb{R} \times \mathbb{R}$ such that $0 \in D(\beta)$ and let $\widetilde{\beta} \subset L^p(\Omega) \times L^p(\Omega)$, $1 \le p < \infty$, be the operator defined by

$$\begin{aligned} &\widetilde{\beta}(u(x)) = \{v \in L^p(\Omega);\ v(x) \in \beta(u(x))),\ \text{a.e. } x \in \Omega\}, \\ &D(\widetilde{\beta}) = \{u \in L^p(\Omega);\ \exists v \in L^p(\Omega) \text{ so that } v(x) \in \beta(u(x)),\ \text{a.e. } x \in \Omega\}. \end{aligned} \tag{3.88}$$

It is easily seen that $\widetilde{\beta}$ is m-accretive in $L^p(\Omega) \times L^p(\Omega)$ and

$$\begin{aligned} &((I + \lambda\widetilde{\beta})^{-1}u) = (1 + \lambda\beta)^{-1}u(x),\ \text{a.e. } x \in \Omega,\ \lambda > 0, \\ &(\widetilde{\beta}_\lambda u)(x) = \beta_\lambda(u(x)),\ \text{a.e. } x \in \Omega,\ \lambda > 0,\ u \in L^p(\Omega). \end{aligned}$$

Very often, this operator $\widetilde{\beta}$ is called the *realization* of the graph $\beta \subset \mathbb{R} \times \mathbb{R}$ in the space $L^p(\Omega) \times L^p(\Omega)$. We have

Theorem 3.16 *Let $A : L^p(\Omega) \to L^p(\Omega)$ be the operator defined by*

$$\begin{aligned} &Au = -\Delta u + \widetilde{\beta}(u), \quad \forall u \in D(A), \\ &D(A) = W_0^{1,p}(\Omega) \cap W^{2,p}(\Omega) \cap D(\widetilde{\beta}) \quad \textit{if } 1 < p < \infty, \\ &D(A) = \{u \in W_0^{1,1}(\Omega);\ \Delta u \in L^1(\Omega)\} \cap D(\widetilde{\beta}) \quad \textit{if } p = 1. \end{aligned} \tag{3.89}$$

Then A is m-accretive and surjective in $L^p(\Omega)$. If $D(\beta) = \mathbb{R}$, then $D(A)$ is dense in $L^p(\Omega)$. Moreover, for each $\lambda > 0$, the operator $(I + \lambda A)^{-1}$ leaves invariant the set $\{f \in L^p(\Omega); f \ge 0,\ \text{a.e. in } \Omega\}$ and

$$\|(I + \lambda A)^{-1} f\|_{L^\infty(\Omega)} \le \|f\|_{L^\infty(\Omega)}, \quad \forall f \in L^\infty(\Omega). \tag{3.90}$$

We note that, for $p = 2$, this result has been proven in Proposition 2.20.

Proof. Let us show first that A is accretive. If $u_1, u_2 \in D(A)$ and $v_1 \in Au_1$, $v_2 \in Au_2$, $1 < p < \infty$, we have

$$\begin{aligned} &\|u_1 - u_2\|_{L^p(\Omega)}^{p-2}(v_1 - v_2, J(u_1 - u_2)) \\ &\quad = -\int_\Omega \Delta(u_1 - u_2)|u_1 - u_2|^{p-2}(u_1 - u_2)dx \\ &\qquad + \int_\Omega (\beta(u_1) - \beta(u_2))(u_1 - u_2)|u_1 - u_2|^{p-2}dx \\ &\quad \le -\int_\Omega \Delta(u_1 - u_2)|u_1 - u_2|^{p-2}(u_1 - u_2)dx, \end{aligned} \tag{3.91}$$

because β is monotone (recall that, by (1.3), $J(u)(x) = |u(x)|^{p-2}u(x)\|u\|^{2-p}_{L^p(\Omega)}$ is the duality mapping of the space $L^p(\Omega)$). (In the previous formula and everywhere in the sequel, by $\beta(u_i)$, $i = 1, 2$, we mean the single-valued sections of $\beta(u_i)$ which arise in the definition of Au_i.)

Let $v = u_1 - u_2$ and $\{v_\varepsilon\} \subset C_0^\infty(\Omega)$ be such that, for $\varepsilon \to 0$,

$$\begin{aligned} v_\varepsilon &\to v \quad \text{in } W_0^{1,p}(\Omega), \\ \Delta v_\varepsilon &\to \Delta v \ \text{ in } W^{2,p}(\Omega). \end{aligned}$$

Then

$$\begin{aligned} \int_\Omega \Delta v |v|^{p-2} v\,dx &= \lim_{\varepsilon\to 0} \int_\Omega \Delta v_\varepsilon (|v_\varepsilon|^2 + \varepsilon)^{\frac{p-2}{2}} v_\varepsilon dx \\ &= -\lim_{\varepsilon\to 0} (p-1) \int_\Omega |\nabla v_\varepsilon|^2 (|v_\varepsilon|^2 + \varepsilon)^{\frac{p-2}{2}} |v_\varepsilon|^2 dx \le 0, \end{aligned}$$

and so, by (3.91), we have $(v_1 - v_2, u_1 - u_2) \ge 0$, as claimed.

The case $p = 1$ is more delicate and to treat it we consider the function $\mathcal{X}_\delta : \mathbb{R} \to \mathbb{R}$ defined by

$$\mathcal{X}_\delta(r) = \begin{cases} 1 & \text{for } r > \delta, \\ \dfrac{r}{\delta} & \text{for } -\delta \le r \le \delta, \\ -1 & \text{for } r < -\delta. \end{cases} \tag{3.92}$$

The function $\mathcal{X}_\delta$ is a smooth monotonically increasing approximation of the signum multivalued function,

$$\operatorname{sign} r = \begin{cases} 1 & \text{for } r > 0, \\ [-1, 1] & \text{for } r = 0, \\ -1 & \text{for } r < 0, \end{cases}$$

we shall invoke frequently in the following. We note that, for each $u \in L^1(\Omega)$, we have

$$\mathcal{X}_\delta(u(x)) \to \frac{u(x)}{|u(x)|}, \quad \text{a.e. on } \{x \in \Omega, |u(x)| \ne 0\}$$

and so $\{\mathcal{X}_\delta\}_\delta$ is weak-star compact in $L^\infty(\Omega)$. Hence, on a subsequence again denoted δ, we have

$$\mathcal{X}_\delta(u) \to \eta \ \text{ weak-star in } L^\infty(\Omega), \eta(x) \in \operatorname{sign} n(x), \quad \text{a.e. } x \in \Omega. \tag{3.93}$$

If $[u_i, v_i] \in A$, $i = 1, 2$, then we have by (3.89)

$$\begin{aligned} \int_\Omega (v_1 - v_2)\mathcal{X}_\delta(u_1 - u_2) dx &= -\int_\Omega \Delta(u_1 - u_2)\mathcal{X}_\delta(u_1 - u_2) dx \\ &\quad + \int_\Omega (\beta(u_1) - \beta(u_2))\mathcal{X}_\delta(u_1 - u_2) dx, \ \forall \delta > 0, \end{aligned} \tag{3.94}$$

where Δ is taken in the sense of distributions on Ω.

Let $\{v_\varepsilon\} \subset C_0^\infty(\Omega)$ be such that, for $\varepsilon \to 0$,

$$\begin{aligned} v_\varepsilon &\to u_1 - u_2 && \text{in } W_0^{1,1}(\Omega), \\ \Delta v_\varepsilon &\to \Delta(u_1 - u_2) && \text{in } L^1(\Omega). \end{aligned} \tag{3.95}$$

Then

$$\int_\Omega \Delta v_\varepsilon \mathcal{X}_\delta(v_\varepsilon)dx = -\int_\Omega |\nabla v_\varepsilon|^2 \mathcal{X}'_\delta(v_\varepsilon)dx \leq 0, \ \forall \varepsilon > 0,$$

and, therefore, by (3.95) we have

$$\int_\Omega \Delta(u_1 - u_2)\mathcal{X}_\delta(u_1 - u_2)dx = \lim_{\varepsilon\to 0} \int_\Omega \Delta v_\varepsilon \mathcal{X}_\delta(v_\varepsilon)dx \leq 0, \ \forall \delta > 0.$$

By (3.94), this yields

$$\int_\Omega (v_1 - v_2)\mathcal{X}_\delta(u_1 - u_2)dx \geq 0, \ \forall \delta > 0,$$

and so, by (3.93), for $\delta \to 0$, $\mathcal{X}_\delta(u_1 - u_2) \to g$ weak-star in $L^\infty(\Omega)$, where $g(x) \in \operatorname{sign} u(x)$, a.e. $x \in \Omega$, $u = u_1 - u_2$, that is (see (1.4)), $g \in \|u\|_{L^1(\Omega)} J_1(u)$, where $J_1 : L^1(\Omega) \to L^\infty(\Omega)$ is the duality mapping of $L^1(\Omega)$. Hence, A is accretive. Now, we prove that $\mathbb{R}(I + A) = L^p(\Omega)$, considering separately the cases $1 < p < \infty$ and $p = 1$.

Case 1. $1 < p < \infty$. Let us denote for $1 < p < \infty$ by A_p the operator $-\Delta$ with the domain $D(A_p) = W_0^{1,p}(\Omega) \cap W^{2,p}(\Omega)$. We have already seen that A_p is accretive in $L^p(\Omega)$. Moreover, by Theorem 1.29, we have that $R(I + A_p) = L^p(\Omega)$ and

$$\|u\|_{W^{2,p}(\Omega)\cap W_0^{1,p}(\Omega)} \leq C\|A_p u\|_{L^p(\Omega)}, \quad \forall u \in D(A_p). \tag{3.96}$$

Hence, A_p is m-accretive $L^p(\Omega)$. Let us prove now that $R(I + A_p + \widetilde{\beta}) = L^p(\Omega)$. Replacing, if necessary, the graph β by $u \to \beta(u) - v_0$, where $v_0 \in \beta(0)$, we may assume that $0 \in \widetilde{\beta}(0)$ and so $\widetilde{\beta}_\lambda(0) = 0$. Assume first that $p \geq 2$. Then, by Green's formula, we have, for all $\lambda > 0$,

$$\begin{aligned} (A_p u, J(\widetilde{\beta}_\lambda u)) &= -\|\widetilde{\beta}_\lambda(u)\|_{L^p(\Omega)}^{2-p} \int_\Omega \Delta u |\beta_\lambda(u)|^{p-2}\beta_\lambda(u)dx \\ &= (p-1)\|\widetilde{\beta}(u)\|_{L^p(\Omega)}^{2-p} \int_\Omega |\nabla u|^2 \frac{d}{du}\, |\beta_\lambda(u)|^{p-2}\beta_\lambda(u)dx \geq 0, \end{aligned} \tag{3.97}$$

and so, by Proposition 2.43, we conclude that $R(I + A_p + \widetilde{\beta}) = L^p(\Omega)$, as claimed.

If $1 < p < 2$, we approximate in (3.97) u by $\{u_\varepsilon\} \subset C_0^\infty(\Omega)$ and let $\varepsilon \to 0$.

To prove the surjectivity of $A_p + \widetilde{\beta}$, consider the equation

$$\varepsilon u + A_p u + \widetilde{\beta}(u) \ni f, \quad \varepsilon > 0, \ f \in L^p(\Omega), \tag{3.98}$$

which, as seen before, has a unique solution u_ε, and $u_\varepsilon = \lim_{\lambda\to 0} u_\lambda^\varepsilon$ in $L^p(\Omega)$, where u_λ^ε is the solution to the approximating equation $\varepsilon u + A_p u + \widetilde{\beta}_\lambda(u) \ni f$. By (3.97), it follows that $\|A_p u_\lambda^\varepsilon\|_{L^p(\Omega)} \leq C$, where C is independent of ε and λ. Hence, letting $\lambda \to 0$, we get $\|A_p u_\varepsilon\|_{L^p(\Omega)} \leq C$, $\forall \varepsilon > 0$, which, by estimate (3.96) implies that $\{u_\varepsilon\}$ is bounded in $W^{1,p}(\Omega) \cap W^{2,p}(\Omega)$. Selecting a subsequence, for simplicity again denoted u_ε, we may assume that, for $\varepsilon \to 0$,

$$\begin{aligned} u_\varepsilon &\to u && \text{weakly in } W^{2,p}(\Omega), \text{ strongly in } L^p(\Omega), \\ A_p u_\varepsilon &\to A_p u && \text{weakly in } L^p(\Omega), \\ \widetilde{\beta}_\varepsilon(u_\varepsilon) &\to g && \text{weakly in } L^p(\Omega). \end{aligned}$$

By Proposition 2.35 we know that $g \in \widetilde{\beta}(u)$, therefore we infer that u is the solution to the equation $A_p u + \widetilde{\beta}(u) \ni f$, that is, $u \in W^{2,p}(\Omega)$ and

$$\begin{cases} -\Delta u + \beta(u) \ni f, & \text{a.e. in } \Omega, \\ u = 0 & \text{on } \partial\Omega. \end{cases} \tag{3.99}$$

Case 2. $p = 1$. We prove directly that $R(A_1 + \widetilde{\beta}) = L^1(\Omega)$, that is, for $f \in L^1(\Omega)$, equation (3.99) has a solution $u \in D(A_1) = \{u \in W_0^{1,1}(\Omega);\ \Delta u \in L^1(\Omega)\}$. (Here, $A_1 = -\Delta$ with the domain $D(A_1)$.)

We fix f in $L^1(\Omega)$ and consider $\{f_n\} \subset L^2(\Omega)$ such that $f_n \to f$ in $L^1(\Omega)$. As seen before, the problem

$$\begin{cases} -\Delta u_n + \beta(u_n) \ni f_n & \text{in } \Omega, \\ u_n = 0 & \text{on } \partial\Omega, \end{cases} \tag{3.100}$$

has a unique solution $u_n \in H_0^1(\Omega) \cap H^2(\Omega)$. Let $v_n(x) = f_n(x) + \Delta u_n(x) \in \beta(u_n(x))$, a.e. $x \in \Omega$. By (3.100) we see that

$$\int_\Omega |v_n(x) - v_m(x)| dx \le \int_\Omega |f_n(x) - f_m(x)| dx, \tag{3.101}$$

because β is monotone and $-\Delta$ is accretive in $L^1(\Omega)$, that is, $\int_\Omega \Delta u \theta\, dx \le 0$, $\forall u \in D(A_1)$, for some $\theta \in L^\infty(\Omega)$ such that $\theta(x) \in \operatorname{sign} u(x)$, a.e. $x \in \Omega$. In fact, multiplying the equation

$$-\Delta(u_n - u_m) + \beta(u_n) - \beta(u_m) = f_n - f_m$$

by $\mathcal{X}_\delta(u_n - u_m)$, where $\mathcal{X}_\delta$ is the function (3.92) and integrating on Ω, we get

$$\int_\Omega (v_n - v_m)\mathcal{X}_\delta(u_n - u_m) dx \le \int_\Omega (f_n - f_m)\mathcal{X}_\delta(u_n - u_m) dx$$

because

$$-\int_\Omega \Delta(u_n - u_m)\mathcal{X}_\delta(u_n - u_m) dx = \int_\Omega |\nabla(u_n - u_m)|^2 \mathcal{X}_\delta'(u_n - u_m) dx \ge 0.$$

Letting $\delta \to 0$, we get

$$\int_\Omega (v_n - v_m)\xi_{n,m}\, dx \le \int_\Omega |f_n - f_m| dx,$$

where $\xi_{n,m} = \lim_{\delta \to 0} \mathcal{X}_\delta(u_n - u_m)$, a.e. in Ω. Clearly, $\xi_{n,m} \in \operatorname{sign}(u_n - u_m)$, a.e. in Ω, and since $\operatorname{sign}(u_n - u_m) = \operatorname{sign}(v_n - v_m)$, a.e. in Ω, it follows (3.101), as claimed. This yields

$$\begin{aligned} v_n &\to v && \text{strongly in } L^1(\Omega), \\ \Delta u_n &\to \xi && \text{strongly in } L^1(\Omega). \end{aligned} \tag{3.102}$$

Now, let $h_i \in L^p(\Omega)$, $i = 0, 1, ..., N$, $p > N$. Then, by a well-known result due to G. Stampacchia [101], *the boundary value problem*

$$\begin{cases} -\Delta\varphi = h_0 + \sum_{i=1}^{N} \dfrac{\partial h_i}{\partial x_i} & in\ \ \Omega, \\ \varphi = 0 & on\ \ \partial\Omega, \end{cases} \tag{3.103}$$

has a unique weak solution $\varphi \in H_0^1(\Omega) \cap L^\infty(\Omega)$ *and*

$$\|\varphi\|_{L^\infty(\Omega)} \le C \sum_{i=0}^{N} \|h_i\|_{L^p(\Omega)}, \quad h_i \in L^p(\Omega). \tag{3.104}$$

This means that

$$\int_\Omega \nabla\varphi \cdot \nabla\psi\, dx = \int_\Omega h_0\psi - \sum_{i=1}^{N} \int_\Omega h_i \frac{\partial\psi}{\partial x_i}\, dx, \quad \forall \psi \in H_0^1(\Omega). \tag{3.105}$$

Substituting $\psi = u_n$ in (3.105), we get, via Green's formula,

$$-\int_\Omega \varphi\Delta u_n dx = \int_\Omega \nabla\varphi \cdot \nabla u_n dx = \int_\Omega h_0 u_n dx - \sum_{i=1}^{N} \int_\Omega h_i \frac{\partial u_n}{\partial x_i}\, dx,$$

and, therefore, by (3.104),

$$\left| \int_\Omega h_0 u_n dx - \sum_{i=1}^{N} h_i \frac{\partial u_n}{\partial x_i}\, dx \right| \le C\|\Delta u_n\|_{L^1(\Omega)} \sum_{i=0}^{N} \|h_i\|_{L^p(\Omega)}.$$

Inasmuch as $\{h_i\}_{i=0}^N \subset (L^p(\Omega))^{N+1}$ are arbitrary, we conclude that the sequence

$$\left\{ \left(u_n, \frac{\partial u_n}{\partial x_1}, ..., \frac{\partial u_n}{\partial x_N} \right) \right\}_{n=1}^{\infty}$$

is bounded in the dual space $(L^q(\Omega))^{N+1}$, $\frac{1}{p} + \frac{1}{q} = 1$. Hence,

$$\|u_n\|_{W_0^{1,q}(\Omega)} \le C\|\Delta u_n\|_{L^1(\Omega)}, \quad \text{where } 1 < q = \frac{p}{p-1} < \frac{N}{N-1}. \tag{3.106}$$

(For $N = 1$, $q = \infty$.) Therefore, $\{u_n\}$ is bounded in $W_0^{1,q}(\Omega)$ and, consequently, compact in $L^1(\Omega)$. Then, extracting a further subsequence if necessary, we may assume that

$$u_n \to u \quad \text{weakly in } W_0^{1,q}(\Omega) \text{ and strongly in } L^1(\Omega). \tag{3.107}$$

Then, by (3.102), it follows that $\xi = \Delta u$, and because the operator $\widetilde{\beta}$ is closed in $L^1(\Omega) \times L^1(\Omega)$, we see by (3.102) and (3.107) that $v(x) \in \beta(u(x))$, a.e. $x \in \Omega$, and $u \in W_0^{1,q}(\Omega)$. Hence $R(A_1) = L^1(\Omega)$ and, therefore, A_1 is m-accretive. If $D(\beta) = \mathbb{R}$, then $C_0^\infty(\Omega) \subset D(A_p)$ and so $\overline{D(A_p)} = L^p(\Omega)$, as claimed.

Now, assume that $f \in L^p(\Omega)$ and $f \ge 0$, a.e. in Ω, where $p > 1$. Then, by multiplying the equation $(I + \lambda A)u = f$ with $u^-|u|^{p-2} \in L^{p'}(\Omega)$, $\frac{1}{p'} + \frac{1}{p} = 1$, and integrating on Ω, we get

$$\int_\Omega |u^-|^p dx \le \int_\Omega \Delta u |u^-|^{p-1} dx = -(p-1) \int_\Omega |\nabla u^-|^2 |u^-|^{p-2} dx \le 0,$$

because $|u^-|^{p-1} \in W_0^{1,p'}(\Omega)$. Hence, $u \geq 0$, a.e. in Ω. The case $p = 1$ follows in a similar way.

Similarly, if $f \in L^\infty(\Omega)$, we have

$$(u - \|f\|_{L^\infty(\Omega)}) - \Delta u + \beta(u) \leq 0 \text{ in } \mathcal{D}'(\Omega)$$

and so, by multiplying with $(u - \|f\|_{L^\infty(\Omega)})$ and integrating on Ω, this yields $u(x) \leq |f|_\infty$, a.e. $x \in \Omega$. By a symmetric argument, we get that $u(x) \geq -|f|_\infty$, a.e. $x \in \Omega$, and so (3.90) follows. □

In particular, we have the following existence result for a semilinear elliptic boundary value problem in $L^1(\Omega)$.

Theorem 3.17 *For every $f \in L^p(\Omega)$, $1 < p < \infty$, the boundary value problem*

$$\begin{cases} -\Delta u + \beta(u) \ni f, & \text{a.e. in } \Omega, \\ u = 0 & \text{on } \partial\Omega, \end{cases} \tag{3.108}$$

has a unique solution $u = u(f) \in W_0^{1,p}(\Omega) \cap W^{2,p}(\Omega)$. If $f \in L^1(\Omega)$, then $u \in W_0^{1,q}(\Omega)$ with $\Delta u \in L^1(\Omega)$, where $1 \leq q < \frac{N}{N-1}$. Moreover, the following estimate holds:

$$\begin{aligned} \|v(f) - v(g)\|_{L^1(\Omega)} \leq \|f - g\|_{L^1(\Omega)}, \ \forall f, g \in L^1(\Omega), \\ v(f) = f + \Delta u(f) \in \beta(u(f)), \ \text{a.e. in } \Omega. \end{aligned} \tag{3.109}$$

$$\|u\|_{W_0^{1,q}(\Omega)} \leq C\|f\|_{L^1(\Omega)}, \quad \forall f \in L^1(\Omega). \tag{3.110}$$

In particular, A_1 is m-accretive in $L^1(\Omega)$, $D(A_1) \subset W_0^{1,q}(\Omega)$, and

$$\|u\|_{W_0^{1,q}(\Omega)} \leq C\|\Delta u\|_{L^1(\Omega)}, \quad \forall u \in D(A_1).$$

Proof. The existence and estimate (3.110) follow as the proof of Theorem 3.16. As regards (3.109), it follows as (3.101), by multiplying the equation $v(f) = v(g) - \Delta(u(f) - u(g)) = f - g$ by $\mathcal{X}_\delta(u(f) - u(g))$, integrating on Ω and letting $\delta \to 0$. (Use here that $\text{sign}(u(f) - u(g)) = \text{sign}(v(f) - v(g))$.) □

Remark 3.18 It is clear from the previous proof that Theorems 3.16 and 3.17 remain true for more general linear second-order elliptic operators A on Ω of the form (2.33).

The semilinear elliptic operator in $L^1(\mathbb{R}^N)$

The previous results partially extend to unbounded domains Ω. Below we treat the case $\Omega = \mathbb{R}^N$.

Let β be a maximal monotone graph in $\mathbb{R} \times \mathbb{R}$ (eventually multivalued) such that $0 \in \beta(0)$ and let $A : L^1(\mathbb{R}^N) \to L^1(\mathbb{R}^N)$ be the operator

$$Au = -\Delta u + \widetilde{\beta}(u), \quad \forall u \in D(A), \quad \text{in } \mathcal{D}'(\mathbb{R}^N), \tag{3.111}$$

where

$$\begin{aligned} D(A) &= \{u \in L^1(\mathbb{R}^N),\ \Delta u \in L^1(\mathbb{R}^N);\ u \in D(\widetilde{\beta})\}, \\ D(\widetilde{\beta}) &= \{u \in L^1(\mathbb{R}^N);\ \exists \eta \in L^1(\mathbb{R}^N),\ \eta(x) \in \beta(u(x)),\ \text{a.e. } x \in \mathbb{R}^N\}, \\ \widetilde{\beta}(u) &= \{\eta \in L^1(\mathbb{R}^N);\ \eta(x) \in \beta(u(x)),\ \text{a.e. } x \in \mathbb{R}^N\}. \end{aligned} \tag{3.112}$$

Here Δu is taken in the sense of Schwartz distributions on $\mathbb{R}^N$, that is,

$$\Delta u(\varphi) = \int_{\mathbb{R}^n} u\Delta\varphi\, dx, \qquad \forall \varphi \in C_0^\infty(\mathbb{R}^N),$$

and so, the equation $Au = f$ is taken in the following distributional sense

$$\int_{\mathbb{R}^N} (-u\Delta\varphi + \eta\varphi)dx = \int_{\mathbb{R}^n} f\varphi\, dx, \qquad \forall \varphi \in C_0^\infty(\mathbb{R}^N),$$

where $\eta \in L^1(\mathbb{R}^N)$ is such that $\eta(x) \in \beta(u(x))$ a.e. $x \in \mathbb{R}^N$.

Theorem 3.19 *The operator A defined by equations* (3.111)–(3.112) *is m-accretive in $L^1(\mathbb{R}^N) \times L^1(\mathbb{R}^N)$. Moreover, if $f \in L^1(\mathbb{R}^N) \cap L^\infty(\mathbb{R}^N)$ and $f \geq 0$, a.e. in $\mathbb{R}^N$, then $u = (I + \lambda A)^{-1} f \geq 0$, a.e. in Ω and $\|u\|_{L^\infty(\mathbb{R}^N)} \leq \|f\|_{L^\infty(\mathbb{R}^N)}$ for all $f \in L^1(\mathbb{R}^N) \cap L^\infty(\mathbb{R}^N)$.*

Proof. We fix $f \in L^1(\mathbb{R}^N)$ and consider the equation $\lambda u + Au \ni f$, that is,

$$\lambda u - \Delta u + \beta(u) \ni f \quad \text{in } \mathbb{R}^N, \tag{3.113}$$

which is taken in the space $\mathcal{D}'(\mathbb{R}^N)$. We shall prove that for each $\lambda > 0$ there is a unique solution $u = u(f)$ and that

$$\|u(f) - u(g)\|_{L^1(\mathbb{R}^n)} \leq \frac{1}{\lambda} \|f - g\|_{L^1(\mathbb{R}^N)}, \quad \forall f, g \in L^1(\mathbb{R}^N), \tag{3.114}$$

and, as seen earlier, this implies the m-accretivity of the operator A in $L^1(\mathbb{R}^N)$.

1°. *Existence in* (3.113). We consider the approximating equation

$$\lambda u_\varepsilon - \Delta u_\varepsilon + \beta_\varepsilon(u_\varepsilon) = f \quad \text{in } \mathcal{D}'(\mathbb{R}^N), \tag{3.115}$$

where $\beta_\varepsilon = \varepsilon^{-1}(1 - (1 + \varepsilon\beta)^{-1})$, $\forall \varepsilon > 0$. Assume first that $f \in L^1(\mathbb{R}^N) \cap L^2(\mathbb{R}^N)$ and rewrite (3.115) as

$$u_\varepsilon - \frac{\varepsilon}{1 + \varepsilon\lambda} \Delta u_\varepsilon = \frac{\varepsilon}{1 + \varepsilon\lambda} f + \frac{1}{1 + \varepsilon\lambda} (1 + \varepsilon\beta)^{-1} u_\varepsilon. \tag{3.116}$$

On the other hand, it is well known (in fact, it follows by the Lax–Millgram Lemma (1.20)) that, for each $g \in L^2(\mathbb{R}^N)$ and any constant $\mu > 0$, the equation

$$v - \mu\,\Delta\, v = g \quad \text{in } \mathcal{D}'(\mathbb{R}^N)$$

has a unique solution $v \in H^2(\mathbb{R}^N)$ and

$$\|v\|_{L^2(\mathbb{R}^N)} \leq \|g\|_{L^2(\mathbb{R}^N)}.$$

If set $v = T_\mu(g)$, we may rewrite (3.116) as

$$u_\varepsilon = T_{\frac{\varepsilon}{1+\varepsilon\lambda}}\left(\frac{\varepsilon}{1+\varepsilon\lambda}f + \frac{1}{1+\varepsilon\lambda}(1+\varepsilon\beta)^{-1}u_\varepsilon\right),$$

and so, by the Banach fixed point theorem, it follows the existence of a unique solution $u_\varepsilon = u_\varepsilon(f) \in L^2(\mathbb{R}^N)$ to (3.116). Moreover, multiplying (3.115) by u_ε, $\beta_\varepsilon(u_\varepsilon)$ and $-\Delta u_\varepsilon$, respectively, and integrating on $\mathbb{R}^N$, we get

$$\lambda\|u_\varepsilon\|_{L^2(\mathbb{R}^N)} + \|\nabla u_\varepsilon\|^2_{L^2(\mathbb{R}^N)} + \|\beta_\varepsilon(u_\varepsilon)\|^2_{L^2(\mathbb{R}^N)} + \|\Delta u_\varepsilon\|^2_{L^2(\mathbb{R}^N)} \leq 3\|f\|^2_{L^2(\mathbb{R}^N)} \quad (3.117)$$

because $\beta_\varepsilon(u_\varepsilon)u_\varepsilon \geq 0$ and, by Green's formula,

$$-\int_{\mathbb{R}^d} \Delta u_\varepsilon \beta_\varepsilon(u_\varepsilon)dx = \int_{\mathbb{R}^N} |\nabla u_\varepsilon|^2 \beta'_\varepsilon(u_\varepsilon)dx \geq 0.$$

Next, we multiply the equation

$$\lambda(u_\varepsilon(f) - u_\varepsilon(g)) - \Delta(u_\varepsilon(f) - u_\varepsilon(g)) + \beta_\varepsilon(u_\varepsilon(f)) - \beta_\varepsilon(u_\varepsilon(g)) = f - g,$$

by $\mathcal{X}_\delta(u_\varepsilon(f) - u_\varepsilon(g))$, where $\mathcal{X}_\delta$ is the function (3.92). We get

$$\begin{aligned}\lambda\int_{\mathbb{R}^N} &(u_\varepsilon(f) - u_\varepsilon(g))\mathcal{X}_\delta(u_\varepsilon(f) - u_\varepsilon(g))dx \\ &+ \int_{\mathbb{R}^N} (\beta_\varepsilon(u_\varepsilon(f) - u_\varepsilon(g)))\mathcal{X}_\delta(u_\varepsilon(f) - u_\varepsilon(g))dx \leq \int_{\mathbb{R}^N} |f - g|dx,\end{aligned} \quad (3.118)$$

because

$$\begin{aligned}-\int_{\mathbb{R}^N} &\Delta(u_\varepsilon(f) - u_\varepsilon(g))\mathcal{X}_\delta(u_\varepsilon(f) - u_\varepsilon(g))dx \\ &= \int_{\mathbb{R}^N} |\nabla(u_\varepsilon(f) - u_\varepsilon(g))|^2 \mathcal{X}'_\delta(u_\varepsilon(f) - u_\varepsilon(g))dx \geq 0.\end{aligned}$$

Then, (3.118) yields

$$\limsup_{\delta\to 0}\int_{[|u_\varepsilon(f)-u_\varepsilon(g)|\geq\delta]} (\lambda|u_\varepsilon(f) - u_\varepsilon(g)| + |\beta_\varepsilon(u_\varepsilon(f) - \beta_\varepsilon|u_\varepsilon(g)|)dx \leq \int_{\mathbb{R}^N} |f-g|dx,$$

and so, we get via Fatou's lemma that $u_\varepsilon(f), u_\varepsilon(g) \in L^1(\mathbb{R}^N)$ and

$$\lambda\|u_\varepsilon(f) - u_\varepsilon(g)\|_{L^1(\mathbb{R}^N)} + \|\beta_\varepsilon(u_\varepsilon(f)) - \beta_\varepsilon(u_\varepsilon(g))\|_{L^1(\mathbb{R}^N)} \leq \|f - g\|_{L^1(\mathbb{R}^N)}. \quad (3.119)$$

In particular, it follows by (3.117) that the sequence $\{u_\varepsilon\}$ is compact in $H^1_{\text{loc}}(\mathbb{R}^N)$ and weakly compact in $H^2_{\text{loc}}(\mathbb{R}^N)$. Moreover, since $\{\beta_\varepsilon(u_\varepsilon) = \frac{1}{\varepsilon}(u_\varepsilon - (1+\varepsilon\beta)^{-1}u_\varepsilon)\}$ is bounded in $L^2(\mathbb{R}^N)$, this implies that $\{(1+\varepsilon\beta)^{-1}u_\varepsilon\}$ is compact in $L^2_{\text{loc}}(\mathbb{R}^N)$. Hence, on a subsequence, again denoted $\{\varepsilon\} \to 0$, we have

$$\begin{aligned}&u_\varepsilon \to u,\ (1+\varepsilon\beta)^{-1}u_\varepsilon \to u && \text{strongly in } L^2_{\text{loc}}(\mathbb{R}^N) \text{ and weakly in } L^2(\mathbb{R}^N) \\ &\beta_\varepsilon(u_\varepsilon) \rightharpoonup \eta && \text{in } L^2(\mathbb{R}^N) \\ &\Delta u_\varepsilon \rightharpoonup \Delta u && \text{in } L^2(\mathbb{R}^N),\end{aligned} \quad (3.120)$$

where $u_\varepsilon = u_\varepsilon(f)$.

Since β is maximal monotone, so is its realization $\widetilde{\beta} \subset L^2(\mathbb{R}^N) \times L^2(\mathbb{R}^N)$, that is, the operator

$$\widetilde{\beta} = \{[u,v] \in L^2(\mathbb{R}^N) \times L^2(\mathbb{R}^N),\ v(x) \in \beta(u(x)),\ \text{ a.e. } x \in \mathbb{R}^N\},$$

and so, by (3.120) it follows that $\eta(x) \in \beta(u(x))$, a.e. $x \in \mathbb{R}^N$ and $\lambda u - \Delta u + \eta = f$ in $\mathcal{D}'(\mathbb{R}^N)$. Then, we infer that $u = u(f) \in L^1(\mathbb{R}^N)$, and letting $\varepsilon \to 0$ in (3.119), we get

$$\|u(f) - u(g)\|_{L^1(\mathbb{R}^N)} \le \frac{1}{\lambda}\ \|f - g\|_{L^1(\mathbb{R}^N)}, \forall f, g \in L^1(\mathbb{R}^N) \cap L^2(\mathbb{R}^N). \qquad (3.121)$$

Hence, $u(f)$ is a solution to (3.113), and (3.114) holds for all $f \in L^1(\mathbb{R}^N) \cap L^2(\mathbb{R}^N)$. If $\eta = \eta(f)$ is given by (3.120), it also follows by (3.119) that

$$\|\eta(f) - \eta(g))\|_{L^1(\mathbb{R}^N)} \le \|f - g\|_{L^1(\mathbb{R}^N)}, \qquad (3.122)$$

where $\lambda u(f) - \Delta u(f) + \eta(f) = f$ in $\mathcal{D}'(\mathbb{R}^N)$, $\eta(f) \in \beta(u(f))$, a.e. in $\mathbb{R}^N$. To summarize, for $f \in L^1(\mathbb{R}^N) \cap L^2(\mathbb{R}^N)$, $u = u(f)$ is a solution to (3.113), which satisfies (3.121)–(3.122). It is also easily seen that it is unique, because if we subtract equations (3.13) for $f, g \in L^1(\mathbb{R}^N) \cap L^2(\mathbb{R}^N)$ and multiply by $u(f) - u(g)$, we get $u(f) = u(g)$.

Now, let $f \in L^1(\mathbb{R}^N)$ be arbitrary but fixed and let $\{f_n\} \subset L^1(\mathbb{R}^N) \cap L^2(\mathbb{R}^N)$ be such that $f_n \to f$ strongly in $L^1(\mathbb{R}^N)$ as $n \to \infty$. We set $u_n = u(f_n)$, $\eta_n = \eta(f_n)$ and note that, by (3.121)–(3.122),

$$\|u_n - u_m\|_{L^1(\mathbb{R}^N)} \le \frac{1}{\lambda}\ \|f_n - f_m\|_{L^1(\mathbb{R}^N)},\ \forall n, m \in \mathbb{N},$$

$$\|\eta_n - \eta_m\|_{L^1(\mathbb{R}^N)} \le \|f_n - f_m\|_{L^1(\mathbb{R}^N)}.$$

Hence, $u_n \to u$ and $\eta_n \to \eta$ strongly in $L^1\mathbb{R}^N)$, and also a.e. in $\mathbb{R}^N$. Since $\eta_n \in \beta(u_n)$, a.e. in $\mathbb{R}^N$ and β is a maximal monotone graph in $\mathbb{R} \times \mathbb{R}$, we have

$$\eta(x) \in \beta(u(x)),\ \text{ a.e. } x \in \mathbb{R}^N.$$

We also have

$$\lambda\eta - \Delta u + \eta = f \text{ in } \mathcal{D}'(\mathbb{R}^N).$$

Hence, $u = u(f)$ is a solution to (3.113). Since

$$\|u(f_n) - u(g_n)\|_{L^1(\mathbb{R}^N)} \le \frac{1}{\lambda}\ \|f_n - g_n\|_{L^1(\mathbb{R}^N)},$$

we see also that $u(f)$ satisfies (3.114).

Assume now that $f \ge 0$, a.e. in $\mathbb{R}^N$. If we multiply equation (3.115) by u_ε^- and integrate on $\mathbb{R}^N$, it follows that $u_\varepsilon^- = 0$, a.e. in $\mathbb{R}^N$, and so $u_\varepsilon \ge 0$, a.e. in $\mathbb{R}^N$. Then, by (3.120), it follows that $u \ge 0$, a.e. in $\mathbb{R}^N$, as claimed.

Now, assume that $f \in L^1(\mathbb{R}^N) \cap L^\infty(\mathbb{R}^N)$ and let $\varphi \in C^\infty[0,\infty)$ be such that $\varphi \ge 0$, $\varphi(k) = 1$ for $0 \le r \le 1$, $\varphi(r) = 0$ for $r \ge 2$, $|\varphi'(r)| \le 1$, multiplying (3.115) by $\varphi\left(\frac{1}{n}(u_\varepsilon - M)^+\right)$, $M = \|f\|_{L^\infty(\mathbb{R}^N)}$ and integrating on $\mathbb{R}^N$, we get

$$\lambda \int_{\mathbb{R}^N} (u_\varepsilon - M)\varphi\left(\frac{1}{n}(u_\varepsilon - M)^+\right) dx + \int_{\mathbb{R}^N} (\beta_\varepsilon(u_\varepsilon) - \beta_\varepsilon(M))\varphi\left(\frac{1}{n}\,(u_\varepsilon - M)^+\right) dx$$
$$+\frac{1}{n}\int_{\mathbb{R}^N} \nabla(u_\varepsilon - M)\cdot\nabla((u_\varepsilon - M)^+)\varphi'\left(\frac{1}{n}(u_\varepsilon - M)^+\right) dx \le 0,$$

and, since β_ε is monotonically nondecreasing, we get

$$\int_{[x;(u_\varepsilon(x)-M)^+\leq n]} (u_\varepsilon - M)^+ dx \leq \frac{C}{n}, \quad \forall n.$$

Hence, $u_\varepsilon \leq M$, a.e. in $\mathbb{R}^N$ and, similarly, it follows that $u_\varepsilon \geq -M$, a.e. in $\mathbb{R}^N$. Then, for $\varepsilon \to 0$ it follows $|u(x)| \leq M$, a.e. $x \in \mathbb{R}^N$, as claimed.

2°. *Uniqueness.* If u_1, u_2 are two solutions to (3.113), we have

$$\lambda(u_1 - u_2) - \Delta(u_1 - u_2) + \eta_1 - \eta_2 = 0 \quad \text{in} \quad \mathcal{D}'(\mathbb{R}^N), \tag{3.123}$$

where $u_i, \eta_i \in L^1(\mathbb{R}^N)$ and $\eta_i \in \beta(u_i)$, a.e. in $\mathbb{R}^N$ for $i = 1, 2$.

We set $u = u_1 - u_2$ and take $u_\varepsilon = u * \rho_\varepsilon$ where ρ_ε is a C_0^∞ mollifier, that is,

$$\rho_\varepsilon(x) = \frac{1}{\varepsilon^N}\,\rho\left(\frac{|x|}{\varepsilon}\right), \quad \rho \in C_0^\infty(\mathbb{R}), \quad \int_{\mathbb{R}^N} \rho_\varepsilon(x)dx = 1,$$

and $*$ stands for the convolution product. We have

$$\lambda u_\varepsilon - \Delta u_\varepsilon + (\eta_1 - \eta_2) * \rho_\varepsilon = 0 \quad \text{in} \quad \mathbb{R}^N. \tag{3.124}$$

It follows, of course, that $u_\varepsilon, (\eta_1 - \eta_2) * \rho_\varepsilon \in L^2(\mathbb{R}^N)$ and $u_\varepsilon \in H^1(\mathbb{R}^N)$ because

$$\begin{aligned} \|u_\varepsilon\|_{L^2(\mathbb{R}^N)} &\leq \|\rho_\varepsilon\|_{L^2(\mathbb{R}^N)} \|u\|_{L^1(\mathbb{R}^N)}, \\ \|\nabla u_\varepsilon\|_{L^2(\mathbb{R}^N)} &\leq \|\nabla \rho_\varepsilon\|_{L^2(\mathbb{R}^N)} \|u\|_{L^1(\mathbb{R}^N)}. \end{aligned} \tag{3.125}$$

Then, multiplying (3.124) by $\mathcal{X}_\delta(u_\varepsilon)$, where $\mathcal{X}_\delta$ is as above and integrating on $\mathbb{R}^N$, we obtain that

$$\lambda \int_{\mathbb{R}^N} u_\varepsilon \mathcal{X}_\delta(u_\varepsilon)dx + \int_{\mathbb{R}^N} ((\eta_1 - \eta_2) * \rho_\varepsilon)\mathcal{X}_\delta(u_\varepsilon)dx \leq 0$$

because, as seen earlier,

$$\int_{\mathbb{R}^N} \Delta u_\varepsilon \mathcal{X}_\delta(u_\varepsilon)dx \leq 0.$$

Then, letting $\delta \to 0$, we get

$$\lambda \int_{\mathbb{R}^N} |u_\varepsilon(x)|dx + \int_{\mathbb{R}^N} ((\eta_1 - \eta_2) * \rho_\varepsilon)\eta_\varepsilon\, dx \leq 0, \quad \forall \delta > 0,$$

where $\eta_\varepsilon(x) \in \text{sign}(u_\varepsilon(x))$, a.e. $x \in \mathbb{R}^N$. Taking into account that by the monotonicity of β, we have that $\text{sign}(\eta_1 - \eta_2) = \text{sign}\, u$, a.e. in $\mathbb{R}^N$, this yields

$$\liminf_{\varepsilon \to 0} \int_{\mathbb{R}^N} ((\eta_1 - \eta_2) * \rho_\varepsilon)(x)\text{sign}\, u_\varepsilon(x)dx \geq 0.$$

Hence, $u_\varepsilon \to 0$ in $L^1(\mathbb{R}^N)$ as $\varepsilon \to 0$ and this implies $u_1 = u_2$, as claimed.

Hence A is m-accretive in $L^1(\mathbb{R}^N)$. Assume now that $D(\beta) = \mathbb{R}$ and prove that $\overline{D(A)} = L^1$. Indeed, if u is arbitrary in $L^1(\mathbb{R}^N)$, there is $\{u_\varepsilon\} \subset C_0^\infty(\mathbb{R}^N)$ such that $u_\varepsilon \to u$ in $L^1(\mathbb{R}^N)$ as $\varepsilon \to 0$. Clearly, $u_\varepsilon \in D(A)$ because β is bounded on $\{x \in \mathbb{R}^N;\ u_\varepsilon(x) \neq 0\}$ and, since it is measurable (as a multivalued function), it has an $L^1(\mathbb{R}^N)$ section. Hence, $u \in \overline{D(A)}$. This completes the proof. □

One might expect that for $\lambda \to 0$ the solution $u = y_\lambda$ to equation (3.113) is convergent (in an appropriate space) to a solution $y \in L^1_{\rm loc}(\mathbb{R}^N)$ to equation

$$-\Delta y + \beta(y) \ni f \quad \text{in } \mathcal{D}'(\mathbb{R}^N). \tag{3.126}$$

It turns out that this is indeed the case and that equation (3.126) has a unique solution. More precisely, one has the following existence result essentially due to Bénilan, Brezis and Crandall [32].

Theorem 3.20 *Assume that $f \in L^1(\mathbb{R}^N)$. Then,*

(i) *If $N = 1$ and $0 \in \operatorname{int}(R(\beta))$, then equation (3.126) has a unique solution $y \in W^{1,\infty}(\mathbb{R})$ with $\Delta y \in L^1(\mathbb{R})$.*
(ii) *If $N = 2$ and $0 \in \operatorname{int}(R(\beta))$, then there is a unique solution $y \in L^1_{\rm loc}(\mathbb{R}^2) \cap W^{1,1}_{\rm loc}(\mathbb{R}^2)$ with $\Delta y \in L^1(\mathbb{R}^2)$ and $\nabla y \in M^2(\mathbb{R}^2)$.*
(iii) *If $N \geq 3$, then there is a unique solution $y \in M^{\frac{N}{N-2}}(\mathbb{R}^N) \cap L^1_{\rm loc}(\mathbb{R}^N)$ with $\Delta y \in L^1(\mathbb{R}^N)$.*

Here, $R(\beta)$ is the range of β and $M^p(\mathbb{R}^N)$, $p > 1$, is the Marcinkiewicz space of order p (see (1.22)).

Proof. We are going to pass to the limit $\lambda \to 0$ in equation (3.113), that is,

$$\lambda y_\lambda - \Delta y_\lambda + \beta(y_\lambda) \ni f. \tag{3.113$'$}$$

The main problem is, however, the boundedness of $\{y_\lambda\}$ in $L^1(\mathbb{R}^N)$ or in $L^1_{\rm loc}(\mathbb{R}^N)$. We set $w_\lambda = \beta(y_\lambda)$ (or, if β is multivalued, the section of β arising in (3.113)$'$). We have

$$\begin{aligned}\lambda(y_\lambda(x+h) - y_\lambda(x)) - \Delta(y_\lambda(x+h) - y_\lambda(x)) + w_\lambda(x+h) - w_\lambda(x)\\ = f(x+h) - f(x), \quad x, h \in \mathbb{R}^N,\end{aligned}$$

and, arguing as above (see, e.g., (3.94)), we get

$$\lambda \int_{\mathbb{R}^N} |y_\lambda(x+h) - y_\lambda(x)|dx + \int_{\mathbb{R}^N} |w_\lambda(x+h) - w_\lambda(x)|dx \leq \int_{\mathbb{R}_N} |f(x+h) - f(x)|dx,$$

and, similarly,

$$\int_{\mathbb{R}^N} |w_\lambda(x)|dx \leq \int_{\mathbb{R}^N} |f(x)|dx. \tag{3.127}$$

Hence, by the Kolmogorov compactness theorem (Theorem 1.31), the set $\{w_\lambda\}$ is compact in $L^1_{\rm loc}(\mathbb{R}^N)$and so, there is $w \in L^1_{\rm loc}(\mathbb{R}^N)$ such that, as $\lambda \to 0$,

$$w_\lambda \to w \quad \text{in } L^1_{\rm loc}(\mathbb{R}^N). \tag{3.128}$$

On the other hand, by (3.127) and by Fatou's lemma, it follows that $w \in L^1(\mathbb{R}^N)$ and $\|w\|_{L^1(\mathbb{R}^N)} \leq \|f\|_{L^1(\mathbb{R}^N)}$.

We recall (see Proposition 1.18) that, for each $g \in L^1(\mathbb{R}^N)$ and $N \geq 3$, the equation

$$-\Delta u = g \quad \text{in } \mathcal{D}'(\mathbb{R}^N),$$

has a unique solution $u \in M^{\frac{N}{N-2}}$ given by

$$u = E * g, \quad E(x) \equiv \omega_N |x|^{2-N}.$$

Moreover, one has the estimate

$$\|u\|_{M^{\frac{N}{N-2}}} + \|\nabla u\|_{M^{\frac{N}{N-1}}} \leq C\|f\|_{L^1(\mathbb{R}^N)}. \tag{3.129}$$

Since $\Delta y_\lambda = \lambda y_\lambda + w_\lambda - f$ is bounded in $L^1(\mathbb{R}^N)$, by (3.129) we have

$$\|y_\lambda\|_{M^{\frac{N}{N-2}}(\mathbb{R}^N)} + \|\nabla y_\lambda\|_{M^{\frac{N}{N-1}}(\mathbb{R}^N)} \leq C, \quad \forall \lambda > 0.$$

In particular, it follows that $\{y_\lambda\}$ is bounded in $W^{1,1}_{\rm loc}(\mathbb{R}^N)$ and so it is compact in $L^1_{\rm loc}(\mathbb{R}^N)$. Then, on a subsequence, $y_\lambda \to y$ in $L^1_{\rm loc}(\mathbb{R}^N)$ and by (3.128), since β is maximal monotone, we infer that $w(x) = \beta(y(x))$, a.e. $x \in \mathbb{R}^N$. Clearly, y is a solution to (3.126) because $\Delta y_\lambda \to \Delta y$ in $\mathcal{D}'(\mathbb{R}^N)$ as $\lambda \to 0$.

Now, we consider separately the cases: $N = 2$ and $N = 1$.

The case $N = 2$. In this case, in order to get the boundedness of $\{y_\lambda\}$, one must assume further that $0 \in \operatorname{int} R(\beta)$. If we denote by $j : \mathbb{R} \to \overline{\mathbb{R}}$ the potential of β (i.e., $\beta = \partial j$), we have that $j(r) \geq c|r|$, for some $c > 0$ and $|r| \geq R_1$. Indeed, as seen earlier (Proposition 1.7), $\operatorname{int} R(\beta) = \operatorname{int} D(\beta^{-1}) = \operatorname{int} D(j^*)$, where j^* is the conjugate of j:

$$j(r) = \sup\{rp - j^*(p);\ p \in \mathbb{R}\}, \ \forall r \in \mathbb{R}.$$

We have, therefore, $|j^*(p)| \leq C$ for all $p \in \mathbb{R}$, $|p| \leq r^*$, where $r^* > 0$ is suitably chosen. This yields

$$j(r) \geq \rho|r| - j^*\left(\rho\,\frac{r}{|r|}\right) \geq \frac{r^*}{2}\,|r| \quad \text{for } |r| \geq 1.$$

Now, we come back to equation (3.113)$'$ and notice that multiplying by $\operatorname{sign} y_\lambda$ (or, more exactly, by $\mathcal{X}_\delta(y_\lambda)$, where $\mathcal{X}_\delta$ is given by (3.92) and let $\delta \to 0$), we get as above

$$\int_{[|y_\lambda|>1]} \frac{\beta(y_\lambda)y_\lambda}{|y_\lambda|}\,dx \leq \int_\Omega |f|dx$$

and so, taking into account that $\beta(y_\lambda)y_\lambda \geq j(y_\lambda) \geq C|y_\lambda|$ on $[|y_\lambda| \geq 1]$ we get

$$\int_{\mathbb{R}^N} |y_\lambda(x)|dx \leq C, \quad \forall \lambda > 0$$

and therefore $\{y_\lambda\}$ is bounded in $L^1(\mathbb{R}^N)$. Then, taking into account that $\Delta y_\lambda = \lambda y_\lambda + w_\lambda - f$ from Proposition 1.28, we infer that $\{\nabla y_\lambda\}$ is bounded in $M^2(\mathbb{R}^2)$ and, therefore, $\{y_\lambda\}$ is bounded in $W^{1,1}_{\rm loc}$. This implies that $y = \lim_{\lambda\downarrow 0} y_\lambda$ exists (on a subsequence) in $L^1_{\rm loc}(\mathbb{R}^2)$ and also that $y \in W^{1,1}_{\rm loc}(\mathbb{R}^2)$. Then, by (3.128), we see that $w(x) \in \beta(y(x))$, a.e. $x \in \Omega$, and so y is the desired solution.

The case $N = 1$. It follows as above that the sets $\{y_\lambda\}$ and $\{\beta_\lambda(y_\lambda)\}$ are bounded in $L^1(\mathbb{R}^N)$ and, because $\{y''_\lambda\}$ is bounded in $L^1(\mathbb{R})$, we also get that $\{y'_\lambda\}$ is bounded in $L^\infty(\mathbb{R})$. In fact, because $\{y'_\lambda\}$ is bounded in $L^1(\mathbb{R})$, then there is at least one $x_0 \in \mathbb{R}$ such that $\{y'_\lambda(x_0)\}$ is bounded and this, clearly, implies that $\{y'_\lambda\}$ is bounded in $L^\infty(\mathbb{R})$. Then we infer, as in the previous cases, that $y = \lim_{\lambda\downarrow 0} y_\lambda$ is the solution to (3.126) and satisfies the required conditions. □

The porous media equation in $L^1(\Omega)$

We have already studied the equation $u - \Delta\beta(u) = f$ in the space $H^{-1}(\Omega)$ and proved (see Proposition 2.23) that, under assumptions (2.50), it has for each $f \in H^{-1}(\Omega)$ a unique solution $u \in L^1(\Omega)$ with $\beta(u) \in H_0^1(\Omega)$. (The energetic approach.) Here, we consider it in the $L^1(\Omega)$-space framework.

In the space $X = L^1(\Omega)$ define the operator

$$\begin{cases} Au = -\Delta\beta(u), \quad \forall u \in D(A), \\ D(A) = \{u \in L^1(\Omega);\ \beta(u) \in W_0^{1,1}(\Omega),\ \Delta\beta(u) \in L^1(\Omega)\}, \end{cases} \tag{3.130}$$

where β is a maximal monotone graph in $\mathbb{R}\times\mathbb{R}$ such that $0 \in \beta(0)$ and Ω is an open bounded subset of $\mathbb{R}^N$ with smooth boundary $\partial\Omega$. More precisely, the operator $A \subset L^1(\Omega) \times L^1(\Omega)$ is defined by

$$A = \{[u, -\Delta\eta],\ u \in L^1(\Omega),\ \eta \in W_0^{1,1}(\Omega),\ \Delta\eta \in L^1(\Omega),\ \eta(x) \in \beta(u(x)),\ \text{a.e. } x \in \Omega\}. \tag{3.131}$$

We have the following.

Theorem 3.21 *The operator A is m-accretive in $L^1(\Omega) \times L^1(\Omega)$ and, for $\lambda > 0$, $(I + \lambda A)^{-1}$ leaves invariant the sets $\{f \in L^1(\Omega);\ f \geq 0,\ a.e.\ in\ \Omega\}$ and $L^\infty(\Omega)$.*

Proof. One should show that, for all $\lambda > 0$, $\mathbb{R}(I+\lambda A) = L^1(\Omega)$ and that (see (2.69))

$$\|u_1 - u_2 + \lambda(f_1 - f_2)\|_{L^1(\Omega)} \geq \|u_1 - u_2\|_{L^1(\Omega)},\ \forall f_i \in Au_i,\ i = 1, 2. \tag{3.132}$$

To this end, we note that, for each $f \in L^1(\Omega)$, the equation

$$u + \lambda Au = f \tag{3.133}$$

can be equivalently written as

$$\beta^{-1}(v) - \lambda\Delta v = f \quad \text{in } \Omega, \quad v \in W_0^{1,1}(\Omega),\ \Delta v \in L^1(\Omega). \tag{3.134}$$

But, according to Theorem 3.17, equation (3.134) has a unique solution $v = v(f) \in W_0^{1,q}(\Omega)$, $\Delta v \in L^1(\Omega)$, $1 < q < \frac{N}{N-1}$ and (see (3.109))

$$\|v(f) - v(g)\|_{L^1(\Omega)} \leq \|f - g\|_{L^1(\Omega)},\ \forall f, g \in L^1(\Omega).$$

Hence,

$$\|(I + \lambda A)^{-1}f - (I - \lambda A)^{-1}g\|_{L^1(\Omega)} \leq \|f - g\|_{L^1(\Omega)},\ \forall f, g \in L^1(\Omega),\ \lambda > 0,$$

which clearly implies (3.132), as claimed.

Now, if $f \in L^\infty(\Omega)$ and $u = (I + \lambda A)^{-1}f$, we have

$$(u - \|f\|_{L^\infty(\Omega)}) - \Delta(\beta(u) - \beta(\|u\|_{L^\infty(\Omega)})) \leq 0 \text{ in } \Omega,$$

and multiplying by $\text{sign}((u - \|f\|_{L^\infty(\Omega)})^+)$ and integrating on Ω, we get as above

$$(u - \|f\|_{L^\infty(\Omega)})^+ = 0, \quad \text{a.e. in } \Omega,$$

and so, $u \leq \|f\|_{L^\infty(\Omega)}$, a.e. in Ω. Similarly, it follows that $u \geq -\|f\|_{L^\infty}$, a.e. in Ω. Hence, $u \in L^\infty(\Omega)$. Finally, if $f \geq 0$ in Ω, then multiplying the equation $u + \lambda Au = f$ by $\text{sign}\, u^-$ (or, more exactly, by $\mathcal{X}_\delta(u^-)$ and let $\delta \to 0$) we get $u^- = 0$, a.e. in Ω. □

Remark 3.22 Assume that β is maximal monotone in $\mathbb{R}\times\mathbb{R}$ and that

$$|\beta^{-1}(r)| \le C_1|r| + C_2, \quad \forall r \in \mathbb{R}. \tag{3.135}$$

Then, for each $f \in L^1(\Omega)$, the equation

$$\Delta\beta(u) = f \text{ in } \Omega; \quad \beta(u) = 0 \text{ on } \partial\Omega, \tag{3.136}$$

has a unique solution $u \in L^1(\Omega)$ such that $\beta(u) \in W^{1,q}(\Omega)$, $1 \le q \le \frac{N}{N-1}$. This follows by (3.134) taking into account that, as seen earlier in Theorem 3.17, the operator $\Delta v = f$ in Ω; $v = 0$ on $\partial\Omega$ has, for each $f \in L^1(\Omega)$, a unique solution $v \in W_0^{1,q}(\Omega)$.

The porous media equation in $\mathbb{R}^N$

Consider the equation

$$\lambda y - \Delta\beta(y) \ni f \quad \text{in } \mathcal{D}'(\mathbb{R}^N), \tag{3.137}$$

where $\lambda > 0$, and β is a maximal monotone graph in $\mathbb{R}\times\mathbb{R}$ such that $0 \in \beta(0)$. By solution y to (3.137) we mean a function $y \in L^1(\mathbb{R}^N)$ such that $\exists \eta \in L^1_{\text{loc}}(\mathbb{R}^N)$, $\eta(x) \in \beta(y(x))$, a.e. $x \in \mathbb{R}^N$, and

$$\lambda y - \Delta\eta = f \quad \text{in } \mathcal{D}'(\mathbb{R}^N), \tag{3.138}$$

that is

$$\int_{\mathbb{R}^N} (\lambda y\varphi - \eta\Delta\varphi - f\varphi)dx = 0, \ \forall \varphi \in C_0^\infty(\mathbb{R}^N). \tag{3.139}$$

Theorem 3.23 *Assume that $f \in L^1(\mathbb{R}^N)$. Then,*

(i) *If $N = 1$ and $0 \in \text{int}(D(\beta))$, then there is a unique solution $y \in L^1(\mathbb{R}^N)$ with $\eta \in L^1_{\text{loc}}(\mathbb{R}) \cap W^{1,\infty}(\mathbb{R})$.*
(ii) *If $N = 2$ and $0 \in \text{int}(D(\beta))$, then there is a unique solution $y \in L^1(\mathbb{R}^N)$ with $\eta \in W^{1,1}_{\text{loc}}(\mathbb{R}^2)$, $|\nabla\eta| \in M^2(\mathbb{R}^2)$.*
(iii) *If $N \ge 3$, then there is a unique solution $y \in L^1(\mathbb{R}^N)$, with $\eta \in M^{\frac{N}{N-2}}(\mathbb{R}^N)$.*

Proof. By substitution, $\beta(y) \to u$, equation (3.137) reduces to equation (3.126) with β^{-1} in the place of β and so, one can apply Theorem 3.20 to derive (i) $\sim$ (iii). □

In the space $L^1(\mathbb{R}^N)$ consider the operator

$$Ay = -\Delta\beta(y), \quad \forall y \in D(A) \tag{3.140}$$

defined by

$$\begin{aligned} D(A) = \{&y \in L^1(\mathbb{R}^N);\ \exists \eta \in L^1_{\text{loc}}(\mathbb{R}^N),\\ &\eta(x) \in \beta(y(x)),\ \text{a.e. } x \in \Omega,\ \Delta\eta \in L^1(\mathbb{R}^N)\} \end{aligned} \tag{3.141}$$

$$Ay = \{-\Delta\eta \in L^1(\mathbb{R}^N);\ \eta \in \beta(y),\ \text{a.e. in } \mathbb{R}^N,\ \eta \in L^1_{\text{loc}}(\mathbb{R}^N),\ y \in L^1(\mathbb{R}^N)\}. \tag{3.142}$$

We have the following.

Theorem 3.24 *Assume that β is a maximal monotone graph in $\mathbb{R} \times \mathbb{R}$ satisfying the conditions of Theorem 3.23. Then the operator A defined by (3.141)–(3.142) is m-accretive in $L^1(\mathbb{R}^N) \times L^1(\mathbb{R}^N)$ and $(I+\lambda A)^{-1}$ leaves invariant the sets $\{f \in L^1(\mathbb{R}^N),\ f \geq 0,\ \text{a.e. in } \mathbb{R}^N\}$ and $L^1(\mathbb{R}^N) \cap L^\infty(\mathbb{R}^N)$. Moreover, if $D(\beta) = \mathbb{R}$, then $\overline{D(A)} = L^1(\mathbb{R}^N)$.*

Proof. There is nothing left to do, except to apply Theorem 3.23 and to notice that by Theorem 3.19 we have also the inequality

$$\|u - v\|_{L^1(\mathbb{R}^N)} \leq \frac{1}{\lambda} \|f - g\|_{L^1(\mathbb{R}^N)}, \quad \forall \lambda > 0,\ f, g \in L^1(\mathbb{R}^N)$$

if u, v are solutions to (3.137) for f and g, respectively. □

3.3 Quasilinear Partial Differential Equations of First Order

Here, we shall study the first-order partial differential operator

$$(Au)(x) = \sum_{i=1}^{N} \frac{\partial}{\partial x_i} a_i(u(x)), \quad x \in \mathbb{R}^N, \tag{3.143}$$

in the space $X = L^1(\mathbb{R}^N)$. We use the notations $a = (a_1, a_2, ..., a_N)$, $\varphi_x = (\varphi_{x_1}, ..., \varphi_{x_N})$, $a(u)_x = \sum_{i=1}^N \frac{\partial}{\partial x_i} a_i(u(x)) = \operatorname{div} a(u)$. This solution is related to the concept of entropy solutions for the conservation of laws equations to be treated in Chapter 4. The function $a : \mathbb{R} \to \mathbb{R}^N$ is assumed to be continuous.

We define the operator A in $L^1(\mathbb{R}^N) \times L^1(\mathbb{R}^N)$ as the closure of the operator $A_0 \subset L^1(\Omega) \times L^1(\Omega)$ defined in the following way.

Definition 3.25 $A_0 = \{[u, v] \in L^1(\mathbb{R}^N) \times L^1(\mathbb{R}^N);\ a(u) \in (L^1(\mathbb{R}^N))^N\}$ and

$$\int_{\mathbb{R}^N} \operatorname{sign}_0(u(x) - k)((a(u(x)) - a(k)) \cdot \varphi_x(x) + v(x)\varphi(x))dx \geq 0, \tag{3.144}$$

for all $\varphi \in C_0^\infty(\mathbb{R}^N)$ such that $\varphi \geq 0$, and all $k \in \mathbb{R}$. Here, $\operatorname{sign}_0 r = \frac{r}{|r|}$ for $r \neq 0$, $\operatorname{sign}_0 0 = 0$.

It is readily seen that, if $a \in C^1(\mathbb{R})$ and $u \in C_0^1(\mathbb{R}^N)$, then $u \in D(A_0)$ and $A_0(u) = a(u)_x$. Indeed, if ρ is a smooth approximation function sign_0, then we have

$$\begin{aligned}\int_{\mathbb{R}^N} \rho(u(x) - k)a(u(x))_x \varphi(x)dx &= \int_{\mathbb{R}^N} dx \left(\int_k^{u(x)} \rho(s-k)a'(s)ds\right)_x \varphi(x)dx \\ &= -\int_{\mathbb{R}^N} dx \left(\int_k^{u(x)} \rho(s-k)a'(s)ds\right) \cdot \varphi_x(s),\end{aligned}$$

where $a' = (a_1', a_2', ..., a_N')$ is the derivative of a. Now, letting ρ tend to sign_0, we get

$$\int_{\mathbb{R}^N} \operatorname{sign}_0(u(x) - k)(a(u(x) - a(k)) \cdot \varphi_x(x) + a(u(x))_x \varphi(x))dx = 0$$

for all $\varphi \in C_0^\infty(\mathbb{R}^N)$. Hence, $u \in D(A_0)$ and $A_0(u) = (a(u))_x$.

Conversely, if $u \in D(A_0) \cap L^\infty(\mathbb{R}^N)$ and $v \in A_0(u)$, then using inequality (3.144) with $k = \|u\|_{L^\infty(\mathbb{R}^N)} + 1$ and $k = -(\|u\|_{L^\infty(\mathbb{R}^N)} + 1)$, we get

$$\int_{\mathbb{R}^N} ((a(u(x)) - a(k)) \cdot \varphi_x(x) + v(x)\varphi(x))dx \leq 0, \quad \forall \varphi \in C_0^\infty(\mathbb{R}^N), \ \varphi \geq 0,$$

respectively,

$$\int_{\mathbb{R}^N} ((a(u(x)) - a(k)) \cdot \varphi_x(x) + v(x)\varphi(x))dx \geq 0, \quad \forall \varphi \in C_0^\infty(\mathbb{R}^N), \ \varphi \geq 0.$$

Hence, $-(a(u))_x + v = 0$ in $\mathcal{D}'(\mathbb{R}^N)$.

Let A be the closure of A_0 in $L^1(\mathbb{R}^N) \times L^1(\mathbb{R}^N)$, that is, $A = \{[u,v] \in L^1(\mathbb{R}^N) \times L^1(\mathbb{R}^N);\ \exists [u_n, v_n] \in A_0,\ u_n \to u,\ v_n \to v \text{ in } L^1(\mathbb{R}^N)\}$.

Theorem 3.26 *Let $a : \mathbb{R} \to \mathbb{R}^N$ be continuous and* $\limsup_{r\to 0} \frac{\|a(r)\|}{|r|} < \infty$. *Then A is m-accretive.*

We prove Theorem 3.26 in several steps but, before proceeding with its proof, we must emphasize that a function u satisfying (3.144) is not a simple distributional solution to equation $(a(u))_x = v$. Its precise meaning becomes clear in the context of the so-called entropy solution to the conservation law equation $u_t + (a(u))_x = v$ which is discussed later on in Chapter 4. We shall first prove the following.

Lemma 3.27 *A is accretive in $L^1(\mathbb{R}^N) \times L^1(\mathbb{R}^N)$.*

Proof. Let $[u,v]$ and $[\bar{u},\bar{v}]$ be two arbitrary elements of A_0. By Definition 3.25, we have, for $k = \bar{u}(y)$, $\varphi(x) = \psi(x,y)$ ($\psi \in C_0^\infty(\mathbb{R}^N \times \mathbb{R}^N)$, $\psi \geq 0$),

$$\begin{aligned}\int_{\mathbb{R}^N \times \mathbb{R}^N} & \operatorname{sign}_0(u(x) - \bar{u}(y))(a(u(x)) - a(\bar{u}(y)) \cdot \psi_x(x,y) \\ & + v(x)\psi(x,y))dx\,dy \geq 0.\end{aligned} \tag{3.145}$$

Now, it is clear that we can interchange u and $\bar{u}$, v and $\bar{v}$, x and y to obtain, by adding it to (3.145), the inequality,

$$\begin{aligned}\int_{\mathbb{R}^N \times \mathbb{R}^N} & \operatorname{sign}_0(u(x) - \bar{u}(y))((a(u(x)) - a(\bar{u}(y)) \cdot (\psi_x(x,y)) \\ & + \psi_y(x,y)) + (v(x) - \bar{v}(y))\psi(x,y))dx\,dy \geq 0,\end{aligned} \tag{3.146}$$

for all $\psi \in C_0^\infty(\mathbb{R}^N \times \mathbb{R}^N)$, $\psi \geq 0$. Now, we take

$$\psi(x,y) = \frac{1}{\varepsilon^n} \varphi(x+y)\rho\left(\frac{x-y}{\varepsilon}\right),$$

where $\varphi \in C_0^\infty(\mathbb{R}^N)$, $\varphi \geq 0$, and $\rho \in C_0(\mathbb{R}^N)$ is such that $\operatorname{supp}\rho \subset \{y;\ \|y\| \leq 1\}$, $\int \rho(y)dy = 1$, $\rho(y) = \rho(-y)$, $\forall y \in \mathbb{R}^N$.

Substituting in (3.146), we get after some calculation that

$$\begin{aligned}&\int_{\mathbb{R}^N \times \mathbb{R}^N} \operatorname{sign}_0(u(y+\varepsilon z) - \bar{u}(y))(2(a(u(y+\varepsilon z)) \\ &-a(\bar{u}(y))) \cdot \nabla\varphi(y+\varepsilon z)) + (v(y+\varepsilon z) - \bar{v}(y))\varphi(y+\varepsilon z))\rho(z)dy\,dz \geq 0.\end{aligned} \tag{3.147}$$

Now, letting ε tend to zero in (3.147), we get

$$\int_{\mathbb{R}^N} \theta(y)(v(y) - \bar{v}(y))\varphi(y)dy + 2\int_{\mathbb{R}^N} \theta(y)(a(u(y)) - a(\bar{u}(y)) \cdot \nabla\varphi(y))dy \geq 0, \tag{3.148}$$

for all $\theta(y) \in \text{sign}(u(y) - \bar{u}(y))$, a.e. $y \in \mathbb{R}^N$. Hence, for every $\varphi \in C_0^\infty(\mathbb{R}^N)$, $\varphi \geq 0$, there exists $\theta \in J(u - \bar{u})$ such that (3.148) holds, where J is the duality mapping of the space $L^1(\Omega)$ (see (1.4)). If in (3.148) we take $\varphi = \alpha(\varepsilon\|y\|^2)$, where $\alpha \in C_0^\infty(\mathbb{R})$, $\alpha \geq 0$, and $\alpha(r) = 1$ for $|r| \leq 1$, and let $\varepsilon \to 0$, we get

$$\int_{\mathbb{R}^N} \theta(y)(v(y) - \bar{v}(y))dy \geq 0$$

for some $\theta \in J(u-\bar{u})$. Hence, A_0 is accretive in $L^1(\mathbb{R}^N)$ and, therefore, so is its closure A. □

In order to prove that A is m-accretive, taking into account that A_0 is accretive, it suffices to show that the range of $I + A_0$ is dense in $L^1(\mathbb{R}^N)$, that is, that the equation $u + a(u)_x = f$ has a solution (in the generalized sense (3.146)) for a sufficiently large class of functions f dense in $L^1(\mathbb{R}^N)$. This means, adopting a terminology used in linear theory, that A_0 is essentially m-accretive. To this end, we shall use the so called *vanishing-viscosity approach.* Namely, we approximate this equation by the following family of elliptic equations

$$u + a(u)_x - \varepsilon\Delta u = f \quad \text{in } \mathbb{R}^N. \tag{3.149}$$

Such a solution is called *viscosity solution* and prove that, for a smooth function $a : \mathbb{R} \to \mathbb{R}^N$ and each $\varepsilon > 0$, equation (3.149) has a unique regular solution. Then, we consider the solution u_ε of (3.149), where a is replaced by a smooth approximating family $\{a_\varepsilon\}$, and show that the corresponding solution u_ε is a solution to $u + \lambda Au = f$. We prove first

Lemma 3.28 *Let $a \in C^1(\mathbb{R}; \mathbb{R}^N)$, with $a' = \{a'_j\}_{j=1}^N$ bounded, and let $\varepsilon > 0$. Then, for each $f \in L^2(\mathbb{R}^N)$, equation* (3.149) *has a solution $u \in H^2(\mathbb{R}^N)$.*

Proof. We shall prove first that, for each $k > 0$, there is a solution $u_k \in H_0^1(\Sigma_k) \cap H^2(\Sigma_k)$, where

$$\Sigma_k = \{r \in \mathbb{R}^N;\ |r| < k\}.$$

Denote by Λ the operator defined in $L^2(\Sigma_k)$ by

$$\Lambda = -\Delta, \quad D(\Lambda) = H_0^2(\Sigma_k)$$

and let $Bu = -a(u)_x$, $\forall u \in D(B) = H_0^1(\Sigma_k)$. The operator $T = (I + \varepsilon\Lambda)^{-1}B$ is continuous and bounded from $H_0^1(\Sigma_k)$ to $H^2(\Sigma_k)$, and therefore it is compact in $H^1(\mathbb{R}^N)$. For a given $f \in L^2(\mathbb{R}^N)$, equation (3.149) is equivalent to

$$u = Tu + (I + \varepsilon\Lambda)^{-1}f. \tag{3.150}$$

Let $D = \{u \in H_0^1(\Sigma_k);\ \|u\|^2_{L^2(\Sigma_k)} + \varepsilon\|\nabla u\|^2_{L^2(\Sigma_k)} < R^2\}$, where $R = \|f\|_{L^2(\Sigma_k)} + 1$. We note that

$$(I + \varepsilon\Lambda)^{-1}f \notin (I - tT)(\partial D), \quad 0 \le t \le 1. \tag{3.151}$$

Indeed, otherwise there is $u \in \partial D$ and $t \in [0, 1]$ such that

$$u - \varepsilon\Delta u + ta(u)_x = f \quad \text{in } \Sigma_k,$$

and we argue from this to a contradiction.

Multiplying the last equation by u and integrating on Σ_k, we get

$$\|u\|^2_{L^2(\Sigma_k)} + \varepsilon\|\nabla u\|^2_{L^2(\Sigma_k)} + t\int_{\Sigma_k} a(u)_x u\, dx = \int_{\Sigma_k} fu\, dx.$$

On the other hand, we have

$$\int_{\Sigma_k} a(u)_x u\, dx = -\int_{\Sigma_k} a(u)\cdot u_x\, dx = -\int_{\Sigma_k} \operatorname{div} b(u) dx = 0, \tag{3.152}$$

where $b(u) = \int_0^u a(s)ds$. Hence,

$$\|u\|^2_{L^2(\Sigma_k)} + \varepsilon\|\nabla u\|^2_{L^2(\Sigma_k)} \le \|f\|_{L^2(\Sigma_k)}\|u\|_{L^2(\Sigma_k)} \le (R-1)R < R^2, \tag{3.153}$$

and so $u \notin \partial D$.

Let us denote by $d(I - tT, D, (I + \varepsilon\Lambda)^{-1}f)$ the Leray–Schauder degree of the map $I - tT$ relative to D at the point $(I + \varepsilon\Lambda)^{-1}f$ (see, e.g., [65], p. 49). By (3.151) and the invariance property of topological degree, it follows that

$$d(I - tT, D, (I + \varepsilon\Lambda)^{-1}f) = d(I, D, (I + \varepsilon\Lambda)^{-1}f)$$

for all $0 \le t \le 1$. Hence,

$$d(I - T, D, (I + \varepsilon\Lambda)^{-1}f) = d(I, D, (I + \varepsilon\Lambda)^{-1}f) = 1$$

because $(I + \varepsilon\Lambda)^{-1}f \in D$. Hence, equation (3.150) has at least one solution $u \in D(\Lambda) = H_0^1(\Sigma_k) \cap H_0^2(\Sigma_k)$.

To complete the proof, we let $k \to \infty$ in the equation

$$u_k + a(u_k)_x - \varepsilon\Delta u_k = f \text{ in } \Sigma_k,$$
$$u_k \in H_0^1(\Sigma_k).$$

By (3.153), we see that $\{u_k\}$ is uniformly bounded in $H^2(\Sigma_k)$ and so, for $k \to \infty$, we have on a subsequence

$$u_k \to u^* \text{ strongly in } H^1_{\text{loc}}(\mathbb{R}^N), \text{ weakly in } H^2_{\text{loc}}(\mathbb{R}^N).$$

Then, clearly, u^* is a solution to (3.149), and so the proof of Lemma 3.28 is complete. □

Lemma 3.29 *Under the assumptions of Lemma* 3.28, *if* $f \in L^p(\mathbb{R}^N) \cap L^2(\mathbb{R}^N)$, $1 \le p \le \infty$, *then* $u \in L^p(\mathbb{R}^N)$ *and*

$$\|u\|_{L^p(\mathbb{R}^N)} \le \|f\|_{L^p(\mathbb{R}^N)}. \tag{3.154}$$

Proof. We first treat the case $1 < p < \infty$. Let $\alpha_n : \mathbb{R} \to \mathbb{R}$ be defined by

$$\alpha_n(r) = \mathcal{X}_{\frac{1}{n}}(r)|r|^{p-2}r,$$

where $\mathcal{X}_\delta$ is given by (3.92). If we multiply equation (3.149) by $\alpha_n(u) \in L^2(\mathbb{R}^N)$ and integrate on $\mathbb{R}^N$, we get

$$\int_{\mathbb{R}^N} \alpha_n(u)u\,dx \le \int_{\mathbb{R}^N} f\alpha_n(u)dx \tag{3.155}$$

because

$$\begin{aligned}\int_{\mathbb{R}^N} a(u)_x\alpha_n(u)dx &= \int_{\mathbb{R}^N} dx\left(\int_0^{u(x)} a'(s)\alpha_n(s)ds\right)_x dx\\ &= \lim_{|x|\to\infty}\int_0^u a(s)\alpha_n(s)ds = 0,\end{aligned}$$

is consequence of the fact that α_n is monotonically nondecreasing and, since $u \in L^2(\mathbb{R}^N)$, we have

$$\int_0^u a(s)\alpha_n(s) \in L^2(\mathbb{R}^N).$$

We also have

$$-\int_{\mathbb{R}^N} \Delta u\alpha_n(u)dx = \int_{\mathbb{R}^N} \alpha_n'(u)|\nabla u|^2dx \ge 0.$$

Note also the inequality

$$\alpha_n(r)r \ge |\alpha_n(r)|^q, \quad \forall r \in \mathbb{R},\ \frac{1}{p} + \frac{1}{q} = 1.$$

Then, using the Hölder inequality in (3.155), we get

$$\int_{\mathbb{R}^N} |\alpha_n(u)|^q dx \le \left(\int_{\mathbb{R}^N} |f|^p dx\right)^{\frac{1}{p}}\left(\int_{\mathbb{R}^N} |\alpha_n(u)|^q dx\right)^{\frac{1}{q}},$$

whence

$$\int_{[|u(x)|\ge n^{-1}]} |u(x)|^p dx \le \|f\|^p_{L^p(\mathbb{R}^N)},$$

which, clearly, implies that $u \in L^p(\mathbb{R}^N)$ and that (3.154) holds. In the case $p = 1$, we multiply equation (3.149) by $\delta_n(u) = \mathcal{X}_{\frac{1}{n}}(u)$. Note that $\delta_n(u) \in L^2(\mathbb{R}^N)$ because $m\{x \in \mathbb{R}^N;\ |u(x)| > n^{-1}\} \le n^2\|u\|^2_{L^2(\mathbb{R}^N)}$. Then, arguing as before, we get

$$\begin{aligned}\int_{[|u(x)|\ge n^{-1}]} |u(x)|dx &\le \int_{\mathbb{R}^N} |f|\,|\delta_n(u)|dx\\ &\le n\int_{[|u|\le n^{-1}]} |f|\,|u|dx + \int_{[|u|>n^{-1}]} |f|dx \le \|f\|_{L^1(\mathbb{R}^N)}.\end{aligned}$$

Then, letting $n \to \infty$, we get (3.154), as claimed.

Finally, in the case $p = \infty$, we set $M = \|f\|_{L^\infty(\mathbb{R}^N)}$. Then, we have

$$u - M + a(u)_x - \varepsilon\Delta(u - M) = f - M \le 0, \quad \text{a.e. in } \mathbb{R}^N.$$

Multiplying this by $(u-M))^+$ (which, as is well known, belongs to $H^1(\mathbb{R}^N)$), we get

$$\int_{\mathbb{R}^N} ((u-M)^+)^2 dx \le 0$$

because

$$\int_{\mathbb{R}^N} a(u)_x (u-M)^+ dx = 0,$$

$$-\int_{\mathbb{R}^N} \Delta(u-M)(u-M)^+ dx = \int_{\mathbb{R}^N} |\nabla(u-M)^+|^2 dx \ge 0.$$

Hence, $u(x) \le M$, a.e. $x \in \mathbb{R}^N$. Now, we multiply the equation

$$u + M + (a(u))_x - \varepsilon\Delta(u+M) = f + M \ge 0$$

by $(u+M)^-$, integrate on $\mathbb{R}^N$ and get as before that $(u+M)^- = 0$, a.e. in $\mathbb{R}^N$. Hence, $u \in L^\infty(\mathbb{R}^N)$ and

$$|u(x)| \le \|f\|_{L^\infty(\mathbb{R}^N)}, \quad \text{a.e. } x \in \mathbb{R}^N,$$

as desired. □

Lemma 3.30 *Under the assumptions of Lemma* 3.28, *let* $f, g \in L^2(\mathbb{R}^N) \cap L^1(\mathbb{R}^N)$ *and let* $u, v \in H^2(\mathbb{R}^N) \cap L^1(\mathbb{R}^N)$ *be the corresponding solutions to equation* (3.149). *Then we have*

$$\|(u-v)^+\|_{L^1(\mathbb{R}^N)} \le \|(f-g)^+\|_{L^1(\mathbb{R}^N)}, \tag{3.156}$$

$$\|u-v\|_{L^1(\mathbb{R}^N)} \le \|(f-g)\|_{L^1(\mathbb{R}^N)}. \tag{3.157}$$

Proof. Because (3.157) is an immediate consequence of (3.156) we confine ourselves to the latter estimate. If we multiply the equation

$$u - v + (a(u) - a(v))_x - \varepsilon\Delta(u-v) = f - g$$

by $\mathcal{X}_\delta((u-v)^+)$, where $\mathcal{X}_\delta$ is the function (3.92)), integrate on $\mathbb{R}^N$, we get

$$\int_{\mathbb{R}^N} (u-v)\mathcal{X}_\delta((u-v)^+)dx + \frac{1}{\delta}\int_{[|u-v|\le\delta]} (a(u)-a(v)) \cdot \nabla(u-v)dx$$
$$\le \int_{\mathbb{R}^N} (f-g)^+ \mathcal{X}_\delta((u-v)^+)dx,$$

and, since $|a(u) - a(v)| \le |a'|_\infty |u-v|$, we have by Proposition 1.16,

$$\lim_{\delta\to 0} \frac{1}{\delta}\int_{[|u-v|\le\delta]} (a(u)-a(v)) \cdot \nabla(u-v)dx = 0$$

and this yields, for $\delta \to 0$,

$$\|(u-v)^+\|_{L^1(\mathbb{R}^N)} \le \|(f-g)^+\|_{L^1(rr^N)}$$

and also (3.157).

Proof of Theorem 3.26 (continued). Let us show first that $L^1(\mathbb{R}^N) \cap L^\infty(\mathbb{R}^N) \subset R(I + A_0)$. To this end, consider a sequence $\{a_\varepsilon\}$ of C^1 functions such that $a_\varepsilon(0) = 0$

and $a_\varepsilon \overset{\varepsilon\to 0}{\longrightarrow} a$ uniformly on compacta. For $f \in L^1(\mathbb{R}^N)\cap L^\infty(\mathbb{R}^N)$, let $u_\varepsilon \in H^1(\mathbb{R}^N)\cap L^1(\mathbb{R}^N) \cap L^\infty(\mathbb{R}^N)$ be the solution to equation (3.149), where $u = u_\varepsilon$. Note the estimates

$$\|u_\varepsilon\|_{L^1(\mathbb{R}^N)} \leq \|f\|_{L^1(\mathbb{R}^N)},\ \|u_\varepsilon\|_{L^\infty(\mathbb{R}^N)} \leq \|f\|_{L^\infty(\mathbb{R}^N)}, \tag{3.158}$$

which were proven earlier in Lemma 3.30. Also, multiplying (3.149) by u_ε and integrating on $\mathbb{R}^N$, we get

$$\|u_\varepsilon\|^2_{L^2(\mathbb{R}^N)} + \varepsilon\|\nabla u_\varepsilon\|^2_{L^2(\mathbb{R}^N)} \leq C\|f\|^2_{L^2(\mathbb{R}^N)}. \tag{3.159}$$

Moreover, applying Lemma 3.29 to the functions $u = u_\varepsilon(x)$ and $v = v_\varepsilon(x+y)$, we get the estimate

$$\int_{\mathbb{R}^N} |u_\varepsilon(x+y) - u_\varepsilon(x)|dx \leq \int_{\mathbb{R}^N} |f(x+y) - f(x)|dx, \quad \forall y \in \mathbb{R}^N.$$

By the Kolmogorov's compactness criterion (Theorem 1.31), these estimates imply that $\{u_\varepsilon\}$ is compact in $L^1_{\rm loc}(\mathbb{R}^N)$ and, therefore, there is a subsequence, again denoted u_ε, such that

$$\begin{aligned} u_\varepsilon &\to u && \text{strongly in every } L^1(\Sigma_R),\ \forall R > 0,\\ u_\varepsilon(x) &\to u(x), && \text{a.e. } x \in \mathbb{R}^N, \end{aligned} \tag{3.160}$$

where $\Sigma_R = \{x;\ |x| \leq R\}$. We show now that $u + A_0 u = f$.

Let $\varphi \in C_0^\infty(\mathbb{R}^N)$, $\varphi \geq 0$, and let $\alpha \in C^1(\mathbb{R})$ be such that $\alpha'' \geq 0$. We multiply equation (3.149) by $\alpha'(u_\varepsilon)\varphi$, and integrate on $\mathbb{R}^N$. Then, the integration by parts yields

$$\begin{aligned} \int_{\mathbb{R}^N} \alpha'(u_\varepsilon)u_\varepsilon\varphi\, dx - \int_{\mathbb{R}^N} (\alpha'(u_\varepsilon)\varphi)_x(a(u_\varepsilon) - a(k))dx + \varepsilon\int_{\mathbb{R}^N} \alpha''(u_\varepsilon)|\nabla u_\varepsilon|^2\varphi\, dx\\ +\varepsilon\int_{\mathbb{R}^N} (\nabla u_\varepsilon\cdot\nabla\varphi)\alpha'(u_\varepsilon)dx = \int_{\mathbb{R}^N} f\alpha'(u_\varepsilon)\varphi\, dx. \end{aligned}$$

This yields

$$\begin{aligned} \int_{\mathbb{R}^N} (\alpha'(u_\varepsilon)u_\varepsilon\varphi + \varepsilon\alpha'(u_\varepsilon)\nabla u_\varepsilon\cdot\nabla\varphi - (\alpha'(u_\varepsilon)\varphi)_x(a(u_\varepsilon) - a(k)))dx\\ \leq \int_{\mathbb{R}^N} f\alpha'(u_\varepsilon)\varphi\, dx. \end{aligned}$$

Now, letting ε tend to zero, it follows by (3.158)–(3.160) that

$$\int_{\mathbb{R}^N} (\alpha'(u)u\varphi - (\alpha'(u)\varphi)_x(a(u) - a(k)))dx \leq \int_{\mathbb{R}^N} f\alpha'(u)\varphi\, dx.$$

Next, we take $\alpha'(s) = \mathcal{X}_\delta(s-k)$, where $\mathcal{X}_\delta$ is given by (3.92).

Then, letting $\mathcal{X}_\delta \to \mathrm{sign}_0$, for $\delta \to 0$, we get the inequality

$$\int_{\mathbb{R}^N} \mathrm{sign}_0(u-k)[u\varphi - (a(u) - a(k))\varphi_x - f\varphi]dx \leq 0.$$

On the other hand, because $\limsup_{|r|\to 0} \frac{\|a(r)\|}{|r|} < \infty$, we have that $a(u) \in L^1(\mathbb{R}^N)$. We have therefore shown that $f \in u + A_0 u$. Now, let $f \in L^1(\mathbb{R}^N)$, and let $f_n \in$

$L^1(\mathbb{R}^N) \cap L^\infty(\mathbb{R}^N)$ be such that $f_n \to f$ in $L^1(\mathbb{R}^N)$ for $n \to \infty$. Let $u_n \in D(A_0)$ be the solution to the equation $u + A_0 u \ni f_n$. Because A_0 is accretive in $L^1(\mathbb{R}^N) \times L^1(\mathbb{R}^N)$, we see that $\{u_n\}$ is convergent in $L^1(\mathbb{R}^N)$. Hence, there is $u \in L^1(\mathbb{R}^N)$ such that

$$u_n \to u, \quad v_n - u_n \to f \quad \text{in} \quad L^1(\mathbb{R}^N), \; v_n \in A_0 u_n.$$

This implies that $f \in u + Au$. □

In particular, we have proved that for every $f \in L^1(\mathbb{R}^N)$ the first-order partial differential equation

$$u - \sum_{i=1}^{N} \frac{\partial}{\partial x_i} a_i(u) = f \quad \text{in} \quad \mathbb{R}^N \tag{3.161}$$

has a unique solution $u \in L^1(\mathbb{R}^N)$ (in the generalized sense of $u{+}Au{=}f$) and the map $f \to u$ is Lipschitz continuous in $L^1(\mathbb{R}^N)$.

3.4 Porous Media Equations with Drift Term

We shall study here the elliptic nonlinear equation

$$u - \lambda\Delta\beta(u) + \lambda\,\mathrm{div}(Db(u)u) = f \text{ in } \mathcal{D}'(\mathbb{R}^N), \quad N \geq 1, \tag{3.162}$$

where $f \in L^1$, $\lambda > 0$ and the functions $\beta : \mathbb{R} \to \mathbb{R}$, $D : \mathbb{R}^N \to \mathbb{R}^N$, $b : \mathbb{R} \to \mathbb{R}$, satisfy the following hypotheses.

(i) $\beta \in C^1$, $\beta(0) = 0$ and

$$\beta'(r) > 0, \; \forall\, r \neq 0, \tag{3.163}$$

$$\beta' \in L^\infty(\mathbb{R}). \tag{3.164}$$

(ii) $D \in (L^2 \cap L^\infty)(\mathbb{R}^N; \mathbb{R}^N)$, $\mathrm{div}\, D \in L^\infty(\mathbb{R}^N)$.
(iii) $b \in C_b(\mathbb{R}) \cap \mathrm{Lip}(\mathbb{R})$.
(iv) *If* $b \equiv 1$, *then* (3.163) *is weaken to* $\beta'(r) \geq 0$, $\forall\, r \in \mathbb{R}$.
(v) *If* $N \geq 3$, *then condition* (3.164) *is dropped.*

In the following, we use the notation $L^p = L^p(\mathbb{R}^N)$, $1 \leq p \leq \infty$, with the norm denoted $|\cdot|_p$, $L^p_{\mathrm{loc}} = L^p_{\mathrm{loc}}(\mathbb{R}^N)$ and $C_0^\infty = C_0^\infty(\mathbb{R}^N)$, while $C_b(\mathbb{R}) = C_b$ is the space of continuous and bounded real valued functions and $C^1(\mathbb{R}) = C^1$ is the space of continuously differentiable functions on $\mathbb{R} = (-\infty, +\infty)$. $\mathrm{Lip}(\mathbb{R})$ is the space of Lipschitzian real valued functions. By H^k, $k = 1, 2$, we denote the standard Sobolev spaces $H^k(\mathbb{R}^N)$ with the norm denoted by $|\cdot|_{H^k}$.

In equation (3.162), the operator $u \to -\Delta\beta(u)$ is the diffusion term, while $u \to \mathrm{div}(Db(u)u)$ is the so called the *drift term*. In the limit case (that is, for $\frac{1}{\lambda} \to 0$) and $f \equiv 0$, we get

$$-\Delta\beta(u) + \mathrm{div}(Db(u)u) = 0 \text{ in } \mathcal{D}'(\mathbb{R}^N). \tag{3.165}$$

Equation (3.165) is known in literature as the stationary (or steady-state) Fokker–Planck equation and is used in statistical mechanics and in the mean field theory to represent the equilibrium states of the corresponding open dynamical system. As seen later on, the operator $Au = -\Delta\beta(u) + \operatorname{div}(Db(u)u)$, defined on an appropriate domain $D(A) \subset L^1$, is the generator of a Fokker–Planck semiflow in the space L^1. Special cases of these equations are the porous media equation (3.137) and the first order partial differential equation (3.161) studied earlier. As a matter of fact, as seen later on in Section 4.6, (3.162) is just the resolvent equation $u+\lambda Au = f$ corresponding to the generator A. The main interest here is the existence of distributional solutions $u \in L^1(\mathbb{R}^d)$, which are density probabilities (see (3.171)–(3.172) below). We have

Theorem 3.31 *Assume that hypotheses* (i)–(v) *hold. Let $f \in L^1$ and $\lambda > 0$. Then there is a solution $u = u(\lambda, f) \in L^1$ to* (3.162) *satisfying*

$$\beta(u) \in L^1_{\text{loc}},\ \Delta\beta(u) - \operatorname{div}(Db(u)u) \in L^1, \tag{3.166}$$

$$\int_{\mathbb{R}^N} (u\varphi - \lambda\beta(u)\Delta\varphi - \lambda b(u)uD \cdot \nabla\varphi)dx = \int_{\mathbb{R}^N} f\varphi\, dx,\ \forall \varphi \in C_0^\infty, \tag{3.167}$$

$$|u(\lambda, f) - u(\lambda, g)|_1 \le |f - g|_1,\ \forall f, g \in L^1,\ \lambda > 0, \tag{3.168}$$

$$u(\lambda_2, f) = u\left(\lambda_1, \frac{\lambda_1}{\lambda_2} f + \left(1 - \frac{\lambda_1}{\lambda_2}\right) u(\lambda_2, f)\right), \quad \forall f \in L^1,\ 0 < \lambda_1, \lambda_2 < \infty, \tag{3.169}$$

$$|u(\lambda, f) - f|_1 \le C\lambda |f|_{H^2},\ \forall f \in C_0^\infty(\mathbb{R}^N). \tag{3.170}$$

Moreover, we have

$$u(\lambda, f) \ge 0,\ \textit{a.e. in } \mathbb{R}^N \textit{ if } f \ge 0,\ \textit{a.e. in } \mathbb{R}^N, \tag{3.171}$$

$$\int_{\mathbb{R}^N} u(\lambda, f)dx = \int_{\mathbb{R}^N} f(x)dx. \tag{3.172}$$

Theorem 3.31 amounts to saying that under assumptions (i)–(iii) there is a family of distributional solutions $\{u = u(\lambda, f)\}_{\lambda > 0}$ to (3.162) satisfying (3.166)–(3.172). In general, $u(\lambda, f) \in L^1$ is not, however, the unique distributional solution to equation (3.162).

We postpone for the time being the proof of Theorem 3.31 and pause briefly to construct an m-accretive operator A in $L^1 \times L^1$ associated with $\{u(\lambda, f)\}_{\lambda>0}$. To this aim, we consider the operator $A_0 : L^1 \to L^1$ defined by

$$\begin{aligned} A_0 y &= -\Delta\beta(y) + \operatorname{div}(Db(y)y),\ \forall y \in D(A_0), \\ D(A_0) &= \{y \in L^1; \beta(y) \in L^1_{\text{loc}}, -\Delta\beta(y) + \operatorname{div}(Db(y)y) \in L^1\}, \end{aligned} \tag{3.173}$$

where the differential operators Δ and div are taken in the sense of Schwartz distributions, that is, in $\mathcal{D}'(\mathbb{R}^N)$.

By Theorem 3.31, it follows that

$$R(I + \lambda A_0) = \mathbb{R}^N,\ \ \forall \lambda > 0, \tag{3.174}$$

$$|J_\lambda(f) - J_\lambda(g)|_1 \le |f-g|_1, \quad \forall \lambda > 0,\ f,g \in L^1, \tag{3.175}$$

where

$$J_\lambda(f) = u(\lambda, f) \in (I+\lambda A_0)^{-1} f, \quad \forall f \in L^1,\ \lambda > 0.$$

In other words, for each $\lambda > 0$, $J_\lambda : L^1 \to L^1$ is a nonexpansive section of the (eventually) multivalued operator $(I+\lambda A_0)^{-1} : L^1 \to L^1$ satisfying (see (3.169) the resolvent equation

$$J_{\lambda_2}(f) = J_{\lambda_1}\left(\frac{\lambda_1}{\lambda_2} f + \left(1 - \frac{\lambda_1}{\lambda_2}\right) J_{\lambda_2}(f)\right), \quad \forall f \in L^1,\ 0 < \lambda_1, \lambda_2 < \infty. \tag{3.176}$$

It should be emphasized that, contrary to appearances, the operator A_0 defined by (3.173) might not be m-accretive because, as mentioned earlier, (3.175) does not imply the accretivity of the operator A_0 on $D(A_0)$, that is,

$$|(u-v)| + \lambda(A_0 u - A_0 v)|_1 \ge |u-v|_1, \quad \forall u, v \in D(A_0). \tag{3.177}$$

In fact, since in general the solution u of (3.166), (3.167) is not unique, the operator $(I+\lambda A_0)^{-1}$ might be multivalued, and so (3.177) fails (see Proposition 2.41). However, as we see below, A_0 has an m-accretive extension A which will be constructed as in Proposition 2.41.

Indeed, for $\lambda_0 > 0$ arbitrary but fixed we define the operator $A : L^1 \to L^1$,

$$Au = A_0 u, \ \forall u \in D(A) = \{u = J_{\lambda_0}(v);\ v \in L^1\} = J_{\lambda_0}(L^1). \tag{3.178}$$

We have

Theorem 3.32 *Under assumptions* (i)–(iii), *the operator A defined by* (3.173), (3.178) *is m-accretive in $L^1 \times L^1$ and*

$$(I+\lambda A)^{-1} = J_\lambda, \quad \forall \lambda > 0, \tag{3.179}$$

Moreover, if $\beta \in C^2(\mathbb{R})$, then

$$\overline{D(A)} = L^1. \tag{3.180}$$

Proof. By (3.178), it follows (3.179) and also that $R(I+\lambda A) = L^1$, $\forall \lambda > 0$.

For $u_i = J_\lambda(v_i) \in D(A)$, $i = 1,2$, we have

$$Au_i = A_0 u_i, \ u_i + \lambda A_0 u_i = v_i, \ i = 1,2,$$

and, therefore, $Au_i = \frac{1}{\lambda}(v_i - u_i)$. Then, by (3.175), this yields the accretivity condition

$$|u_1 - u_2|_1 \le |v_1 - v_2|_1 = |u_1 - u_2 + \lambda(Au_1 - Au_2)|_1.$$

Hence, A is m-accretive, as claimed. By Proposition 2.41, it also follows that A is independent of λ_0 and (3.179) holds.

Finally, by (3.170) it follows (3.180) because $J_\lambda(f) \in D(A)$, $\forall \lambda > 0$. □

The set $D(A)$ is in general different and, obviously, smaller than the original (distributional) domain $D(A_0)$ of A_0. However, in the special case of porous media equation, we have $D(A) = D(A_0)$ (see Theorem 3.24).

Proof of Theorem 3.31. For $f \in L^1$, consider the equation

$$u + \lambda A_0 u = f. \tag{3.181}$$

Equivalently,

$$\begin{aligned} &u - \lambda\Delta\beta(u) + \lambda\,\mathrm{div}(Db(u)u) = f \text{ in } \mathcal{D}'(\mathbb{R}^N), \\ &u \in L^1, \quad \beta(u) \in L^1_{\mathrm{loc}}, \ -\Delta\beta(u) + \mathrm{div}(Db(u)u) \in L^1. \end{aligned} \tag{3.182}$$

We assume first $f \in L^1 \cap L^2$ and consider the approximating equation

$$u - \lambda\Delta(\beta(u) + \varepsilon u) + \varepsilon\lambda\beta(u) + \lambda\ \mathrm{div}(Db^*_\varepsilon(u)) = f \text{ in } \mathcal{D}'(\mathbb{R}^N). \tag{3.183}$$

Similarly,

$$\begin{aligned} (\varepsilon I - \Delta)^{-1}u + \lambda\beta(u) + \lambda(\varepsilon I - \Delta)^{-1}(\mathrm{div}(Db^*_\varepsilon(u)) - \varepsilon\Delta u) \\ = (\varepsilon I - \Delta)^{-1} f \text{ in } L^2, \end{aligned} \tag{3.184}$$

where $b^*_\varepsilon(u) = \frac{b(u)u}{1+\varepsilon|u|}$. We set

$$\begin{aligned} &F_\varepsilon(u) = (\varepsilon I - \Delta)^{-1}u, \ G(u) = \lambda\beta(u), \ u \in L^2, \\ &G_\varepsilon(u) = \lambda(\varepsilon I - \Delta)^{-1}(\mathrm{div}(Db^*_\varepsilon(u)) - \varepsilon\Delta u), \ u \in L^2, \end{aligned}$$

and note that F_ε and G are accretive and continuous in L^2.

We also have by assumptions (ii)–(iii) that G_ε is continuous in L^2 and

$$\begin{aligned} &\int_{\mathbb{R}^N} (G_\varepsilon(u) - G_\varepsilon(\bar{u}))(u - \bar{u})dx \\ &\quad = -\lambda \int_{\mathbb{R}^N} ((b^*_\varepsilon(u) - b^*_\varepsilon(\bar{u}))D \cdot \nabla(\varepsilon I - \Delta)^{-1}(u - \bar{u}) + \varepsilon|u - \bar{u}|^2 \\ &\quad\quad + \varepsilon^2(\varepsilon I - \Delta)^{-1}(u - \bar{u})(u - \bar{u}))dx \\ &\quad \geq -C_\varepsilon\lambda|u - \bar{u}|_2|\nabla(\varepsilon I - \Delta)^{-1}(u - \bar{u})|_2, \ \forall u, \bar{u} \in L^2(\mathbb{R}^N), \end{aligned}$$

for some $C_\varepsilon > 0$. Moreover, we have

$$\int_{\mathbb{R}^N} (\varepsilon I - \Delta)^{-1}u u\,dx = \varepsilon|(\varepsilon I - \Delta)^{-1}u|_2^2 + |\nabla(\varepsilon I - \Delta)^{-1}u|_2, \ \forall u \in L^2.$$

We see that, for $u^* = u - \bar{u}$, we have

$$\begin{aligned} &(F_\varepsilon(u^*) + G_\varepsilon(u) - G_\varepsilon(\bar{u}) + G(u) - G(\bar{u}), u^*)_2 \\ &\geq \lambda\gamma|u^*|_2^2 + |\nabla(\varepsilon I - \Delta)^{-1}u^*|_2^2 + \varepsilon|(\varepsilon I - \Delta)^{-1}u^*|_2^2 - C_\varepsilon\lambda|u^*|_2|\nabla(\varepsilon I - \Delta)^{-1}u^*|_2. \end{aligned}$$

This implies that $F_\varepsilon + G_\varepsilon + G$ is accretive and coercive on L^2 for $\lambda < \lambda_\varepsilon$, where λ_ε is sufficiently small. Since this operator is continuous and accretive, it follows that it is m-accretive in L^2 and, therefore, surjective (see Theorem 2.42). Hence, for each $f \in L^2 \cap L^1$ and $\lambda < \lambda_\varepsilon$, equation (3.184) has a unique solution $u_\varepsilon \in L^2$.

Since $u_\varepsilon \in L^2$, $|b^*_\varepsilon(r)| \leq C_\varepsilon|r|$, $r \in \mathbb{R}$, and $D \in L^\infty$, we see that by hypothesis (3.164), which implies $|\beta(u_\varepsilon)| \leq C|u_\varepsilon|$, it follows that $u_\varepsilon, \beta(u_\varepsilon) \in H^1(\mathbb{R}^N)$. Indeed, multiplying (3.183) where $u = u_\varepsilon$ by u_ε and $\beta(u_\varepsilon)$, respectively, and integrating over $\mathbb{R}^N$ we get after some calculation that, for $\lambda < \lambda_\varepsilon$,

$$|u_\varepsilon|_2^2 + \lambda|\nabla\beta(u_\varepsilon)|_2^2 + \lambda\varepsilon|\nabla u_\varepsilon|_2^2 + \varepsilon\lambda|\beta(u_\varepsilon)|_2^2 \leq C|f|_2^2, \tag{3.185}$$

where C is independent of ε.

Assume now that $N \geq 3$ and drop condition (3.164). Then, by Theorem 1.11 we have

$$|\beta(u_\varepsilon)|_{p^*} \leq C|\nabla\beta(u_\varepsilon)|_2, \ \frac{1}{p^*} = \frac{1}{2} - \frac{1}{N}$$

and so, by the Hölder inequality,

$$\left|\int_{\mathbb{R}^N} \beta(u_\varepsilon) f\, dx\right| \leq |\beta(u_\varepsilon)|_{p^*} |f|_{q^*}, \ \frac{1}{p^*} + \frac{1}{q^*} = 1.$$

Then, multiplying (3.183), where $u = u_\varepsilon$, by $\beta(u_\varepsilon)$ and integrating as above on $\mathbb{R}^N$, we get instead of (3.185) the estimate

$$|u_\varepsilon|_2^2 + \lambda|\nabla\beta(u_\varepsilon)|_2^2 + \lambda\varepsilon|\nabla u_\varepsilon|_2^2 + \varepsilon\lambda|\beta(u_\varepsilon)|_2^2 \leq C|f|_{q^*}^2, \tag{3.186}$$

for all $0 < \lambda < \lambda_\varepsilon$.

We also note that we have, by (3.183), the resolvent relation

$$\begin{gathered} u_\varepsilon(\lambda_2, f) = u_\varepsilon\left(\lambda_1, \frac{\lambda_1}{\lambda_2} f + \left(1 - \frac{\lambda_1}{\lambda_2}\right) u_\varepsilon(\lambda_2, f)\right), \\ 0 < \lambda_1,\ \lambda_2 < \lambda_\varepsilon,\ f \in L^1 \cap L^2. \end{gathered} \tag{3.187}$$

We assume first that condition (3.164) holds and we denote by $u_\varepsilon(\lambda, f) \in H^1(\mathbb{R}^N)$ the solution to (3.183) for $f \in L^2 \cap L^1$ and prove that

$$|u_\varepsilon(f_1) - u_\varepsilon(f_2)|_1 \leq |f_1 - f_2|_1, \ \forall f_1, f_2 \in L^1 \cap L^2. \tag{3.188}$$

We set $u = u_1 - u_2$, $f = f_1 - f_2$. We have, for $u_i = u_\varepsilon(\lambda, f_i)$, $i = 1, 2$,

$$\begin{aligned} u - \lambda\Delta(\beta(u_1) - \beta(u_2)) - \lambda\varepsilon\Delta(u_1 - u_2) + \varepsilon\lambda(\beta(u_1) - \beta(u_2)) \\ +\lambda \operatorname{div}(D(b_\varepsilon^*(u_1) - b_\varepsilon^*(u_2))) = f \ \text{ in } L^2. \end{aligned} \tag{3.189}$$

We consider the function $\mathcal{X}_\delta : \mathbb{R} \to \mathbb{R}$, defined by (3.92), we set

$$F_\varepsilon = \lambda\nabla(\beta(u_1) - \beta(u_2)) + \lambda\varepsilon\nabla(u_1 - u_2) - \lambda D(b_\varepsilon^*(u_1) - b_\varepsilon^*(u_2))$$

and rewrite (3.189) as

$$u = \operatorname{div} F_\varepsilon - \varepsilon\lambda(\beta(u_1) - \beta(u_2)) + f.$$

Let $\Lambda_\delta = \mathcal{X}_\delta(\widetilde{\beta}(u_1) - \widetilde{\beta}(u_2))$, where $\widetilde{\beta}(u) = \beta(u) + \varepsilon u$. Since $\Lambda_\delta \in H^1(\mathbb{R}^N)$, it follows by (3.183) that $\Lambda_\delta \operatorname{div} F_\varepsilon \in L^1$, and so we have

$$\begin{aligned} \int_{\mathbb{R}^N} u\Lambda_\delta dx &= -\int_{\mathbb{R}^N} F_\varepsilon \cdot \nabla\Lambda_\delta dx \\ &\quad - \varepsilon\lambda\int_{\mathbb{R}^N} (\beta(u_1) - \beta(u_2))\Lambda_\delta dx + \int_{\mathbb{R}^N} f\Lambda_\delta dx \\ &= -\int_{\mathbb{R}^N} (F_\varepsilon \cdot \nabla(\widetilde{\beta}(u_1) - \widetilde{\beta}(u_2))\mathcal{X}_\delta'(\widetilde{\beta}(u_1) - \widetilde{\beta}(u_2))dx \\ &\quad - \varepsilon\lambda\int_{\mathbb{R}^N} (\beta(u_1) - \beta(u_2))\mathcal{X}_\delta(\widetilde{\beta}(u_1) - \widetilde{\beta}(u_2))dx + \int_{\mathbb{R}^N} f\Lambda_\delta dx. \end{aligned} \tag{3.190}$$

We set

$$\begin{aligned} I_\delta^1 &= \int_{\mathbb{R}^N} D(b_\varepsilon^*(u_1) - b_\varepsilon^*(u_2)) \cdot \nabla \Lambda_\delta dx \\ &= \int_{\mathbb{R}^N} D(b_\varepsilon^*(u_1) - b_\varepsilon^*(u_2)) \cdot \nabla(\widetilde{\beta}(u_1) - \widetilde{\beta}(u_2)) \mathcal{X}_\delta'(\widetilde{\beta}(u_1) - \widetilde{\beta}(u_2)) dx \qquad (3.191) \\ &= \frac{1}{\delta} \int_{[|\widetilde{\beta}(u_1) - \widetilde{\beta}(u_2)| \le \delta]} D(b_\varepsilon^*(u_1) - b_\varepsilon^*(u_2)) \cdot \nabla(\widetilde{\beta}(u_1) - \widetilde{\beta}(u_2)) dx. \end{aligned}$$

Since $|D| \in L^\infty \cap L^2$ and

$$|b_\varepsilon^*(u_1) - b_\varepsilon^*(u_2)| \le \mathrm{Lip}(b_\varepsilon^*)|u_1 - u_2| \le \gamma^{-1} \mathrm{Lip}(b_\varepsilon^*)|\widetilde{\beta}(u_1) - \widetilde{\beta}(u_2)|,$$

by assumption (i) it follows that

$$\begin{aligned} &\lim_{\delta \to 0} \frac{1}{\delta} \int_{[|(\widetilde{\beta}(u_1) - \widetilde{\beta}(u_2))| \le \delta]} |D(b_\varepsilon^*(u_1) - b_\varepsilon^*(u_2)) \cdot \nabla(\widetilde{\beta}(u_1) - \widetilde{\beta}(u_2))| dx \\ &\le \gamma^{-1} \mathrm{Lip}(b_\varepsilon^*)|D|_2 \lim_{\delta \to 0} \left(\int_{[|\widetilde{\beta}(u_1) - \widetilde{\beta}(u_2)| \le \delta]} |\nabla(\widetilde{\beta}(u_1) - \widetilde{\beta}(u_2))|^2 dx \right)^{\frac{1}{2}} = 0. \end{aligned}$$

Then, by Proposition 1.16, we get

$$\lim_{\delta \to 0} I_\delta^1 = 0. \qquad (3.192)$$

On the other hand, since $\mathcal{X}_\delta' \ge 0$, we have

$$\int_{\mathbb{R}^N} \nabla(\widetilde{\beta}(u_1) - \widetilde{\beta}(u_2)) \cdot \nabla(\widetilde{\beta}(u_1) - \widetilde{\beta}(u_2)) \mathcal{X}_\delta'(\widetilde{\beta}(u_1) - \widetilde{\beta}(u_2))\, dx \ge 0. \qquad (3.193)$$

By (3.190)–(3.193), since $|\Lambda_\delta| \le 1$, we get

$$\lim_{\delta \to 0} \int_{\mathbb{R}^N} u \mathcal{X}_\delta(\widetilde{\beta}(u_1) - \widetilde{\beta}(u_2)) dx \le \int_{\mathbb{R}^N} |f|\, dx$$

and, since $u\mathcal{X}_\delta(\widetilde{\beta}(u_1) - \widetilde{\beta}(u_2)) \ge 0$ and $\mathcal{X}_\delta \to \mathrm{sign}$ as $\delta \to 0$, by Fatou's lemma this yields $|u|_1 \le |f|_1$, as claimed. Hence, for $n \to \infty$, we have $u_\varepsilon^n \to u_\varepsilon(\lambda, f)$ in L^1. Moreover, we have

$$|u_\varepsilon(\lambda, f_1) - u_\varepsilon(\lambda, f_2)|_1 \le |f_1 - f_2|_1, \ \forall f_1, f_2 \in L^1 \cap L^2, \ 0 < \lambda \le \lambda_\varepsilon. \qquad (3.194)$$

Now, by (3.187) we see that $u_\varepsilon(\lambda, f)$ can be extended as solution to (3.183) for all $\varepsilon > 0$. Indeed, for a fixed $\lambda \in (\lambda_\varepsilon, \infty)$ we consider the equation

$$v = u_\varepsilon \left(\lambda_\varepsilon, \frac{\lambda_\varepsilon}{\lambda} f + \left(1 - \frac{\lambda_\varepsilon}{\lambda}\right) v \right) = T_\varepsilon(v), \ v \in L^1.$$

By (3.188), the operator T_ε satisfies

$$|T_\varepsilon(v_1) - T_\varepsilon(v_2)| \le \left(1 - \frac{\lambda_\varepsilon}{\lambda}\right) |v_1 - v_2|,$$

and, therefore, it is a contraction for $\lambda > \lambda_\varepsilon$. Hence, there is $v_\lambda = v \in L^1$ such that $T_\varepsilon(v_\lambda) = v_\lambda$. Of course, by (3.187) we see that $v_\lambda = u_\varepsilon(\lambda, f)$ for $0 < \lambda \le \lambda_\varepsilon$. Hence,

$v_\lambda = v_\lambda(f)$ is an extension of the solution $u_\varepsilon(\lambda, f)$ for all $\lambda > 0$ and, clearly, (3.187) extends to all $\lambda_1, \lambda_2 > 0$. Moreover, by

$$u_\varepsilon(\lambda, f) = u_\lambda\left(\lambda_\varepsilon, \frac{\lambda_\varepsilon}{\lambda} f + \left(1 - \frac{\lambda_\varepsilon}{\lambda}\right) u_\varepsilon(\lambda, f)\right),$$

it follows that $u_\varepsilon(\lambda, f)$ satisfies (3.183), as claimed.

Fix $\lambda > 0$ and let $f \in L^1 \cap L^2$. Assume first that $1 \leq N < \infty$ and that (3.164) holds. If $u_\varepsilon = u_\varepsilon(\lambda, f)$, by (3.185), it follows that $\{u_\varepsilon\}$ and $\{\beta(u_\varepsilon)\}$ are bounded in $L^2(\mathbb{R}^N)$ and $H^1(\mathbb{R}^N)$, respectively. Hence, along a subsequence, again denoted $\{\varepsilon\} \to 0$ because $H^1_{\rm loc}(\mathbb{R}^N)$ is compact in $L^2_{\rm loc}(\mathbb{R}^N)$, we have

$$\begin{aligned} u_\varepsilon &\longrightarrow u \quad \text{weakly in } L^2(\mathbb{R}^N),\\ \beta(u_\varepsilon) &\longrightarrow \eta \quad \text{weakly in } H^1(\mathbb{R}^N) \text{ and strongly in } L^2_{\rm loc}(\mathbb{R}^N),\\ \Delta\beta(u_\varepsilon) &\longrightarrow \Delta\eta \text{ weakly in } H^{-1}(\mathbb{R}^N). \end{aligned} \tag{3.195}$$

Assume now that $N \geq 3$ and that condition (3.164) is dropped. Then, by (3.186) it follows that, for $f \in L^1 \cap L^{q^*} \cap L^2$, instead of (3.195) we have

$$\begin{aligned} u_\varepsilon &\longrightarrow u \quad \text{weakly in } L^2(\mathbb{R}^N),\\ \beta(u_\varepsilon) &\longrightarrow \eta \quad \text{weakly in } L^{p^*}(\mathbb{R}^N) \cap H^1_{\rm loc}(\mathbb{R}^N),\\ \Delta\beta(u_\varepsilon) &\longrightarrow \Delta\eta \text{ weakly in } H^{-1}_{\rm loc}(\mathbb{R}^N). \end{aligned}$$

In particular, it follows also in this case that $\{\beta(u_\varepsilon)\}$ is compact in $L^2_{\rm loc}(\mathbb{R}^N)$.

Since the graph $u \to \beta(u)$ is maximal monotone in each $L^2(\Omega)$, $\Omega \subset \mathbb{R}^N$, it is weakly-strongly closed and so by (3.195) it follows that $\eta(x) = \beta(u(x))$, a.e. $x \in \mathbb{R}^N$.

Assume first that (i) holds, that is, $b \not\equiv 1$. Since β is strictly monotone on $\mathbb{R} \setminus \{0\}$, it follows by (3.195) that $u_\varepsilon \to u$, a.e. in $\mathbb{R}^N$ as $\varepsilon \to 0$ and by virtue of (iii) this implies that, for $\varepsilon \to 0$,

$$b^*_\varepsilon(u_\varepsilon(x)) \to b^*(u(x)), \quad \text{a.e. } x \in \mathbb{R}^N. \tag{3.196}$$

Taking into account that

$$|b^*_\varepsilon(u_\varepsilon(x))| \leq C|u_\varepsilon(x)|, \quad \text{a.e. } x \in \mathbb{R}^N,$$

we also infer that $\{b^*_\varepsilon(u_\varepsilon)\}$ is weakly compact in $L^2(\mathbb{R}^N)$. Let $\varphi \in C^\infty_0(\mathbb{R}^N)$ and let $\Omega = support\, \varphi$. By (3.196) and the Egorov theorem, it follows that, for each $\delta > 0$, there is $\Omega_\delta \subset \Omega$ such that $\text{meas}(\Omega \setminus \Omega_\delta) \leq \delta$ and

$$b^*_\varepsilon(u_\varepsilon) \to b^*(u) \text{ uniformly on } \Omega_\delta.$$

Then, we have

$$\begin{aligned} \int_{\mathbb{R}^N} \text{div}(Db^*_\varepsilon(u_\varepsilon))\varphi\, dx &= -\int_\Omega (D \cdot \nabla\varphi) b^*_\varepsilon(u_\varepsilon) dx\\ &= -\int_{\Omega_\delta} (D \cdot \nabla\varphi) b^*_\varepsilon(u_\varepsilon) dx - \int_{\Omega\setminus\Omega_\delta} (D \cdot \nabla\varphi) b^*_\varepsilon(u_\varepsilon) dx,\\ &\qquad \forall \varphi \in C^\infty_0(\Omega), \end{aligned}$$

and, therefore,

$$\lim_{\varepsilon\to 0}\left|\left(\int_{\mathbb{R}^N}\operatorname{div}(Db^*_\varepsilon(u_\varepsilon))\varphi\,dx+\int_{\mathbb{R}^N}(D\cdot\nabla\varphi)b^*(u)dx\right)\right|$$
$$\le\int_{\Omega\setminus\Omega_\delta}(|D\cdot\nabla\varphi|(|b^*_\varepsilon(u_\varepsilon)|+|b^*(u)|)\le C(m(\Omega\setminus\Omega_\delta))^{\frac{1}{2}}|u|_2\le C\delta^{\frac{1}{2}}|u|_2,$$
$$\forall\varphi\in C_0^\infty(\Omega).$$

Hence

$$\operatorname{div}(Db^*_\varepsilon(u_\varepsilon))\to\operatorname{div}(Db(u)u)\ \text{ in } \mathcal{D}'(\mathbb{R}^N) \tag{3.197}$$

as $\varepsilon\to 0$.

If $b\equiv 1$, then by (3.195) it follows directly that (3.197) holds because $u_\varepsilon\to u$ weakly in L^2.

Then, passing to the limit in (3.183), we obtain by (3.195)

$$u-\lambda\Delta\beta(u)+\lambda\operatorname{div}(Db(u)u)=f \text{ in } \mathcal{D}'(\mathbb{R}^N), \tag{3.198}$$

where $u=u(f)\in L^1(\mathbb{R}^N)$. By (3.194) and (3.195), it follows that

$$|u(f_1)-u(f_2)|_1\le|f_2-f_2|_1,\ \forall f_1,f_2\in L^2\cap L^1, \tag{3.199}$$

and, respectively, for all $f_1,f_2\in L^1\cap L^2\cap L^{q^*}$ if $N\ge 3$. Indeed, by (3.195) it follows that, for $\varepsilon\to 0$,

$$|u_\varepsilon(\lambda,f_1)-u_\varepsilon(\lambda,f_2)|\to|u(\lambda,f_1)-u(\lambda,f_2)|,\ \text{a.e. in } \mathbb{R}^N,$$

and so, by the Fatou lemma,

$$|u(\lambda,f_1)-u(\lambda,f_2)|_1\le\liminf_{\varepsilon\to 0}|u_\varepsilon(\lambda,f_1)-u_\varepsilon(\lambda,f_2)|_1$$

and so, by (3.194) it follows (3.199), as claimed. Hence $u(\lambda,f)\in D(A_0)$, and $u+\lambda A_0(u)=f$ for $0<\lambda<\lambda_0$. By (3.199), it follows also (3.168).

Now, let us prove that (3.198) and (3.199) extend to all $f\in L^1$ and to $f_1,f_2\in L^1$, respectively. Here is the argument.

For $f\in L^1$ we select $\{f_n\}\subset L^1\cap L^2$, respectively $f_n\in L^1\cap L^2\cap L^{q^*}$ if $N\ge 3$, such that $f_n\to f$ in L^1 as $n\to\infty$. We set $u_n=u(\lambda,f_n)$ and by (3.199) we have

$$|u_n-u_m|_1\le|f_n-f_m|_1,\quad\forall n,m\in\mathbb{N}.$$

Hence, $\{u_n\}$ is convergent in L^1 to $u\in L^1$ which, clearly, satisfy (3.198). Let us show that u is a solution to (3.198). To this purpose we are going to pass to limit in the equation

$$u_n-\lambda\Delta\beta(u_n)+\lambda\operatorname{div}(Db(u_n)u_n)=f\ \text{ in } \mathcal{D}'(\mathbb{R}^d), \tag{3.200}$$

that is,

$$\int_{\mathbb{R}^d}(u_n\varphi-\lambda\beta(u_n)\Delta\varphi-\lambda(D\cdot\nabla\varphi)b(u_n)u_n)dx=\int_{\mathbb{R}^d}f\varphi\,dx. \tag{3.201}$$

Since $u_n\to u$ in L^1 and $b(u_n)u_n\to b(u)u$ in L^1, it remains to show that

$$\beta(u_n)\to\beta(u)\ \text{ in } L^1_{\text{loc}}(\mathbb{R}^d). \tag{3.202}$$

To this end we write, by (3.199),

$$\begin{aligned}\beta(u_n) &= \frac{1}{\lambda}(-\Delta)^{-1}(f - u_n - \lambda\,\mathrm{div}(Db(u_n)u_n))\\ &= \frac{1}{\lambda}E_N * (f - u_n - \lambda\,\mathrm{div}(Db(u_n)u_n))\\ &= \frac{1}{\lambda}E_N * (f - u_n) - \mathrm{div}(E_N * (Db(u_n)u_n)),\end{aligned}$$

where E_N is the fundamental solution to $-\Delta$ in $\mathbb{R}^N$ (see (1.33), (1.34)).

Then, by (1.34) we get, for $N \geq 3$,

$$\|\beta(u_n)\|_{M^{\frac{N}{N-2}}} \leq C(|f|_1 + |u_n|_1)$$

and, since $M^{\frac{N}{N-2}} \subset L^q_{\mathrm{loc}}$ for $1 \leq q < \frac{N}{N-2}$, we infer that $\{\beta(u_n)\}$ is bounded and, consequently, weakly compact in $L^{\frac{N}{N-2}}_{\mathrm{loc}}$. Hence, on a subsequence, we have

$$\beta(u_n) \to \eta \ \text{ weakly in } L^{\frac{N}{N-2}}(K)$$

on each compact set. Since $\{u_n\}$ is strongly convergent to u in L^1, we infer that $\eta(x) = \beta(u(x))$, a.e. $x \in \mathbb{R}^N$ and so we may pass to limit in (3.201) to get (3.198).

We also have

$$u_\varepsilon(\lambda, f) \to u(\lambda, f) = J_\lambda(f) \ \text{ strongly in } L^1, \tag{3.203}$$

but we omit the proof. Then, letting $\varepsilon \to 0$ in (3.199), we get (3.175). Note also that (3.171)–(3.172) follow by the approximating equation (3.183). Indeed, if $f \geq 0$, a.e. in $\mathbb{R}^N$, then multiplying (3.183) by u_ε^-, we get

$$-\int_{\mathbb{R}^N}|u_\varepsilon^-|^2dx - \lambda\int_{\mathbb{R}^N}\beta'(u_\varepsilon)|\nabla u_\varepsilon^-|^2dx + \lambda\int_{\mathbb{R}^N}(D\cdot\nabla u_\varepsilon^-)b(u_\varepsilon)u_\varepsilon^-dx = \int_{\mathbb{R}^N}fu_\varepsilon^-dx.$$

This yields

$$|u_\varepsilon^-|_2^2 + \alpha\lambda|\nabla u_\varepsilon^-|_2^2dx \leq \lambda|D|_\infty|u_\varepsilon^-|^2|\nabla u_\varepsilon^-|_2,$$

and, therefore, $|u_\varepsilon^-|_2 = 0$ for $0 < \lambda \leq \lambda_1$ sufficiently small, hence $u_\varepsilon \geq 0$, a.e. in $\mathbb{R}^N$, for $0 \leq \lambda \leq \lambda_1$. By (3.187), the latter extends to all $\lambda > 0$, as claimed. Finally, for $\varepsilon \to 0$, it follows by (3.203) that (3.189) holds. By (3.183), where $u = u_\varepsilon$, we have

$$\int_{\mathbb{R}^N}u_\varepsilon(x)dx = \int_{\mathbb{R}^N}f(x)dx. \tag{3.204}$$

Formally, (3.204) follows by integrating (3.183) on $\mathbb{R}^N$, but a rigorous argument should take into account that u_ε satisfies (3.183) in the following weak sense

$$\int_{\mathbb{R}^N}u_\varepsilon\psi dx + \lambda\int_{\mathbb{R}^N}(\nabla(\beta(u_\varepsilon) + \varepsilon u_\varepsilon) - Db_\varepsilon^*(u_\varepsilon))\cdot\nabla\psi dx = \int_{\mathbb{R}^N}f\psi dx, \quad \forall\psi \in H^1. \tag{3.205}$$

Now, we choose $\psi(x) = \varphi\left(\frac{|x|^2}{n}\right)$, where $\varphi \in C^1([0,\infty))$ is such that $\varphi(r) = 1$ for $0 \le r \le 1$, $\varphi(r) = 0$ for $r \ge 2$. Then, $\nabla\psi(x) = \frac{2}{n}\, x\varphi'\left(\frac{|x|^2}{n}\right)$ and, by substituting in (3.205), we see that

$$\int_{[|x|\le\sqrt{n}]} (u_\varepsilon(x) - f(x))dx = -\int_{[|x|>\sqrt{n}]} |u_\varepsilon(x) - f(x)|dx + \alpha_n,$$

where $\alpha_n = 0\left(\frac{1}{n}\right)$ as $n \to \infty$. Since $u_\varepsilon, f \in L^1$, we infer that

$$\lim_{n\to\infty} \int_{[|x|\le\sqrt{n}]} (u_\varepsilon(x) - f(x))dx = 0$$

and so (3.204) follows.

For $\varepsilon \to 0$, (3.204) yields (3.190) and letting $\varepsilon \to 0$ in (3.187), we get (3.169).

Finally, if $f \in C_0^\infty$, we have by (3.181),

$$f + \lambda A_0 f = -\lambda\Delta\beta(f) + \lambda\,\mathrm{div}(Db(f)f)$$

and this yields

$$|u(\lambda, f) - f|_1 \le \lambda| - \Delta\beta(f) + \mathrm{div}(Db(f)f)| \le C\lambda\|f\|_{H^2},$$

where C is independent of λ and f, as claimed. □

Remark 3.33 Under assumptions (i)–(v), Theorem 3.31 remains true for the elliptic equation

$$u - \lambda \sum_{i,j=1}^{N} a_{ij} D_{ij}^2(\beta(u)) + \lambda\,\mathrm{div}(Db(u)u) = f \ \text{ in } \mathcal{D}'(\mathbb{R}^N), \tag{3.206}$$

where $D_{ij}^2 = \frac{\partial^2}{\partial x_i \partial x_j}$, $a_{ij} = a_{ji}$, $\forall i, j = 1, ..., N$, and for some $\alpha < 0$

$$\sum_{i,j=1}^{N} a_{ij}\xi_i\xi_j \ge \alpha|\xi|^2, \quad \forall \xi \in \mathbb{R}^N.$$

Then, the corresponding operator A defined by (3.178) is m-accretive in $L^1(\mathbb{R}^N)$. The proof is exactly the same, but we omit the details.

Remark 3.34 Keeping in mind Theorem 3.26, one may speculate that also in this case one could take instead A defined by (3.178) the closure $\overline{A}_0$ in L^1 of the operator (3.173). Apparently, this is the natural definition for realization of the operator A_0 in L^1, but it is not clear if this multivalued operator is m-accretive under the general conditions considered here. However, one might use the construction (3.178) to the treatment of equation (3.161), but this remains an open problem. As seen earlier, this happens if the equation $u + \lambda A_0 u = f$ has a unique solution and this is true, for instance, if $Db(u)u \in L^2(\mathbb{R}^d)$ for $\forall u \in L^1(\mathbb{R}^d)$, but we do not give details.

Comments. The existence theory of semilinear elliptic equations in L^1 presented here is essentially due to Bénilan, Brezis, and Crandall [32], and Brezis and Strauss [49].

The m-accretivity of operator associated with first-order linear equation in $\mathbb{R}^N$ (Theorem 3.26) was proven by Crandall [58] in connection with the conservation law equation which will be discussed later on in Chapter 4. As regards Theorem 3.31 to be used in Chapter 4 to construct the Fokker–Planck semiflows in $L^1(\mathbb{R}^N)$, it was proven in [23] under appropriate assumptions (see also [25]). More general equations of this form were studied in [22], in connections with nonlinear Fokker–Planck equations corresponding to stochastic differential equations.

Chapter 4

Nonlinear Dissipative Dynamics

One cannot escape the feeling that these mathematical formulas have an independent existence and an intelligence of their own, that they are wiser than we are, wiser even their discovers. Heinrich Rudolf Hertz

I consider that I understand an equation when I can predict the properties of its solutions without actually solve it. Paul Dirac

In this chapter we present several applications of general theory to nonlinear dynamics governed by partial differential equations of dissipative type illustrating the ideas and general existence theory developed in the previous section. Most of the significant well posed dynamics described by partial differential equations can be written in the abstract form (2.96) with an appropriate quasi-m-accretive operator A in a Banach space X. The boundary value conditions are incorporated in the domain of A. Usually, X is a space of functions on an open subset Ω of the Euclidean space $\mathbb{R}^N$ and A is a nonlinear differential operator in space variables. In this approach, the time variable t has a privileged role. The whole existence strategy for a given dynamics is to find the appropriate operator A and to prove that it is quasi-m-accretive. The basic result to be frequently used here is the existence theorem for the Cauchy problem and its consequences which provides existence and uniqueness of a mild (or strong) solution as limit of a convergent finite difference scheme. The main emphasis here is put on parabolic-like boundary value problems although the area of problems covered by the general theory is much larger and includes also nonlinear hyperbolic problems. In most cases, the operator A is single-valued but there are, however, notable situations where A is multivalued and this happens, for instance, in the case of problems with free boundary or with discontinuous coefficients.

4.1 Semilinear Parabolic Equations

The classical linear heat (or diffusion) equation perturbed by a nonlinear potential $\beta = \beta(y)$, where y is the state of system, is the simplest form of semilinear parabolic equation arising in applications and will be treated in some details below. The nonlinear potential β might describe exogeneous driving forces intervening over diffusion process or might induce unilateral state constraints.

The principal motivation for choosing multivalued functions β in the examples which follow is not the mathematical generality of the problem, but the necessity to treat problems with a free (or moving) boundary as well as problems with discontinuous monotone nonlinearities. In the latter case, by filling the jumps $[\beta(r_0-0),\beta(r_0+0)]$ of function β, we get a maximal monotone multivalued graph $\widetilde{\beta}\subset\mathbb{R}\times\mathbb{R}$, where $\widetilde{\beta}(r)=\beta(r)$, $\forall r\neq r_0$, for which the general existence theory applies. To be more specific, assume that β is a maximal monotone graph such that $0\in D(\beta)$, and Ω is an open and bounded subset of $\mathbb{R}^N$ with a sufficiently smooth boundary $\partial\Omega$ (for instance, of class C^2). To be more specific, consider the parabolic boundary value problem

$$\begin{cases} \dfrac{\partial y}{\partial t}-\Delta y+\beta(y)\ni f & \text{in } (0,T)\times\Omega=Q,\\ y(0,x)=y_0(x) & \forall x\in\Omega,\\ y=0 & \text{on } (0,T)\times\partial\Omega=\Sigma, \end{cases} \tag{4.1}$$

where $y_0\in L^2(\Omega)$ and $f\in L^2(\Omega)$ are given.

We may represent the parabolic problem (4.1) as a nonlinear differential equation in the space $H=L^2(\Omega)$:

$$\begin{cases} \dfrac{dy}{dt}(t)+Ay(t)\ni f(t), & t\in[0,T],\\ y(0)=y_0, \end{cases} \tag{4.2}$$

where $A:L^2(\Omega)\to L^2(\Omega)$ is the multivalued operator defined by

$$\begin{aligned} Ay&=\{z\in L^2(\Omega);\ z=-\Delta y+w,\ w(x)\in\beta(y(x)),\ \text{a.e. } x\in\Omega\},\\ D(A)&=\{y\in H_0^1(\Omega)\cap H^2(\Omega);\ \exists w\in L^2(\Omega),\ w(x)\in\beta(y(x)),\ \text{a.e. } x\in\Omega\}. \end{aligned} \tag{4.3}$$

Here, $(\frac{d}{dt})y$ is the strong derivative of $y:[0,T]\to L^2(\Omega)$ and the Laplace operator Δ is considered in the sense of distributions on Ω. As a matter of fact, it is readily seen that if y is absolutely continuous from $[a,b]$ to $L^1(\Omega)$, then $\frac{dy}{dt}=\frac{\partial y}{\partial t}$ in $\mathcal{D}'((a,b);\ L^1(\Omega))$, and so a strong solution to equation (4.2) satisfies this equation in the sense of distributions in $(0,T)\times\Omega$. For this reason, whenever there is no any danger of confusion we write $\frac{\partial y}{\partial t}$ instead of $\frac{dy}{dt}$.

Recall (see Proposition 2.20) that A is maximal monotone (i.e., m-accretive) in $L^2(\Omega)\times L^2(\Omega)$ and $A=\partial\varphi$, where

$$\varphi(y)=\begin{cases} \dfrac{1}{2}\displaystyle\int_\Omega|\nabla y|^2dx+\int_\Omega g(y)dx, & \text{if } y\in H_0^1(\Omega),\ g(y)\in L^1(\Omega),\\ +\infty, & \text{otherwise}, \end{cases}$$

and $\partial g=\beta$. Moreover, we have

$$\|y\|_{H^2(\Omega)}+\|y\|_{H_0^1(\Omega)}\le C(\|A^0y\|_{L^2(\Omega)}+1),\ \ \forall y\in D(A). \tag{4.4}$$

Writing equation (4.1) in the form (4.2), we view its solution y as a function of t from $[0,T]$ to $L^2(\Omega)$. The boundary conditions that appear in (4.1) are implicitly incorporated into problem (4.2) through the condition $y(t)\in D(A)$, $\forall t\in[0,T]$.

The function $y : \Omega \times [0, T] \to \mathbb{R}$ is called a *strong solution* to problem (4.1) if $y : [0, T] \to L^2(\Omega)$ is continuous on $[0, T]$, absolutely continuous on $(0, T)$, and satisfies the equation

$$\begin{cases} \dfrac{d}{dt} y(t,x) - \Delta y(t,x) + \beta(y(t,x)) \ni f(t,x), & \text{a.e. } t \in (0,T),\ x \in \Omega, \\ y(0,x) = y_0(x), & \text{a.e. } x \in \Omega, \\ y(t,x) = 0, & \text{a.e. } x \in \partial\Omega,\ t \in (0,T). \end{cases} \tag{4.5}$$

Proposition 4.1 *Let $y_0 \in L^2(\Omega)$ and $f \in L^2(0, T; L^2(\Omega)) = L^2(Q)$ be such that $y_0(x) \in \overline{D(\beta)}$, a.e. $x \in \Omega$. Then, problem* (4.1) *has a unique strong solution*

$$y \in C([0,T]; L^2(\Omega)) \cap W^{1,1}((0,T]; L^2(\Omega))$$

that satisfies

$$t^{\frac{1}{2}} y \in L^2(0,T; H_0^1(\Omega) \cap H^2(\Omega)),\ t^{\frac{1}{2}} \frac{dy}{dt} \in L^2(0,T; L^2(\Omega)). \tag{4.6}$$

If, in addition, $f \in W^{1,1}([0,T]; L^2(\Omega))$, then $y(t) \in H_0^1(\Omega) \cap H^2(\Omega)$ for every $t \in (0, T]$ and

$$t\, \frac{dy}{dt} \in L^\infty(0,T; L^2(\Omega)). \tag{4.7}$$

If $y_0 \in H_0^1(\Omega)$, $g(y_0) \in L^1(\Omega)$, and $f \in L^2(0,T; L^2(\Omega))$, then

$$\frac{dy}{dt} \in L^2(0,T; L^2(\Omega)),\quad y \in L^\infty(0,T; H_0^1(\Omega)) \cap L^2(0,T; H^2(\Omega)). \tag{4.8}$$

Finally, if $y_0 \in D(A)$ and $f \in W^{1,1}([0,T]; L^2(\Omega))$, then

$$\frac{dy}{dt} \in L^\infty(0,T; L^2(\Omega)),\quad y \in L^\infty(0,T; H^2(\Omega) \cap H_0^1(\Omega)) \tag{4.9}$$

and

$$\frac{d^+}{dt} y(t) + (-\Delta y(t) + \beta(y(t)) - f(t))^0 = 0, \qquad \forall t \in [0,T]. \tag{4.10}$$

Proof. This is a direct consequence of Theorem 2.69, because, as seen in Proposition 2.20, we have

$$\overline{D(A)} = \{u \in L^2(\Omega);\ u(x) \in \overline{D(\beta)}, \quad \text{a.e. } x \in \Omega\}.$$

In particular, it follows that for $y_0 \in H_0^1(\Omega)$, $g(y_0) \in L^1(\Omega)$, and $f \in L^2((0,T) \times \Omega)$, the solution y to problem (4.1) belongs to the space

$$H^{2,1}(Q) = \left\{ y \in L^2(0,T; H^2(\Omega)),\ \frac{\partial y}{\partial t} \in L^2(Q) \right\},\quad Q = (0,T) \times \Omega.$$

Problem (4.1) can be studied in the L^p setting, $1 \le p < \infty$ as well, if one takes the operator $A : L^p(\Omega) \to L^p(\Omega)$ as

$$Ay = \{z \in L^p(\Omega);\ z = -\Delta y + w,\ w(x) \in \beta(y)),\ \text{a.e. } x \in \Omega\}, \tag{4.11}$$

$$D(A) = \{y \in W_0^{1,p}(\Omega) \cap W^{2,p}(\Omega);\ w \in L^p(\Omega) \text{ such that } w(x) \in \beta(y(x)),\ \text{a.e. } x \in \Omega\} \quad \text{if } p > 1, \tag{4.12}$$

$$D(A) = \{y \in W_0^{1,1}(\Omega);\ \Delta y \in L^1(\Omega),\ \exists w \in L^1(\Omega) \text{ such that } w(x) \in \beta(y(x)),\ \text{a.e. } x \in \Omega\} \quad \text{if } p = 1. \tag{4.13}$$

As seen earlier in Theorem 3.16, the operator A is m-accretive in $L^p(\Omega) \times L^p(\Omega)$ and so, also in this case, the general existence theory is applicable. Namely, we have

Proposition 4.2 *Let $y_0 \in D(A)$ and $f \in W^{1,1}([0,T]; L^p(\Omega))$, $1 < p < \infty$. Then, problem (4.1) has a unique strong solution $y \in C([0,T]; L^p(\Omega))$, that satisfies*

$$\frac{d}{dt} y \in L^\infty(0,T; L^p(\Omega)),\ y \in L^\infty(0,T; W_0^{1,p}(\Omega) \cap W^{2,p}(\Omega)) \tag{4.14}$$

$$\frac{d^+}{dt} y(t) + (-\Delta y(t) + \beta(y(t)) - f(t))^0 = 0,\ \forall t \in [0,T]. \tag{4.15}$$

If $y_0 \in \overline{D(A)}$ and $f \in L^1(0,T; L^p(\Omega))$, then (4.1) has a unique mild solution

$$y \in C([0,T]; L^p(\Omega)).$$

Proposition 4.2 follows by Theorem 2.61 (recall that $X = L^p(\Omega)$ is uniformly convex for $1 < p < \infty$). □

By Theorem 2.49 (see (2.128)), we have the following.

Proposition 4.3 *Assume $p = 1$. Then, for each $y_0 \in \overline{D(A)}$ and $f \in L^1(0,T; L^1(\Omega))$, problem (4.1) has a unique mild solution $y \in C([0,T]; L^1(\Omega))$, that is,*

$$y(t) = \lim_{h \to 0} y_h(t),$$

where $y_h : [0,T] \to L^1(\Omega)$ is the step function given by

$$y_h^{i+1} = y_h^i + h\Delta y_h^{i+1} - h\beta(y_h^{i+1}) + \int_{ih}^{(i+1)h} f(t)dt \ \ in\ \ \Omega,\ i = 0,1,...,m = \left[\tfrac{T}{h}\right] + 1,$$

$$y_h^0 = y_0,\ \ y_h^{i+1} \in H_0^1(\Omega)$$

$$y_h(t) = y_h^i \ \ for\ \ t \in [ih,(i+1)h),\ i = 0,1,...,m.$$

Because the space $X = L^1(\Omega)$ is not reflexive, according to the general existence theory (Theorem 2.49), the mild solution to the Cauchy problem (4.2) in $L^1(\Omega)$ is only continuous as a function of t, even if y_0 and f are regular. However, also in this case we have a regularity property of mild solutions, that is, a smoothing effect on initial data, which resembles the case $p = 2$. Namely, we have

Proposition 4.4 *Let $\beta : \mathbb{R} \to \mathbb{R}$ be a maximal monotone graph, $0 \in D(\beta)$, and $\beta = \partial g$. Let $f \in L^2(0,T;L^\infty(\Omega))$ and $y_0 \in L^1(\Omega)$ be such that $y_0(x) \in \overline{D(\beta)}$, a.e. $x \in \Omega$. Then, the mild solution $y \in C([0,T];L^1(\Omega))$ to problem (4.1) satisfies*

$$\|y(t)\|_{L^\infty(\Omega)} \le C\left(t^{-\frac{N}{2}}\|y_0\|_{L^1(\Omega)} + \int_0^t \|f(s)\|_{L^\infty(\Omega)}ds\right), \tag{4.16}$$

$$\begin{aligned}&\int_0^T\int_\Omega (t^{\frac{N+4}{2}}y_t^2 + t^{\frac{N+2}{2}}|\nabla y|^2)dx\,dt + T^{\frac{N+4}{2}}\int_\Omega |\nabla y(T,x)|^2dx\\ &\le C\left(\left(\|y_0\|_{L^1(\Omega)}^{\frac{4}{N+2}} + \int_0^T\int_\Omega |f|dx\,dt\right)^{\frac{N+2}{2}} + T^{\frac{N+4}{2}}\int_0^T\int_\Omega f^2dx\,dt\right).\end{aligned} \tag{4.17}$$

Proof. Without loss of generality, we may assume that $0 \in \beta(0)$. Also, let us assume first that $y_0 \in H_0^1(\Omega) \cap H^2(\Omega)$. Then, as seen in Proposition 4.1, problem (4.1) has a unique strong solution such that $t^{\frac{1}{2}}y_t \in L^2(Q)$, $t^{\frac{1}{2}}y \in L^2(0,T;H_0^1(\Omega) \cap H^2(\Omega))$:

$$\begin{cases}\dfrac{\partial y}{\partial t}(t,x) - \Delta y(t,x) + \beta(y(t,x)) \ni f(t,x), & \text{a.e. } (t,x) \in Q,\\ y(0,x) = y_0(x), & x \in \Omega,\\ y = 0, & \text{on } \Sigma.\end{cases} \tag{4.18}$$

Consider the linear problem

$$\begin{cases}\dfrac{\partial z}{\partial t} - \Delta z = \|f(t)\|_{L^\infty(\Omega)} & \text{in } Q,\\ z(0,x) = |y_0(x)|, & x \in \Omega,\\ z = 0, & \text{on } \Sigma.\end{cases} \tag{4.19}$$

Subtracting these two equations and multiplying the resulting equation by $(y-z)^+$, and integrating on Ω we get

$$\frac{1}{2}\frac{d}{dt}\|(y-z)^+\|_{L^2(\Omega)}^2 + \int_\Omega |\nabla(y-z)^+|^2dx \le 0, \text{ a.e. } t \in (0,T),$$

$$(y-z)^+(0) \le 0 \qquad \text{in } \Omega,$$

because $z \ge 0$, β is monotonically increasing, and so $\beta(z) \ge 0$. Hence, $y(t,x) \le z(t,x)$, a.e. in Q and so $|y(t,x)| \le z(t,x)$, a.e. $(t,x) \in Q$. On the other hand, the solution z to problem (4.19) can be represented as

$$z(t,x) = S(t)(|y_0|)(x) + \int_0^t S(t-s)(\|f(s)\|_{L^\infty(\Omega)})ds, \quad \text{a.e. } (t,x) \in Q,$$

where $S(t)$ is the semigroup generated on $L^1(\Omega)$ by $-\Delta$ with Dirichlet homogeneous conditions on $\partial\Omega$. We know, by the regularity theory of $S(t)$ (see also Theorem 4.21 below), that

$$\|S(t)u_0\|_{L^\infty(\Omega)} \le Ct^{-\frac{N}{2}}\|u_0\|_{L^1(\Omega)}, \quad \forall u_0 \in L^1(\Omega),\ t > 0.$$

Hence,

$$|y(t,x)| \le Ct^{-\frac{N}{2}}\|y_0\|_{L^1(\Omega)} + \int_0^t \|f(s)\|_{L^\infty(\Omega)}ds, \quad (t,x)\in Q. \tag{4.20}$$

Now, for an arbitrary $y_0 \in L^1(\Omega)$ such that $y_0 \in \overline{D(\beta)}$, a.e. in Ω, we choose a sequence $\{y_0^n\} \subset H_0^1(\Omega)\cap H^2(\Omega)$, $y_0^n \in \overline{D(\beta)}$, a.e. in Q, such that $y_0^n \to y_0$ in $L^1(\Omega)$ as $n\to\infty$. (We may take, for instance, $y_0^n = S(n^{-1})(1+n^{-1}\beta)^{-1}y_0$.) If y_n is the corresponding solution to problem (4.1), then we know that $y_n \to y$ strongly in $C([0,T];L^1(\Omega))$, where y is the solution with the initial value y_0. By (4.20), it follows that y satisfies estimate (4.16).

Because $y(t) \in L^\infty(\Omega) \subset L^2(\Omega)$ for all $t>0$, it follows by Proposition 4.1 that $y \in W^{1,2}([\delta,T];L^2(\Omega)) \cap L^2(\delta,T;H_0^1(\Omega)\cap H^2(\Omega))$ for all $0<\delta<T$ and it satisfies equation (4.18), a.e. in $Q=(0,T)\times\Omega$. (Arguing as before, we may assume that $y_0 \in H_0^1(\Omega)\cap H^2(\Omega)$ and so $y_t, y \in L^2(0,T;L^2(\Omega))$.) To get the desired estimate (4.17), we multiply equation (4.18) by $y_t t^{k+2}$ and integrate on Q to get

$$\int_0^T\int_\Omega t^{k+2}y_t^2\,dx\,dt + \frac{1}{2}\int_0^T\int_\Omega t^{k+2}|\nabla y|_t^2\,dx\,dt + \int_0^T\int_\Omega t^{k+2}\frac{\partial}{\partial t}g(y)dx\,dt$$
$$= \int_0^T\int_\Omega t^{k+2}y_t f\,dx\,dt,$$

where $y_t = \frac{\partial y}{\partial t}$ and $\partial g = \beta$. This yields

$$\int_Q t^{k+2}y_t^2dx\,dt + \frac{T^{k+2}}{2}\int_\Omega |\nabla y(T,x)|^2dx + T^{k+2}\int_\Omega g(y(T,x))dx$$
$$\le \frac{k+2}{2}\int_Q t^{k+1}|\nabla y|^2dx\,dt + (k+2)\int_Q t^{k+1}g(y)dx\,dt$$
$$+\frac{1}{2}\int_0^T t^{k+2}y_t^2dx\,dt + \frac{1}{2}\int_Q t^{k+2}f^2dx\,dt.$$

Hence,

$$\int_Q t^{k+2}y_t^2dx\,dt + T^{k+2}\int_\Omega |\nabla y(T,x)|^2dx$$
$$\le (k+2)\int_Q t^{k+1}|\nabla y|^2dx\,dt + 2(k+2)\int_Q t^{k+1}\beta(y)dx + T^{k+2}\int_Q f^2dx\,dt.$$

(If β is multivalued, then $\beta(y)$ is of course the section of $\beta(y)$ arising in (4.18).)

Finally, writing $\beta(y)y$ as $(f+\Delta y - y_t)y$ and using Green's formula, we get

$$\begin{aligned}
&\int_Q t^{k+2}y_t^2dx\,dt + T^{k+2}\int_\Omega |\nabla y(T,x)|^2dx + \int_Q t^{k+1}|\nabla y|^2dx\,dt\\
&\le (k+2)(k+1)\int_Q y^2t^kdx\,dt\\
&\quad + T^{k+2}\int_Q f^2dx\,dt + 2(k+2)\int_Q t^{k+1}|f|\,|y|dx\,dt\\
&\le C\left(\int_Q t^ky^2dx\,dt + T^{k+2}\int_Q f^2dx\,dt\right).
\end{aligned} \tag{4.21}$$

Next, we have, by the Hölder inequality

$$\int_\Omega y^2 dx \le \|y\|_{L^p(\Omega)}^{\frac{N-2}{N+2}} \|y\|_{L^1(\Omega)}^{\frac{4}{N+2}}$$

for $p = 2N(N-2)^{-1}$. Then, by the Sobolev embedding theorem,

$$\int_\Omega |y(t,x)|^2 dx \le \left(\int_\Omega |\nabla y(t,x)|^2 dx\right)^{\frac{2N}{N+2}} \left(\int_\Omega |y(t,x)| dx\right)^{\frac{4}{N+2}}. \tag{4.22}$$

On the other hand, multiplying equation (4.18) by $\mathcal{X}_\delta(y(t,x))$ where $\mathcal{X}_\delta$ is given by (3.92) and integrating on $\Omega \times (0,t)$, we get, after letting $\delta \to 0$,

$$\|y(t)\|_{L^1(\Omega)} \le \|y_0\|_{L^1(\Omega)} + \int_0^t \int_\Omega |f(s,x)| dx\, ds, \qquad t \ge 0,$$

because, as seen earlier in Section 3.2, if $y \in W_0^{1,1}(\Omega)$ and $\Delta y \in L^1(\Omega)$, then

$$\int_\Omega \Delta y\ \mathcal{X}_\delta\, dx \le 0.$$

Then, by estimates (4.21) and (4.22), we get

$$\begin{aligned} &\int_Q t^{k+2} y_t^2 dx\, dt + T^{k+2} \int_\Omega |\nabla y(T,x)|^2 dx + \int_Q t^{k+1} |\nabla y(t,x)|^2 dx\, dt \\ &\quad \le C\left(\left(\|y_0\|_{L^1(\Omega)}^{\frac{4}{N+2}} + \int_0^T \int_\Omega |f(t,x)| dx\, dt\right)\right. \\ &\quad\quad \left. \times \int_0^t t^k \|\nabla y(t)\|_{L^2(\Omega)}^{\frac{2N}{N+2}} dt + T^{k+2} \int_Q f^2 dx\, dt\right). \end{aligned}$$

On the other hand, we have, for $k = \frac{N}{2}$,

$$\int_0^T t^k |\nabla y(t)|^{\frac{2N}{N+2}} dt \le \left(\int_0^T t^{k+1} |\nabla y(t)|^2 dt\right)^{\frac{N}{N+2}} T^{\frac{2}{N+2}}.$$

Substituting in the latter inequality, we get after some calculation involving the Hölder inequality

$$\begin{aligned} &\int_Q t^{\frac{N+4}{2}} y_t^2(t,x) dx\, dt + \int_Q t^{\frac{N+2}{2}} |\nabla y(t,x)|^2 dx\, dt + T^{\frac{N+4}{2}} \int_\Omega |\nabla y(t,x)|^2 dx \\ &\quad \le C_1 \left(\|y_0\|_{L^1(\Omega)}^{\frac{4}{N+2}} + \int_Q |f(t,x)| dx\, dt\right)^{\frac{N+2}{2}} + C_2 T^{\frac{N+4}{2}} \int_Q f^2(t,x) dx\, dt, \end{aligned}$$

as claimed. □

It should be mentioned that the above class of parabolic problems covers a large spectrum of applied models including the heat conduction, population dynamics, chemical reaction or the Hodgkin–Huxley model for the nerve axon.

The smoothing effect on initial data of nonlinear heat and diffusion processes modeled by equation (4.1) is a direct consequence of the entropy growth and the irreversibility of these processes.

The interest for $L^1(\Omega)$ initial data is motivated because, in specific situations, the state y is a concentration or density and also by the fact that in the case where y_0 is a positive measure on Ω, for instance the Dirac measure δ, it can be approximated by a sequence $\{y_0^n\} \subset L^1(\Omega)$ and get by the previous estimate the well posedness of problem for measure initial data under appropriate conditions on β (see Brezis and Friedman [46]).

In particular, it follows by Proposition 4.4 that the semigroup $S(t)$ generated by the operator A (defined by (4.11) and (4.13) on $L^1(\Omega)$) has a regularizing effect on L^1-initial data, that is, for all $t > 0$ it maps $L^1(\Omega)$ into $D(A)$ and is a.e. differentiable on $(0, \infty)$.

In the special case, where β is the multivalued function

$$\beta(r) = \begin{cases} 0 & \text{if } r > 0, \\ \mathbb{R}^- & \text{if } r = 0, \end{cases}$$

problem (4.1) reduces to the parabolic variational inequality (the obstacle problem)

$$\begin{cases} \dfrac{\partial y}{\partial t} - \Delta y = f & \text{in } \{(t,x);\ y(t,x) > 0\}, \\ y \geq 0,\ \dfrac{\partial y}{\partial t} - \Delta y \geq f & \text{in } Q, \\ y(0,x) = y_0(x) & \text{in } \Omega, \quad y = 0 \quad \text{on } (0,T) \times \partial\Omega = \Sigma. \end{cases} \tag{4.23}$$

This is a problem with free (moving) boundary which formally can be represented as

$$\begin{aligned} &\frac{\partial y}{\partial t} - \Delta y = f \text{ in } [y > 0], \\ &y \geq 0 \text{ in } Q; \quad f \leq 0 \text{ in } [y = 0], \\ &y(0,x) = y_0(x) \text{ in } \Omega, \quad y = 0 \text{ on } \Sigma, \end{aligned} \tag{4.24}$$

and is used as a model for the one-phase Stefan problem or diffusion of oxigen in tissue (see, e.g., [10]).

We also point out that Proposition 4.1 remains true for equations of the form

$$\begin{cases} \dfrac{\partial y}{\partial t} - \Delta y + \beta(x,y) \ni f & \text{in } Q, \\ y(0,x) = y_0(x) & \text{in } \Omega, \\ y = 0 & \text{on } \Sigma, \end{cases}$$

where $\beta : \Omega \times \mathbb{R} \to 2^{\mathbb{R}}$ is of the form $\beta(x,y) = \partial_y g(x,y)$ and $g : \Omega \times \mathbb{R} \to \mathbb{R}$ is a normal convex integrand on $\Omega \times \mathbb{R}$ sufficiently regular in X and with appropriate polynomial growth with respect to y. The details are left to the reader.

Now, we consider the equation

$$\begin{cases} \dfrac{\partial y}{\partial t} - \Delta y = f & \text{in } (0,T) \times \Omega = Q, \\ \dfrac{\partial}{\partial n} y + \beta(y) \ni 0 & \text{on } \Sigma, \\ y(0,x) = y_0(x) & \text{in } \Omega, \end{cases} \tag{4.25}$$

where $\beta \subset \mathbb{R} \times \mathbb{R}$ is a maximal monotone graph, $0 \in D(\beta)$, $y_0 \in L^2(\Omega)$, and $f \in L^2(Q)$. As seen earlier (Proposition 2.22), we may write (4.25) as

$$\begin{cases} \dfrac{dy}{dt}(t) + Ay(t) = f(t) \quad \text{in } (0,T), \\ y(0) = y_0, \end{cases}$$

where $Ay = -\Delta y$, $\forall y \in D(A) = \{y \in H^2(\Omega);\ 0 \in \frac{\partial y}{\partial n} + \beta(y), \text{ a.e. on } \partial\Omega\}$.

More precisely, $A = \partial\varphi$, where $\varphi : L^2(\Omega) \to \overline{\mathbb{R}}$ is defined by

$$\varphi(y) = \frac{1}{2}\int_\Omega |\nabla y|^2 dx + \int_{\partial\Omega} j(y)d\sigma, \quad \forall y \in L^2(\Omega),$$

and $\partial j = \beta$. (Here, $d\sigma$ is the Lebesgue measure on $\partial\Omega$.) Then, again applying Theorem 2.69, we get the following.

Proposition 4.5 *Let $y_0 \in \overline{D(A)}$ and $f \in L^2(Q)$. Then, problem (4.25) has a unique strong solution $y \in C([0,T]; L^2(\Omega))$ such that*

$$t^{\frac{1}{2}}\frac{dy}{dt} \in L^2(0,T; L^2(\Omega)), \qquad t^{\frac{1}{2}}y \in L^2(0,T; H^2(\Omega)).$$

If $y_0 \in H^1(\Omega)$ and $j(y_0) \in L^1(\Omega)$, then

$$\frac{dy}{dt} \in L^2(0,T; L^2(\Omega)), \quad y \in L^2(0,T; H^2(\Omega)) \cap L^\infty(0,T; H^1(\Omega)).$$

Finally, if $y_0 \in D(A)$ and $f, \frac{\partial f}{\partial t} \in L^2(\Omega)$, then

$$\frac{dy}{dt} \in L^\infty(0,T; L^2(\Omega)), \quad y \in L^\infty(0,T; H^2(\Omega))$$

and

$$\frac{d^+}{dt}y(t) - \Delta y(t) = f(t), \quad \forall t \in [0,T).$$

Note also that, if $y_0 \geq 0$, a.e. in Ω, and $f \geq 0$, a.e. in Q, then $y \geq 0$, a.e. in Q. It should be mentioned that one uses here the estimate (see (2.48))

$$\|u\|_{H^2(\Omega)} \leq C(\|u - \Delta u\|_{L^2(\Omega)} + 1), \quad \forall u \in D(A).$$

An important special case is

$$\beta(r) = \begin{cases} 0 & \text{if } r > 0, \\ (-\infty, 0] & \text{if } r = 0. \end{cases}$$

Then, problem (4.25) reads as

$$\begin{cases} \dfrac{\partial y}{\partial t} - \Delta y = f & \text{in } Q, \\ y\dfrac{\partial y}{\partial n} = 0,\ y \geq 0,\ \dfrac{\partial y}{\partial n} \geq 0 & \text{on } \Sigma, \\ y(0,x) = y_0(x) & \text{in } \Omega. \end{cases} \tag{4.26}$$

A problem of this type arises in the control of a heat field. More precisely, the thermostat control process is modeled by equation (4.26), where

$$\beta(r)=\begin{cases}a_1(r-\theta_1) & \text{if } -\infty<r<\theta_1,\\ 0 & \text{if } \theta_1\le r\le\theta_2,\\ a_2(r-\theta_2) & \text{if } \theta_2<r<\infty,\end{cases}$$

$a_i \ge 0,\ \theta_1 \in \mathbb{R},\ i=1,2$. In the limit case, we obtain (4.26).

The black body radiation heat emission on $\partial\Omega$ is described by equation (4.26), where β is given by (the Stefan–Boltzman law)

$$\beta(r)=\begin{cases}\alpha(r^4-y_1^4) & \text{for } r\ge 0,\\ -\alpha y_1^4 & \text{for } r<0,\end{cases}$$

and, in the case of convection heat transfer,

$$\beta(r)=\begin{cases}ar^{\frac{5}{4}} & \text{for } r\ge 0,\\ 0 & \text{for } r<0.\end{cases}$$

Note, also, that the Michaelis–Menten dynamic model of enzyme diffusion reaction is described by equation (4.1) (or (4.25)), where

$$\beta(r)=\begin{cases}\dfrac{r}{\lambda(r+k)} & \text{for } r>0,\\ (-\infty,0] & \text{for } r=0,\\ \emptyset & \text{for } r<0,\end{cases}$$

where λ, k are positive constants.

We note that more general boundary value problems of the form

$$\begin{cases}\dfrac{\partial y}{\partial t}-\Delta y+\gamma(y)\ni f & \text{in } Q,\\ y(0,x)=y_0(x) & \text{in } \Omega,\\ \dfrac{\partial y}{\partial n}+\beta(y)\ni 0 & \text{on } \Sigma,\end{cases}$$

where β and γ are maximal monotone graphs in $\mathbb{R}\times\mathbb{R}$ such that $0\in D(\beta)$, $0\in D(\gamma)$ can be written in the form (4.2) where $A=\partial\varphi$ and $\varphi: L^2(\Omega)\to\overline{\mathbb{R}}$ is defined by

$$\varphi(y)=\begin{cases}\dfrac{1}{2}\displaystyle\int_\Omega|\nabla y|^2dx+\int_\Omega g(y)dx+\int_{\partial\Omega}j(y)d\sigma & \text{if } y\in H^1(\Omega),\\ +\infty & \text{otherwise,}\end{cases}$$

and $\partial g=\gamma,\ \partial j=\beta$.

We may conclude, therefore, that for $f\in L^2(\Omega)$ and $y_0\in H^1(\Omega)$ such that $g(y_0)\in L^1(\Omega)$, $j(y_0)\in L^1(\partial\Omega)$ the preceding problem has a unique solution $y\in W^{1,2}([0,T];L^2(\Omega))\cap L^2(0,T;H^2(\Omega))$.

On the other hand, semilinear parabolic problems of the form (4.1) or (4.25) arise very often as feedback systems associated with the linear heat equation. For instance, the feedback relay control

$$u = -\rho \operatorname{sign} y, \tag{4.27}$$

where

$$\operatorname{sign} r = \begin{cases} \dfrac{r}{|r|} & \text{if } r \neq 0, \\ [-1,1] & \text{if } r = 0, \end{cases}$$

applied to the controlled heat equation

$$\begin{cases} \dfrac{\partial y}{\partial t} - \Delta y = u & \text{in } \mathbb{R}^+ \times \Omega, \\ y = 0 & \text{on } \mathbb{R}^+ \times \partial\Omega, \\ y(0,x) = y_0(x) & \text{in } \Omega \end{cases} \tag{4.28}$$

transforms it into a nonlinear equation of the form (4.1), that is,

$$\begin{cases} \dfrac{\partial y}{\partial t} - \Delta y + \rho \operatorname{sign} y \ni 0 & \text{in } \mathbb{R}^+ \times \Omega, \\ y = 0 & \text{on } \mathbb{R}^+ \times \partial\Omega, \\ y(0,x) = y_0(x) & \text{in } \Omega. \end{cases} \tag{4.29}$$

This is the closed-loop system associated with the feedback law (4.27) and, according to Proposition 4.4, for every $y_0 \in L^1(\Omega)$, it has a unique strong solution $y \in C(\mathbb{R}^+; L^2(\Omega))$ satisfying

$$\begin{aligned} &y(t) \in L^\infty(\Omega), \ \forall t > 0, \\ &t^{\frac{N+4}{4}} y_t \in L^2_{\text{loc}}(\mathbb{R}^+; L^2(\Omega)), \ t^{\frac{N+2}{4}} y \in L^2_{\text{loc}}(\mathbb{R}^+; H^1(\Omega)). \end{aligned}$$

(Of course, if $y_0 \in L^2(\Omega)$, then y has sharper properties provided by Proposition 4.1.) Let us observe that the feedback control (4.27) belongs to the constraint set $\{u \in L^\infty(\mathbb{R}^+ \times \Omega); \|u\|_{L^\infty(\mathbb{R}^+\times\Omega)} \leq \rho\}$ and steers the initial state y_0 into the origin in a finite time T. Here is the argument. We assume first that $y_0 \in L^\infty(\Omega)$ and consider the function $w(t,x) = \|y_0\|_{L^\infty(\Omega)} - \rho t$. On the domain $\Omega \times (0, \rho^{-1}\|y_0\|_{L^\infty(\Omega)}) = Q_0$, we have

$$\begin{cases} \dfrac{\partial w}{\partial t} - \Delta w + \rho \operatorname{sign} w \ni 0 & \text{in } Q_0, \\ w(0) = \|y_0\|_{L^\infty(\Omega)} & \text{in } \Omega, \\ w \geq 0 & \text{on } \partial\Omega \times (0, \rho^{-1}\|y_0\|_{L^\infty(\Omega)}). \end{cases} \tag{4.30}$$

Then, subtracting equations (4.29) and (4.30) and multiplying by $(y-w)^+$ and integrating on Q_0 (or, simply, applying the maximum principle), we get

$$(y - w)^+ \leq 0 \quad \text{in } Q_0.$$

Hence, $y \leq w$ in Q_0. Similarly, it follows that $y \geq -w$ in Q_0 and, therefore,

$$|y(t,x)| \leq \|y_0\|_{L^\infty(\Omega)} - \rho t, \quad \forall (t,x) \in Q_0.$$

Hence, $y(t) \equiv 0$ for all $t \geq T = \rho^{-1}\|y_0\|_{L^\infty(\Omega)}$. Now, if $y_0 \in L^1(\Omega)$, then inserting in system (4.28) the feedback control

$$u(t) = \begin{cases} 0 & \text{for } \ 0 \leq t \leq \varepsilon, \\ -\rho \operatorname{sign} y(t) & \text{for } \ t > \varepsilon, \end{cases}$$

we get a trajectory $y(t)$ that steers y_0 into the origin in the time

$$T(y_0) < \varepsilon + \rho^{-1}\|y(\varepsilon)\|_{L^\infty(\Omega)} \leq \varepsilon + C(\rho\varepsilon^{\frac{N}{2}})^{-1}\|y_0\|_{L^1(\Omega)},$$

where C is independent of ε and y_0 (see estimate (4.16)). If we choose $\varepsilon > 0$ that minimizes the right-hand side of the latter inequality, then we get

$$T(y_0) \leq \left(\frac{CN}{2\rho}\|y_0\|_{L^1(\Omega)}\right)^{\frac{2}{N+2}} + \left(\frac{N}{2}\right)^{-\frac{N}{N+2}} \left(\frac{C}{\rho}\|y_0\|_{L^1(\Omega)}\right)^{\frac{2}{N+2}}.$$

We have, therefore, proved the following null controllability result for system (4.28).

Proposition 4.6 *For any $y_0 \in L^1(\Omega)$ and $\rho > 0$ there is $u \in L^\infty(\mathbb{R}^+ \times \Omega)$, $\|u\|_{L^\infty(\mathbb{R}^+\times\Omega)} < \rho$, that steers y_0 into the origin in a finite time $T(y_0)$.*

Remark 4.7 Consider the nonlinear parabolic equation

$$\begin{cases} \dfrac{\partial y}{\partial t} - \Delta y + |y|^{p-1}y = 0, & \text{in } \mathbb{R}^+ \times \Omega, \\ y(0,x) = y_0(x), & x \in \Omega, \\ y = 0, & \text{on } \mathbb{R}^+ \times \partial\Omega, \end{cases} \tag{4.31}$$

where $0 < p < \frac{N+2}{N}$ and $y_0 \in L^1(\Omega)$. By Proposition 4.4, we know that the solution y satisfies the estimates

$$\|y(t)\|_{L^\infty(\Omega)} \leq Ct^{-\frac{N}{2}}|y_0\|_{L^1(\Omega)},$$

$$\|y(t)\|_{L^1(\Omega)} \leq C\|y_0\|_{L^1(\Omega)},$$

for all $t > 0$.

Now, if y_0 is a bounded Radon measure on Ω, that is, $y_0 \in M(\Omega) = (C_0(\overline{\Omega}))^*$ ($C_0(\overline{\Omega})$ is the space of continuous functions on $\overline{\Omega}$ that vanish on $\partial\Omega$), there is a sequence $\{y_0^j\} \subset C_0(\Omega)$ such that $\|y_0^j\|_{L^1(\Omega)} \leq C$ and $y_0^j \to y_0$ weak-star in $M(\Omega)$. Then, if y^j is the corresponding solution to equation (4.31) it follows from the previous estimates, for $j \to \infty$ that (see Brezis and Friedman [46]),

$$\begin{aligned} y^j &\to y & &\text{in } L^q(Q), \ 1 < q < \tfrac{N+2}{N}, \\ |y^j|^{p-1}y^j &\to |y|^{p-1}y & &\text{in } L^1(Q). \end{aligned}$$

Moreover, $u \in C^{2,1}(\overline{\Omega} \times (0,\infty))$ is a classical solution to (4.31) on $\Omega \times (0,\infty)$ and

$$\lim_{t\to 0} \int_{\mathbb{R}^N} y(t,x)\varphi(x)dx = y_0(\varphi), \ \varphi \in C_0(\overline{\Omega}).$$

If $p > \frac{N+2}{N}$, there is no solution to (4.31).

Remark 4.8 Consider the semilinear parabolic equation (4.1), where β is a continuous monotonically increasing function, $f \in L^p(Q)$, $p > 1$, and $y_0 \in W_0^{p,2-\frac{2}{p}}(\Omega)$, $g(y_0) \in L^1(\Omega)$, $g(r) = \int_0^r |\beta(s)|^{p-2}\beta(s)ds$. Then, the solution y to problem (4.1) belongs to $W_p^{2,1}(Q)$ and

$$\|y\|^p_{W_p^{2,1}(Q)} \leq C\left(\|f\|^p_{L^p(\Omega)} + \|y_0\|^p_{W_0^{p,2-\frac{2}{p}}(\Omega)} + \int_\Omega g(y_0)dx\right). \tag{4.32}$$

(Here, $W_p^{2,1}(Q)$ is the space

$$\left\{y \in L^p(Q);\ \frac{\partial^{r+s}}{\partial t^r \partial x^s}\, y \in L^p(Q),\ 2r + s \leq 2\right\}.$$

For $p = 2$, we have $W_2^{2,1}(Q) = H^{2,1}(Q)$.)

Indeed, if we multiply equation (4.1) by $|\beta(y)|^{p-2}\beta(y)$ and integrate on Q, we get the estimate (as seen earlier in Proposition 4.1, for f and y_0 smooth enough this problem has a unique solution $y \in W^{1,\infty}([0,T]; L^p(\Omega))$, $y \in L^\infty(0,T; W^{2,p}(\Omega))$)

$$\begin{aligned}
&\int_\Omega g(y(t,x))dx + \int_0^t \int_\Omega |\beta(y(s,x))|^p dx\, ds \\
&\quad \leq \int_0^t \int_\Omega |\beta(y(s,x))|^{p-1} |f(s,x)| dx\, ds + \int_\Omega g(y_0(x))dx \\
&\quad \leq \left(\int_0^t \int_\Omega |\beta(y(s,x))|^p dx\, ds\right)^{\frac{1}{q}} \left(\int_0^t \int_\Omega |f(s,x)|^p dx\, ds\right)^{\frac{1}{p}} + \int_\Omega g(y_0(x))dx,
\end{aligned}$$

where $\frac{1}{p} + \frac{1}{q} = 1$. In particular, this implies that

$$\|\beta(y)\|_{L^p(Q)} \leq C(\|f\|_{L^p(Q)} + \|g(y_0)\|_{L^1(\Omega)})$$

and so, by the standard L^p estimates for linear parabolic equations (see, e.g., Ladyzenskaya, Solonnikov, and Ural'ceva [76]), we find the estimate (4.32), which clearly extends to all $f \in L^p(Q)$ and $y_0 \in W_0^{p,2-\frac{2}{p}}(\Omega)$, $g(y_0) \in L^1(\Omega)$.

Nonlinear parabolic equations of divergence type

Several physical diffusion processes are described by the continuity equation

$$\frac{\partial y}{\partial t} + \operatorname{div} \vec{q} = f,$$

where the flux $\vec{q}$ of the diffusive material is a nonlinear function β of local density gradient ∇y. Such an equation models nonlinear interaction phenomena in material science and in particular in mathematical models of crystal growth as well as in heat propagation, nonlinear diffusion and image processing. This class of problems can be written as

$$\begin{cases}
\dfrac{\partial y}{\partial t}(t,x) - \operatorname{div} \beta(\nabla(y(t,x))) \ni f(t,x), & x \in \Omega,\ t \in (0,T), \\
y = 0 & \text{on } (0,T) \times \partial\Omega, \\
y(0,x) = y_0(x), & x \in \Omega,
\end{cases} \tag{4.33}$$

where $\beta : \mathbb{R}^N \to \mathbb{R}^N$ is a maximal monotone graph satisfying conditions (3.10) and (3.11) (or, in particular, conditions (3.6) and (3.7) of Theorem 3.2). Here and everywhere in the following, the differential operators div and ∇ are taken with respect to the spatial variable $x \in \Omega$.

In the space $X = L^2(\Omega)$, consider the operator A defined by (3.27) in order to represent (4.33) as a Cauchy problem in X, that is,

$$\begin{cases} \frac{dy}{dt}(t) + Ay(t) \ni f(t), \quad t \in (0,T), \\ y(0) = y_0. \end{cases} \tag{4.34}$$

In Section 3.1, we have studied in detail the stationary version of (4.33), i.e., $Ay = f$ and we have proven (Theorem 3.5) that A is maximal monotone (m-accretive) in $L^2(\Omega)$ and so, by Theorem 2.61, we obtain the following existence result.

Proposition 4.9 *Let $f \in W^{1,1}([0,T]; L^2(\Omega))$, $y_0 \in W_0^{1,p}(\Omega)$ be such that $\operatorname{div} \eta_0 \in L^2(\Omega)$ for some $\eta_0 \in (L^q(\Omega))^N$, $\eta_0 \in \beta(\nabla y_0)$, a.e. in Ω. Then, there is a unique strong solution y to (4.33) (equivalently to (4.34)) such that*

$$y \in L^\infty(0,T; W_0^{1,p}(\Omega)) \cap W^{1,\infty}([0,T]; L^2(\Omega))$$

$$\frac{d^+}{dt} y(t) - \operatorname{div} \eta(t) = f(t), \quad \forall t \in [0,T],$$

where $\eta \in L^\infty(0,T; L^2(\Omega))$, $\eta(t,x) \in \beta(\nabla y(t,x))$, a.e. $(t,x) \in (0,T) \times \Omega = Q$. Moreover, if $\beta = \partial j$, then the strong solution y exists for all $y_0 \in L^2(\Omega)$ and $f \in L^2(Q)$. If $y_0 \geq 0$, a.e. in Ω, then

$$y(t,x) \geq 0, \quad a.e.\ (t,x) \in (0,T) \times \Omega. \tag{4.35}$$

(Here and everywhere in the following, $\frac{dy}{dt}$ is the derivative of the function $t \to y(t,\cdot)$ from $[0,T]$ to $X = L^2(\Omega)$.)

The last part of Proposition 4.9 follows by Theorem 2.69, because, as seen earlier in Theorem 3.5, in this latter case $A = \partial\varphi$. As regards (4.35), it follows by the exponential formula (2.112) because, as easily seen, $(I + \lambda A)^{-1}$ leaves invariant the set of nonnegative $L^2(\Omega)$-functions. In fact, if $f \geq 0$, a.e. in Ω, then multiplying the equation

$$y - \lambda \operatorname{div} \beta(\nabla y) = f \text{ in } \Omega; \qquad y = 0 \text{ on } \partial\Omega,$$

by y^- and integrating on Ω, we see that $y^- = 0$.

Now, if we refer to Theorem 3.7 and Remark 2.25 we see that Proposition 4.9 remains true under conditions $\beta = \partial j$ and (3.33) and (3.34). We have, therefore, the following.

Proposition 4.10 *Let β satisfy conditions* (3.33) and (3.34). *Then, for each $y_0 \in L^2(\Omega)$ and $f \in L^2(0,T; L^2(\Omega))$ there is a unique strong solution to* (4.33) *or to the* *equation with Neumann boundary conditions* $\beta(\nabla y(x)) \cdot \nu(x) = 0$, *in the following weak sense,*

$$\frac{d}{dt}\int_\Omega y(t,x)v(x)dx + \int_\Omega \eta(t,x)\cdot\nabla v(x)dx = \int_\Omega f(t,x)v(x)dx,\ \forall v \in C^1(\overline{\Omega}),$$
$$\eta(t,x) \in \beta(\nabla y(t,x)),\quad a.e.\ (t,x) \in (0,T)\times\Omega,$$
$$y(0,x) = y_0(x).$$

Now, we consider the singular diffusion boundary value problem

$$\begin{cases} \dfrac{\partial y}{\partial t} - \mathrm{div}(\mathrm{sign}\ (\nabla y)) \ni f & \text{in } (0,T)\times\Omega, \\ y = 0 & \text{on } (0,T)\times\partial\Omega, \\ y(0,x) = y_0(x), & x \in \Omega, \end{cases} \tag{4.36}$$

where $\mathrm{sign}\, u \equiv u|u|^{-1}$, $\forall u \in \mathbb{R}^N$, $u \neq 0$, $\mathrm{sign}(0) = \{u \in \mathbb{R}^N;\ |u| \leq 1\}$.

As seen earlier in Section 3.2, the operator $y \to \mathrm{div}(\mathrm{sign}(\nabla y))$ with the domain $\{y \in W_0^{1,1}(\Omega); \mathrm{div}(\mathrm{sign}(\nabla y)) \in L^2\}$ is not m-accretive in the space $L^2(\Omega)$ (neither in every $L^p(\Omega)$, $1 \leq p \leq \infty$, as well, but this happens if we replace it by its realization in the space $BV(\Omega)$. Namely, the operator $A = \partial\Phi : L^2(\Omega) \to L^2(\Omega)$, where its potential $\Phi : BV(\Omega) \to]-\infty, +\infty]$ is the total variation function defined by (3.60), is maximal monotone (equivalently, m-accretive) in $L^2(\Omega) \times L^2(\Omega)$. Then, the Cauchy problem (the *total variation flow equation*)

$$\begin{aligned} &\frac{dy}{dt} + \partial\Phi(y) \ni 0, \quad \text{a.e. } t \in (0,T), \\ &y(0) = y_0 \end{aligned} \tag{4.37}$$

is well posed and is the natural extension of (4.36). In fact, by Theorem 2.49, it follows that, for $y_0 \in D(\Phi) = BV(\Omega)$, equation (4.37) has a unique solution (the *total variation flow* on Ω)

$$y \in C([0,T]; L^2(\Omega)) \cap W^{1,2}([0,T]; L^2(\Omega)),$$
$$\Phi(y(t)) \leq \Phi(y_0), \quad \forall t \in [0,T].$$

In other words, $y(t) = S(t)y_0$, where $S(t)$ is the semigroup on $L^2(\Omega)$ generated by $\partial\Phi$, that is,

$$S(t)y_0 = e^{-t\partial\Phi}y_0,\ \forall t \geq 0.$$

The operator $\partial\Phi$ is hard to describe explicitly, and so problem (4.37) as well, but a convenient approximation of $\partial\Phi$ already studied earlier is the operator $A_\varepsilon : L^2(\Omega) \to L^2(\Omega)$ defined by

$$A_\varepsilon u = -\varepsilon\Delta u - \mathrm{div}(\mathrm{sign}(\nabla u)),\ \forall u \in D(A_\varepsilon) = H_0^1(\Omega) \cap H^2(\Omega),\ \forall \varepsilon > 0.$$

More precisely, we have

$$A_\varepsilon u = \{-\varepsilon\Delta u - \mathrm{div}\ \eta;\ \eta \in L^2(\Omega),\ \eta \in \mathrm{sign}\,\nabla u,\ \text{a.e. in } \Omega\},\ \forall u \in D(A_\varepsilon).$$

By Theorem 2.15, we know that, if Ω is a bounded, convex and open set of $\mathbb{R}^N$ with smooth boundary $\partial\Omega$, then, for each $\varepsilon > 0$, the operator A_ε is m-accretive in $L^2(\Omega) \times L^2(\Omega)$. Moreover, for $f \in L^2(\Omega)$, we have for the solution $y_\varepsilon \in H_0^1(\Omega) \cap H^2(\Omega)$ to

$$y_\varepsilon - \varepsilon\Delta y_\varepsilon - \operatorname{div}(\operatorname{sign}(\nabla y_\varepsilon)) = f, \tag{4.38}$$

that as $\varepsilon \to 0$,

$$y_\varepsilon \to y \ \text{ weakly in } L^2(\Omega) \text{ and strongly in } L^p(\Omega) \text{ for } 1 \le p < \tfrac{N}{N-1}. \tag{4.39}$$

In particular, by Proposition 4.5 it follows that, for each $y_0 \in L^2(\Omega)$ and $f \in L^2((0,T) \times \Omega)$, the problem

$$\begin{aligned} &\frac{\partial y_\varepsilon}{\partial t} - \varepsilon\Delta y_\varepsilon - \operatorname{div}(\operatorname{sign}(\nabla y_\varepsilon)) = f \ \text{ in } (0,T) \times \Omega, \\ &y_\varepsilon = 0 \qquad\qquad\qquad\qquad\qquad\qquad \text{on } (0,T) \times \partial\Omega, \\ &y_\varepsilon(0,x) = y_0, \qquad\qquad\qquad\qquad\quad x \in \Omega, \end{aligned} \tag{4.40}$$

has a unique solution

$$y_\varepsilon \in C([0,T]; L^2(\Omega)) \cap W^{1,2}([\delta,T]; L^2(\Omega)) \cap L^2([\delta,T]; H_0^1(\Omega) \cap H^2(\Omega)), \quad \forall \delta \in (0,T),$$

while, for $y_0 \in H_0^1(\Omega)$, we have

$$y_\varepsilon \in W^{1,2}([0,T]; L^2(\Omega)) \cap L^2(0,T; H_0^1(\Omega) \cap H^2(\Omega)).$$

Moreover, it turns out that the total variation flow $y(t) = e^{-t\partial\Phi}y_0$ defined by (4.37) can be approximated by the solution y_ε to (4.40). Namely, we have

Proposition 4.11 *Let $y_0 \in L^2(\Omega)$, $f \equiv 0$, and let y be the corresponding solution to* (4.37). *Then, for $\varepsilon \to 0$,*

$$y_\varepsilon \to y \quad \text{in } C([0,T]; L^2(\Omega)).$$

Proof. This follows by the Trotter–Kato theorem (Theorem 2.76), because by Theorem 3.10 we know that, for each $f \in L^2(\Omega)$, we have

$$\lim_{\varepsilon\to 0}(I + \varepsilon A_\varepsilon)^{-1} f = (I + \varepsilon\partial\Phi)^{-1} f \text{ strongly in } L^2(\Omega). \tag{4.41}$$

Remark 4.12 Proposition 4.11 remains true if in (4.36) the Dirichlet boundary condition is replaced by the Neumann homogeneous boundary condition. In this case, the function Φ is replaced by the function Ψ defined by (3.61).

Equation (4.40) can be viewed as a viscosity approach to the total variation flow generated by (4.37) and it is a convenient way to approximate the total variation flow $S(t) = e^{-t\partial\Phi}$ by a smooth one, namely,

$$S_\varepsilon(t) = e^{-tA_\varepsilon}.$$

One should also note that, by Propositions 3.11 and 3.12, it follows that

$$\|S(t)y_0\|_{W_0^{1,p}(\Omega)} \le \|y_0\|_{W_0^{1,p}(\Omega)},\ \forall y_0 \in W_0^{1,p}(\Omega) \cap L^2(\Omega),$$

for all $1 \le p < \infty$, and so, for $p = 1$ we get

$$\Phi(S(t)y_0) \le \Phi(y_0),\ \forall y_0 \in BV(\Omega) \cap L^2(\Omega),\ t \ge 0.$$

Then, by Theorem 1.29, we infer that the trajectory $\gamma(y_0) = \{S(t)y_0,\ t \ge 0\}$ is compact in $L^p(\Omega)$ for $1 \le p < \frac{N}{N-1}$. Then, for $N = 1$, the image limit set $\omega(y_0)$ is nonempty and so Theorem 2.72 applies. However, also for $N > 1$ one might apply Theorem 2.71 to get the weak convergence for $t \to \infty$ of $S(t)y_0$ in $L^2(\Omega)$ because $A = \partial\Phi$.

The PDE image restoring model

The nonlinear evolution equation

$$\begin{aligned} &\frac{\partial y}{\partial t} - \operatorname{div}(\zeta(\nabla y)) = 0,\ t \ge 0,\ x \in \Omega, \\ &y(t,x) = y_0(x), \qquad\qquad x \in \Omega, \end{aligned} \tag{4.42}$$

with the Dirichlet bolundary condition

$$y(t,x) = 0,\ \ \forall\, t \ge 0,\ x \in \partial\Omega,$$

or the Neumann boundary condition

$$\zeta(\nabla y(t,x)) \cdot n = 0,\ \ \forall\, t \ge 0,\ x \in \partial\Omega,$$

if is well posed, can be used as a nonlinear filter for restoring a degraded image $\Omega \subset \mathbb{R}^2$. The restored image from the original blurred one $y_0(x)$ at time t is $y(t,x)$. (The finite difference scheme

$$y_{j+1} - h\ \operatorname{div}(\zeta(\nabla y_{j+1}(x))) = y_j(x),\ j = 0, 1, \dots, x \in \Omega, \tag{4.43}$$

associated with problem (4.42) is just the PDE elliptic procedure to image restoring briefly described by equation (3.55) in Section 3.2.)

In the special case $\zeta(r) \equiv \operatorname{sign} r$, equation (4.42) leads to the total variation flow approach to image restoring (see [5, 53, 99])

$$\begin{aligned} &\frac{dy}{dt} + \partial\Phi(y) = 0,\ \ t > 0, \\ &y(0) = y_0. \end{aligned} \tag{4.44}$$

As seen above, for a blurred image $y_0 \in L^2(\Omega)$, the solution $y(\cdot,t)$ at a fixed time t (the restored image) is in $BV(\Omega)$. This means that the restored image $y(\cdot,t) \in BV(\Omega)$ is sufficiently smooth in x, but it is not of class C^1 or absolutely continuous as happens with other filters. This fact implies that $y(\cdot,t)$ detects edges and is continuous on rectifiable curves which is an important feature for a recovered image. Of course, equation (4.42) is hard to solve (even numerically) and so, in specific cases, it is approximated by more regular equations. One of these approximations is equation (4.40) which is discussed in Proposition 4.11. One might suspect, however, that the best results are not obtained for large times t because, as follows by Theorem 2.70, we have

$$\lim_{t\to\infty} y(t) = y_\infty = (\partial\Phi)^{-1}(0), \text{ weakly in } L^2(\Omega),$$

and so the *restored* image might be far away from the original one.

A very popular filter of the form (4.42) was proposed by Perona and Malik [91] and is described by the equation

$$\begin{aligned} &\frac{\partial y}{\partial t} - \operatorname{div}(g(|\nabla y|^2)\nabla y) = 0 \text{ in } && (0,T)\times\Omega, \\ &\nabla y \cdot n = 0 && \text{on } (0,T)\times\partial\Omega, \\ &y(0) = y_0 && \text{in } \Omega, \end{aligned} \tag{4.45}$$

where $g(s) = \frac{\alpha^2}{\alpha^2+s}$, $\forall s \geq 0$. Though the Perona–Malik model is successful in many specific situations and it is at origin of a large variety of other models, however, from the mathematical point of view a severe limitation of this model is that problem (4.45) is, in general, not well posed. Indeed, by a simple computation it follows that the mapping $r \to g(|r|^2)r$ is monotonically decreasing in $\mathbb{R}^N$ for $|r| > \alpha$ which, as is well-known, implies that the problem is ill-posed in this region. However, we have the following local existence result

Proposition 4.13 *Assume that Ω is a bounded convex set with smooth boundary. Let $y_0 \in W^{1,\infty}(\Omega)$. Then, for $\|y_0\|_{W^{1,\infty}(\Omega)} < \alpha$, there is a unique solution y to* (4.45) *satisfying*

$$y \in C([0,T];L^2(\Omega)) \cap L^\infty(0,T;H_0^1(\Omega)),$$

$$\frac{\partial y}{\partial t},\ \operatorname{div}(g(|\nabla y|^2)\nabla y) \in L^\infty(\delta,T;L^2(\Omega)),\ \forall \delta \in (0,T),$$

$$|\nabla y(t,x)| \leq \alpha,\ a.e.\ (t,x) \in (0,T)\times\Omega,$$

$$\frac{\partial^+}{\partial t} y(t,x) - \operatorname{div}(g(|\nabla y(t,x)|^2)\nabla y(t,x)) = 0,\ \forall t \in (0,T),\ x \in \Omega.$$

Moreover, if $y_0 \geq 0$, a.e. in Ω, then $y(t,x) \geq 0$, a.e. $(t,x) \in (0,T)\times\Omega$.

Taking into account that y_0 is the blurred image, condition $y_0 \in W^{1,\infty}(\Omega)$ is, apparently, too restrictive. However, the above result shows that the Perona–Malik model is well-posed in the class of 'moderately' degraded original images y_0.

Proof. We set

$$g_\alpha(s) = \begin{cases} g(s) = \dfrac{\alpha^2}{s+\alpha^2} & \text{for } 0 \leq s \leq \alpha, \\ \dfrac{\alpha}{\alpha+1} & \text{for } s > \alpha, \end{cases}$$

and consider the operator $A_\alpha^\varepsilon : D(A_\alpha^\varepsilon) \subset L^2(\Omega) \to L^2(\Omega)$, $\varepsilon > 0$, defined by

$$A_\alpha^\varepsilon u = -\operatorname{div}(g_\alpha(|\nabla u|^2)\nabla u) - \varepsilon\Delta u,\ u \in D(A_\alpha^\varepsilon),$$

where $D(A_\alpha^\varepsilon) = H_0^1(\Omega) \cap H^2(\Omega)$. We have

Lemma 4.14 *For each $\varepsilon > 0$, the operator A_α^ε is maximal monotone (m-accretive) in $L^2(\Omega) \times L^2(\Omega)$.*

Proof. Consider the operator $\widetilde{A}_\alpha^\varepsilon : H_0^1(\Omega) \to H^{-1}(\Omega)$ defined by

$$_{H^{-1}(\Omega)}\left\langle \widetilde{A}_\alpha^\varepsilon u, v \right\rangle_{H_0^1(\Omega)} = \int_\Omega g_\alpha(|\nabla u|^2)\nabla u \cdot \nabla v\, dx + \varepsilon \int_\Omega \nabla u \cdot \nabla v\, dx,\ \forall v \in H_0^1(\Omega).$$

It is easily seen the mapping $r \to g_\alpha(|r|^2)r$ is monotone and continuous in $\mathbb{R}^N$. This implies that the operator $\widetilde{A}_\alpha^\varepsilon$ is monotone, demicontinuous (that is, strongly-weakly continuous) and coercive. Then, according to Theorem 2.4 (see also Corollary 3.15), it is maximal monotone and surjective. Hence, its restriction A_α^ε to $L^2(\Omega)$ is maximal monotone, that is, $R(I + A_\alpha^\varepsilon) = L^2(\Omega)$. This means that, for each $y_0 \in L^2(\Omega)$, the Cauchy problem

$$\begin{aligned} &\frac{dy}{dt} + A_\alpha^\varepsilon y = 0, \quad t \in (0,T), \\ &y(0) = y_0, \end{aligned} \tag{4.46}$$

has a generalized (mild) solution $y_\varepsilon \in C([0,T]; L^2(\Omega))$. Moreover, as easily seen, $A_\alpha^\varepsilon = \nabla \varphi_\alpha^\varepsilon$, where

$$\varphi_\alpha^\varepsilon(y) = \int_\Omega (j_\alpha(|\nabla y|) + \frac{\varepsilon}{2}|\nabla y|^2)dx,$$

$$j_\alpha(s) = \begin{cases} \dfrac{\alpha^2}{2}\log(s^2+\alpha^2) & \text{for } 0 \le s \le \alpha, \\ \dfrac{s^2}{4} + C(\alpha) & \text{for } s > \alpha, \end{cases}$$

where $C(\alpha) = \frac{\alpha^2}{4}(\log\alpha + 2\log 2 - 1)$. (We note that j_α is convex, continuous and $g_\alpha = \nabla j_\alpha$.) Then, by Theorem 2.69, it follows that y_ε is a strong solution of (4.6) on $(0,T)$ and

$$\frac{dy_\varepsilon}{dt} \in L^\infty(\delta, T; L^2(\Omega)),\ y_\varepsilon \in L^\infty(\delta, T; H_0^1(\Omega) \cap H^2(\Omega)),\ \forall \delta \in (0,T). \tag{4.47}$$

Consider now the closed convex subset K of $L^2(\Omega)$,

$$K = \{y \in W^{1,\infty}(\Omega) \cap H_0^1(\Omega); |\nabla y|_\infty \le \alpha\}.$$

(Here and everywhere in the following, we shall denote by $|\cdot|_p$ the norm of $L^p(\Omega)$, $1 \le p \le \infty$.) We have

Lemma 4.15 *If $y_0 \in K$, then $y_\varepsilon(t) \in K$, $\forall t \in [0,T]$.*

Proof. According to the exponential formula (2.112), it suffices to show that, for each $\lambda > 0$, the resolvent $(I + \lambda A_\alpha^\varepsilon)^{-1}$ leaves invariant the set K. In other words, for each $f \in K$, the solution $v \in H_0^1(\Omega)$ to the equation

$$v - \lambda\,\mathrm{div}(g_\alpha(|\nabla v|^2)\nabla v) - \lambda\varepsilon\Delta v = f \text{ in } \Omega,$$

belongs to K. Since Ω is convex, this follows by (3.79). $\square$

Proof of Proposition 4.13 (*continued*). We come back to equation (4.46) and show that, if $y_0 \in K$, then, for $\varepsilon \to 0$, y_ε is convergent to a solution u to (4.45). Since $g_\alpha = g$ on K, we have

$$\begin{aligned}&\frac{\partial u_\varepsilon}{\partial t} - \operatorname{div}(g(|\nabla y_\varepsilon|^2)\nabla y_\varepsilon) - \varepsilon\Delta y_\varepsilon = 0 \text{ in } (0,T)\times\Omega,\\ &y_\varepsilon(0) = y_0 \text{ in } \Omega,\ \ y_\varepsilon = 0 \text{ on } (0,T)\times\partial\Omega.\end{aligned} \tag{4.48}$$

We note that, taking into account that $g_\alpha(|\nabla y|^2)\nabla y \equiv \partial j_\alpha(|\nabla y|)$, we have (see Corollary 3.15)

$$\int_\Omega \operatorname{div}(g_\alpha(|\nabla v|^2)\nabla v)\Delta v\,dx \le 0,\ \forall v \in H_0^1(\Omega)\cap H^2(\Omega).$$

Then, by multiplying (4.48) with $y_\varepsilon, \Delta y_\varepsilon$ and $\frac{\partial y_\varepsilon}{\partial t}$, and integrating on $(0,T)\times\Omega$, we get the apriori estimates

$$\int_0^T\int_\Omega \left|\frac{\partial y_\varepsilon}{\partial t}\right|^2 dt\,dx + \varphi_\alpha^\varepsilon(y_\varepsilon(t)) \le \varphi_\alpha^\varepsilon(y_0) \le C,\ \forall t\in[0,T],$$

$$\frac{1}{2}\,|y_\varepsilon(t)|_2 + \varepsilon\int_0^T |\Delta y_\varepsilon(t)|_2^2 dt \le \frac{1}{2}|y_0|_2^2,\ \ \forall\, t\in[0,T],$$

and, by Lemma 4.15,

$$|\nabla y_\varepsilon(t)|_\infty \le \alpha,\ \ \text{a.e. } t\in(0,T).$$

In particular, by the Arzelà–Ascoli theorem this implies that $\{y_\varepsilon\}$ is compact in $C([0,T];L^2(\Omega))$ and, therefore, there is $y \in C([0,T];L^2(\Omega))\cap W^{1,2}([0,T];L^2(\Omega))$ such that, for $\varepsilon\to 0$,

$$\begin{aligned} y_\varepsilon &\longrightarrow y \ \ \text{strongly in } C([0,T];L^2(\Omega))\\ &\qquad\quad\ \text{weakly in } L^2(0,T;H_0^1(\Omega)),\\ \nabla y_\varepsilon &\longrightarrow \nabla y \text{ weak-star in } L^\infty((0,T)\times\Omega),\\ \varepsilon\Delta y_\varepsilon &\longrightarrow 0 \ \ \text{strongly in } L^2(0,T;L^2(\Omega)),\\ \operatorname{div} g(|\nabla y_\varepsilon|^2)\nabla y_\varepsilon) &\longrightarrow \eta \ \ \text{weakly in } L^2(0,T;L^2(\Omega)),\end{aligned}$$

$$\begin{aligned}&\frac{\partial y}{\partial t} - \eta = 0, \quad \text{a.e. } (x,t)\in(0,T)\times\Omega,\\ &y(0) = y_0 \qquad \text{in } \Omega,\\ &|\nabla y(t,x)| \le \alpha, \text{ a.e. } (t,x)\in(0,T)\times\Omega.\end{aligned}$$

To complete the proof, it suffices to show that

$$\eta = \operatorname{div}(g(|\nabla y|^2)\nabla y) \text{ in } \mathcal{D}'((0,T)\times\Omega). \tag{4.49}$$

To this end, we recall that $A_\alpha^\varepsilon = \nabla\varphi_\alpha^\varepsilon$ and, since $\nabla g_\alpha = j_\alpha$, this implies that, for all $v\in L^2(0,T;H_0^{-1}(\Omega))$,

$$-\int_0^T\int_\Omega \operatorname{div}(g_\alpha(|\nabla y_\varepsilon|^2)\nabla y_\varepsilon)(y_\varepsilon - v)dx\,dt \ge \int_0^T\int_\Omega (j_\alpha(|\nabla y_\varepsilon|) - j_\alpha(|\nabla v|))dx\,dt.$$

Letting $\varepsilon \to 0$, we obtain, for all $v \in L^2(0,T;H_0^1(\Omega))$,

$$-\int_0^T \int_\Omega \eta(y-v)dx\,dt \geq \int_0^T \int_\Omega (j_\alpha(|\nabla y|) - j_\alpha(|\nabla v|))dx\,dt,$$

because the function $y \to \int_0^T \int_\Omega j_\alpha(|\nabla y|)dx\,dt$ is convex, lower semicontinuous and, therefore, weakly lower semicontinuous in the space $L^2(0,T;H_0^1(\Omega))$. By (4.14), it follows that $\eta \in \partial\varphi_\alpha(y)$, where $\varphi_\alpha : L^2(\Omega) \to \overline{\mathbb{R}}$ is defined by

$$\varphi_\alpha(y) = \int_\Omega j_\alpha(|\nabla y|)dx, \ y \in H_0^1(\Omega),$$

and this implies (4.49), as claimed. The uniqueness of the solution follows by the $L^2(\Omega)$ monotonicity of the operator

$$y \xrightarrow{A} -\text{div}(g(|\nabla y|^2)\nabla y)$$

on the set $\{y \in H_0^1(\Omega);\ |\nabla y|_\infty \leq \alpha\}$.

Proposition 4.11 implies that equation (4.45) is well posed on K and defines a semiflow $t \to y(t,u_0)$ on the set K, which can be computed via the finite difference scheme (4.43), that is,

$$\begin{aligned} &y_{i+1} - h\,\text{div}(g(|\nabla y_{i+1}|^2)\nabla y_{i+1}) = y_i \text{ in } \ \Omega,\ i = 0,1,..., \\ &y_{i+1} = 0 \qquad\qquad\qquad\qquad\qquad\quad \text{on } \ \partial\Omega, \\ &y_1 = y_0 \qquad\qquad\qquad\qquad\qquad\quad\ \text{in } \ \Omega. \end{aligned} \tag{4.50}$$

As a matter of facts, at each step i, (4.50) is just the variational restoring procedure given by the minimization problem (3.53).

In order to extend the above algorithm to all $y_0 \in L^2(\Omega)$, that is, to general blurred images y_0, we consider the projection $P_K : L^2(\Omega) \to K$ and replace (4.50) by

$$\begin{aligned} &y_{i+1} - h \text{ div } g(|\nabla y_{i+1}|^2 \nabla y_{i+1}) = y_i \ \text{ in } \Omega,\ i = 1,..., \\ &y_1 = P_K u_0,\ \ y_{i+1} = 0 \ \text{ on } \partial\Omega. \end{aligned}$$

Recalling that $P_K = (I + N_K)^{-1}$, where N_K is the normal cone to K in $L^2(\Omega)$, we see that the projection $v = P_K f$, for $f \in L^2(\Omega)$, is given by

$$\begin{aligned} &v - \text{div } \beta(\nabla v) \ni f \ \text{ in } \Omega, \\ &v = 0 \qquad\qquad\qquad \text{on } \partial\Omega, \end{aligned} \tag{4.51}$$

where $\beta : \mathbb{R}^N \to 2^{\mathbb{R}^N}$ is the normal cone to the ball $\{r \in \mathbb{R}^N;\ |r| \leq \alpha\}$, that is,

$$\beta(r) = \begin{cases} \lambda \dfrac{r}{|r|},\ \lambda > 0, & \text{for } |r| = \alpha, \\ \qquad 0 & \text{for } |r| < \alpha. \end{cases}$$

Equation (4.51), which is well-posed in $H_0^1(\Omega)$, is equivalent with the minimization problem

$$\text{Min}\left\{\int_\Omega \frac{1}{2}(v-f)^2dx;|\ \nabla v(x)| \leq \alpha, \text{ a.e. } x \in \Omega\right\},$$

and can be approximated by the elliptic problem

$$v_\varepsilon - \text{div } \beta_\varepsilon(\nabla v_\varepsilon) = f \text{ in } \Omega, \ \ v_\varepsilon = 0 \text{ on } \partial\Omega,$$

where $\beta_\varepsilon(v) = \varepsilon^{-1}(I - (I + \varepsilon\beta)^{-1}) = \frac{1}{\varepsilon} \frac{v}{|v|} (|v| - \alpha)^+$, $\forall v \in \mathbb{R}^N$. Then, (4.51) leads to the following denoising algorithm

$$y_{i+1}^\varepsilon - h \text{ div}(g(|\nabla y_{i+1}^\varepsilon|^2)\nabla y_{i+1}^\varepsilon) = y_i^\varepsilon \text{ in } \Omega, \ i = 1, 2, ...,$$

$$y_{i+1}^\varepsilon = 0 \text{ on } \partial\Omega,$$

$$y_1^\varepsilon - \text{div}(\beta_\varepsilon(\nabla y_1^\varepsilon)) = u_0 \text{ in } \Omega, \ \ y_1^\varepsilon = 0 \text{ on } \partial\Omega,$$

which, as mentioned above, is well posed for all i and $\varepsilon > 0$. □

Semilinear parabolic equation in $\mathbb{R}^N$

We consider here equation (4.1) in $\Omega = \mathbb{R}^N$, that is,

$$\begin{cases} \frac{\partial y}{\partial t} - \Delta y + \beta(y) \ni f & \text{in } (0,T) \times \mathbb{R}^N, \\ y(0,x) = y_0(x) & x \in \mathbb{R}^N, \\ y(\cdot, 0) \in L^1(\mathbb{R}^N) & \forall t \in (0,T). \end{cases} \tag{4.52}$$

With respect to the case of bounded domain Ω previously studied, this problem presents some peculiarities and the more convenient functional space to study it is $L^1(\mathbb{R}^N)$. We write (4.52) as a differential equation in $X = L^1(\mathbb{R}^N)$ of the form

$$\begin{cases} \frac{dy}{dt}(t) + Ay(t) \ni f(t), & t \in (0,T), \\ y(0) = y_0, \end{cases}$$

where $A : D(A) \subset L^1(\mathbb{R}^N) \to \mathbb{R}^N$ is defined by

$$Ay = \{z \in L^1(\mathbb{R}^N); \ z = -\Delta y + w, \ w \in \beta(y), \text{ a.e. in } \mathbb{R}^N\},$$

$$D(A) = \{y \in L^1(\mathbb{R}^N); \ \Delta y \in L^1(\mathbb{R}^N), \ \exists w \in L^1(\mathbb{R}^N),$$
$$\text{such that } w(x) \in \beta(y(x)), \text{ a.e. } x \in \mathbb{R}^N\}.$$

By Theorem 3.19 we know that, if $N = 1, 2, 3$, then A is m-accretive in $L^1(\mathbb{R}^N) \times L^1(\mathbb{R}^N)$. Then, by Theorem 2.49, which neatly applies to this situation, we get the following existence result.

Proposition 4.16 *Let $y_0 \in L^1(\mathbb{R}^N)$ and $f \in L^1(0,T;\mathbb{R}^N)$ be such that $\Delta y_0 \in L^1(\mathbb{R}^N)$ and $\exists w \in L^1(\mathbb{R}^N)$, $w(x) \in \beta(y_0(x))$, a.e. $x \in \mathbb{R}^N$. Then, problem (4.52) has a unique mild solution $y \in C([0,T]; L^1(\mathbb{R}^N))$. In other words,*

$$y(t) = \lim_{h\to 0} y_h(t) \quad \textit{strongly in } L^1(\mathbb{R}^n) \textit{ for each } t \in [0,T], \tag{4.53}$$

where y_h is the solution to the finite difference scheme

$$y_h(t) = y_h^i \quad \textit{for } t \in [ih, (i+1)h), \ \ i = 0, 1, ..., M,$$

$$y_h^{i+1} - y_h^i - h\Delta y_h^{i+1} + h\beta(y_h^{i+1}) \ni \int_{ih}^{(i+1)h} f(t)dt \quad \textit{in } \mathbb{R}^n, \tag{4.54}$$

$$y_h^i \in L^1(\mathbb{R}^N), \ \ i = 0, 1, ..., M = \left[\tfrac{T}{h}\right].$$

Of course, this result extends to all of y_0 in the closure of the set $\{y_0 \in L^1(\mathbb{R}^N); \Delta y_0 \in L^1(\mathbb{R}^N), \beta(y_0) \in L^1(\mathbb{R}^N)\}$ in $L^1(\mathbb{R}^N)$.

4.2 The Porous Media Equation

The nonlinear diffusion equation models the dynamic of density in a substance undergoing anomalous diffusive behavior diffusion as that in porous or granular media. However, the class of mathematical models described by this class of equations is much larger and includes phase transition dynamics (the Stefan problem) or other physical processes that are of diffusion type (heat propagation, filtration, or dynamics of biological groups). Such an equation can be schematically written as

$$\begin{cases} \dfrac{\partial y}{\partial t} - \Delta\beta(y) \ni f & \text{in } (0,T)\times\Omega = Q, \\ \beta(y) = 0 & \text{on } (0,T)\times\partial\Omega = \Sigma, \\ y(0,x) = y_0(x) & \text{in } \Omega, \end{cases} \tag{4.55}$$

where Ω is a bounded and open subset of $\mathbb{R}^N$ with smooth boundary, and $\beta : \mathbb{R} \to 2^{\mathbb{R}}$ is a maximal monotone graph in $\mathbb{R}\times\mathbb{R}$ such that $0 \in D(\beta)$. The steady-state equation associated with (4.55) is just the stationary porous media equation studied in Sections 3.3, Remark 3.8. The function $y \in C([0,T]; L^1(\Omega))$ is called a generalized solution to problem (4.55) if $\beta(y) \in L^1(Q)$, $\beta(y(t)) \in W_0^{1,1}(\Omega)$, a.e. $t \in (0,T)$ and satisfies (4.55) in the sense of distributions, that is,

$$\int_Q (y\varphi_t + \beta(y)\Delta\varphi)dx\,dt + \int_Q f\varphi\,dx\,dt + \int_\Omega y_0(x)\varphi(0,x)dx = 0 \tag{4.56}$$

for all $\varphi \in C_0^\infty([0,T)\times\Omega)$. In the existence theory, for this equation we shall use also other (equivalent) definitions (mild solution, for instance).

We shall treat here equation (4.55) via the general existence theory presented in Chapter 2 for the Cauchy problem for two functional spaces: $H^{-1}(\Omega)$ and $L^1(\Omega)$ and should be mentioned that only in these spaces the equation can be represented as a problem of the form (2.96) with the operator A m-accretive. The $H^{-1}(\Omega)$-approach called the *energetic approach* has the advantage that equation (4.42) can be represented as a gradient system of the form (2.126) and so the solution is regular in $H^{-1}(\Omega)$.

Let us first briefly describe some specific diffusive-like problems that lead to equations of this type.

The flow of gases in porous media

Let y be the density of a gas that flows through a porous medium that occupies a domain $\Omega \subset \mathbb{R}^3$ and let $\bar{v}$ be the pore velocity. If p denotes the pressure, we have $p = p_0 y^\alpha$ for $\alpha \geq 1$. Then, the conservation law equation

$$k_1\,\frac{\partial y}{\partial t} + \operatorname{div}(y\,\bar{v}) = 0$$

combined with Darcy's law, $\gamma\bar{v} = -k_2\nabla p$, where k_1 is the porosity of the medium, k_2 the permeability and γ the viscosity, yields the *porous medium equation*

$$\frac{\partial y}{\partial t} - \delta\Delta y^{\alpha+1} = 0 \quad \text{in } Q. \tag{4.57}$$

(Here $\delta = k_2 p_0 (k_1(\alpha+1)\gamma)^{-1}$.) Equation (4.57) is also relevant in the study of other mathematical models, such as population dynamics. The case where $-1 < \alpha < 0$ is that of fast diffusion processes arising in physics of plasma. In particular, the case

$$\beta(x) = \begin{cases} \log x & \text{for } x > 0 \\ -\infty & \text{for } x \leq 0 \end{cases}$$

emerges from the central limit approximation to Carleman's model of Boltzman equations. Nonlinear diffusion equations of the form (4.55) perturbed by a drift (transport term), that is,

$$\frac{\partial y}{\partial t} - \Delta\beta(y) + \operatorname{div} K(y) \ni f$$

with appropriate boundary conditions arise in the dynamics of underground water flows and are known in the literature as the Richards equation. The special case

$$\beta(y) = \begin{cases} \beta_0(y) & \text{for } y < y_s, \\ [\beta_0(y_s), +\infty) & \text{for } y = y_s, \\ \emptyset & \text{for } y > y_s, \end{cases}$$

where $\beta_0 : \mathbb{R} \to \mathbb{R}$ is a continuous and monotonically increasing function, models the dynamics of saturated–unsaturated underground water flows.

The Stefan problem

This problem describes the conduction of heat in a medium involving a phase charge. To be more specific, consider a unit volume of ice Ω at temperature $\theta < 0$. If a uniform heat source of intensity F is applied, then the temperature increases at rate $\frac{E}{C_1}$ until it reaches the melting point $\theta = 0$. Then, the temperature remains at zero until ρ units of heat have been supplied to transform the ice into water (ρ is the latent heat). After all the ice has melted the temperature begins to increase at the rate $\frac{h}{C_2}$ (C_1 and C_2 are specific heats of ice and water, respectively). During the process, the variation of the internal energy $e(t)$ is, therefore, given by

$$e(t) = C(\theta(t)) + \rho H(\theta(t)),$$

where

$$C(\theta) = \begin{cases} C_1\theta & \text{for } \theta \leq 0, \\ C_2\theta & \text{for } \theta > 0, \end{cases}$$

and H is the Heaviside graph

$$H(\theta) = \begin{cases} 1 & \theta > 0, \\ [0, 1] & \theta = 0, \\ 0 & \theta < 0. \end{cases}$$

In other words, we have

$$e = \gamma(\theta) = \begin{cases} C_1\theta & \text{if } \theta < 0, \\ [0, \rho] & \text{if } \theta = 0, \\ C_2\theta + \rho & \text{if } \theta > 0. \end{cases} \tag{4.58}$$

The function γ is called the *enthalpy* of the system.

Now, let $Q = \Omega \times (0, \infty)$ and denote by Q_-, Q_+, Q_0 the regions of Q, where $\theta < 0$, $\theta > 0$, and $\theta = 0$, respectively. We set $S_+ = \partial Q_+$, $S_- = \partial Q_-$, and $S = S_+ \cup S_-$.

If $\theta = \theta(t, x)$ is the temperature distribution in Q and $q = q(t, x)$ the heat flux, then, according to the Fourier law,

$$q(t, x) = -k\nabla\theta(t, x), \tag{4.59}$$

where k is the thermal conductivity. Consider the function

$$K(\theta) = \begin{cases} k_1\theta & \text{if } \theta < 0, \\ k_2\theta & \text{if } \theta > 0, \end{cases}$$

where k_1, k_2 are the thermal conductivity of the ice and water, respectively.

If f is the external heat source, then the conservation law yields

$$\frac{d}{dt}\int_{\Omega^*} e(t, x)dx = -\int_{\partial\Omega^*} (q(t, x), \nu)d\sigma + \int_{\Omega^*} F(t, x)dx$$

for any subdomain $\Omega^* \times (t_1, t_2) \subset Q$ (ν is the normal to $\partial\Omega^*$) if e and q are smooth. Equivalently,

$$\begin{aligned} &\int_{\Omega^*} e_t(t, x)dx + \int_{S\cap\Omega^*} [|e(t)|]V(t)dt \\ &\quad = -\int_{\Omega^*} \operatorname{div} q(t, x)dx + \int_{\partial\Omega^*\cap S} [|(q(t), \nu)|]d\sigma + \int_{\Omega^*} F(t, x)dx, \end{aligned}$$

where $V(t) = -N_t\|N_t\|$ is the true velocity of the interface S ($N = (N_1, N_2)$ is the unit normal to S) and $[|\cdot|]$ is the jump along S.

The previous inequality yields

$$\begin{aligned} &\frac{\partial}{\partial t} e(t, x) + \operatorname{div} q(t, x) = F(t, x) \quad \text{in } Q \setminus S, \\ &[|e(t)|]N_x + [|(q(t), N_t|] = 0 \qquad \text{on } S. \end{aligned} \tag{4.60}$$

If we represent the interface S as $\{(t, x);\ t = \sigma(x)\}$, we get the system

$$\begin{cases} C_1\theta_t - k_1\Delta\theta = f & \text{in } Q_- = \{(t, x);\ \theta(t, x) < 0\} \\ C_2\theta_t - k_2\Delta\theta = f & \text{in } Q_+ = \{(t, x);\ \theta(t, x) > 0\}, \\ (k_2\nabla\theta^+ - k_1\nabla\theta^-) \cdot \nabla\sigma(x) = -\rho & \text{on } S. \end{cases} \tag{4.61}$$

This is the classical two-phase Stefan problem. We may write system (4.61) as

$$\frac{\partial}{\partial t}\gamma(\theta) - \Delta K(\theta) \ni f \quad \text{in } Q, \tag{4.62}$$

where $\gamma : \mathbb{R} \to 2^{\mathbb{R}}$ is given by (4.58). Indeed, for every test function $\varphi \in C_0^\infty(Q)$ we have

$$\begin{aligned}
&\left(\frac{\partial}{\partial t}\gamma(\theta) - \Delta K(\theta)\right)(\varphi) = -\int_Q (\gamma(\theta)\varphi_t + K(\theta)\Delta\varphi)dx\,dt \\
&= C_1\int_{Q_-} \theta_t\varphi\,dx\,dt + C_2\int_{Q_+} \theta_t\,dx\,dt - k_1\int_{Q_-} \varphi\Delta\theta\,dx\,dt \\
&-k_2\int_{Q_+} \varphi\Delta\theta\,dx\,dt + \int_S \left(k_2\frac{\partial\theta^+}{\partial n} - k_1\frac{\partial\theta^-}{\partial n}\right)\varphi\,ds - \rho\int_{Q_+} \varphi_t dx\,dt \qquad (4.63)\\
&= \int_{Q_-} (C_1\theta_t - k_1\Delta\theta)\varphi\,dx\,dt + \int_{Q_+} (C_2\theta_t - k_2\Delta\theta)\varphi\,dx\,dt \\
&+ \int_S ((k_2\nabla\theta^+ - k_1\nabla\theta^-)\cdot\nabla\sigma + \rho)dx = 0.
\end{aligned}$$

If we denote by β the function $\gamma^{-1}K$, that is,

$$\beta(r) = \begin{cases} k_1C_1^{-1}r & \text{for } r < 0, \\ 0 & \text{for } 0 \le r < \rho, \\ k_2C_2^{-1}(r-\rho) & \text{for } r \ge \rho, \end{cases} \tag{4.64}$$

we may rewrite (4.62) in the form (4.55). Problem (4.55) can be treated as a nonlinear accretive Cauchy problem in two functional spaces only; either in $H^{-1}(\Omega)$ or in $L^1(\Omega)$.

In the special case of the *one-phase Stefan problem*, system (4.61) reduces to

$$\begin{aligned}
&C_1\theta_t - k_1\Delta\theta = f \quad \text{in } \{(t,x);\ \theta(t,x) < 0\}, \\
&k_1\nabla\theta\cdot\nabla\sigma = \rho \qquad \text{in } \{(t,x) \in S\},
\end{aligned}$$

which can be written as a parabolic variational inequality of the form (4.23) (see, e.g., [10]).

The energetic approach of porous media equations

In the space $H^{-1}(\Omega)$, consider the operator

$$\begin{aligned}
A = \{\ [y,w] \in (H^{-1}(\Omega)\cap L^1(\Omega)) \times H^{-1}(\Omega);\ w = -\Delta v, \\
v \in H_0^1(\Omega),\ v(x) \in \beta(y(x)), \quad \text{a.e. } x \in \Omega\}.
\end{aligned}$$

We assume that

$$\beta^{-1} : \mathbb{R} \to \mathbb{R} \text{ is everywhere defined and bounded on bounded subsets.} \tag{4.65}$$

Then, by Proposition 2.23, A is maximal monotone in $H^{-1}(\Omega) \times H^{-1}(\Omega)$. More precisely, $A = \partial\varphi$, where $\varphi : H^{-1}(\Omega) \to \overline{\mathbb{R}}$ is defined by

$$\varphi(y) = \begin{cases} \displaystyle\int_\Omega j(y(x))dx & \text{if } y \in L^1(\Omega)\cap H^{-1}(\Omega),\ j(y) \in L^1(\Omega), \\ +\infty & \text{otherwise,} \end{cases}$$

where $\partial j = \beta$. Then, we may write problem (4.55) as

$$\begin{aligned} &\frac{dy}{dt} + Ay \ni f \qquad \text{in } \ (0,T), \\ &y(0) = y_0, \end{aligned} \tag{4.66}$$

and so, by Theorem 2.69, we obtain the following existence result.

Theorem 4.17 *Let β be a maximal monotone graph in $\mathbb{R} \times \mathbb{R}$ satisfying condition (4.65). Let $f \in L^1(0,T;H^{-1}(\Omega))$ and let $y_0 \in H^{-1}(\Omega) \cap L^1(\Omega)$ be such that $y_0(x) \in \overline{D(\beta)}$, a.e. $x \in \Omega$. Then, there is a unique pair of functions $y \in C([0,T];H^{-1}(\Omega)) \cap W^{1,2}(0,T;H^{-1}(\Omega))$ and $v : (0,T) \times \Omega \to \mathbb{R}$, such that $v(t) \in H_0^1(\Omega)$, $\forall t \in [0,T]$ satisfying*

$$\begin{cases} \dfrac{\partial y}{\partial t} - \Delta v = f, & a.e. \ in \ Q = (0,T) \times \Omega, \\ v(t,x) \in \beta(y(t,x)), & a.e. \ (t,x) \in Q, \\ y(0,x) = y_0(x), & a.e. \ in \ \Omega, \end{cases} \tag{4.67}$$

$$t^{\frac{1}{2}} \, \frac{\partial y}{\partial t} \in L^2(0,T;H^{-1}(\Omega)), \quad t^{\frac{1}{2}} v \in L^2(0,T;H_0^1(\Omega)). \tag{4.68}$$

Moreover, if $j(y_0) \in L^1(\Omega)$, then

$$\frac{\partial y}{\partial t} \in L^2(0,T;H^{-1}(\Omega)), \quad v \in L^2(0,T;H_0^1(\Omega)). \tag{4.69}$$

If $y_0 \in D(A)$ and $f \in W^{1,1}([0,T];H^{-1}(\Omega))$, then

$$\frac{\partial y}{\partial t} \in L^\infty(0,T;H^{-1}(\Omega)), \quad v \in L^\infty(0,T;H_0^1(\Omega)). \tag{4.70}$$

We note that the derivative $\frac{\partial y}{\partial t}$ in (4.67) is the strong derivative $\frac{dy}{dt}$ of the function $t \to y(\cdot,t)$ from $[0,T]$ into $H^{-1}(\Omega)$, and it coincides with the derivative $\frac{\partial y}{\partial t}$ in the sense of distributions on Q. It is readily seen that the solution y (see Theorem 4.17) is a generalized solution to (4.55) in the sense of definition (4.56).

Remark 4.18 Taking into account Remark 2.26, Theorem 4.17 extends to equations of the form

$$\begin{aligned} &\frac{dy}{dt} - \Delta\beta(x,y) \ni f, \\ &y(0,x) = y_0(x), \end{aligned}$$

where $\beta(x,y) \equiv \partial j(x,r)$ and $j : \Omega \times \mathbb{R} \to \overline{\mathbb{R}}$ satisfies condition (2.59).

The L^1-approach

In the space $X = L^1(\Omega)$, consider the operator

$$\begin{aligned} &A = \{[y,w] \in L^1(\Omega) \times L^1(\Omega); \ w = -\Delta v\}, \\ &v \in W_0^{1,1}(\Omega)\}, \ v(x) \in \beta(y(x)), \quad \text{a.e. } x \in \Omega. \end{aligned} \tag{4.71}$$

We have seen earlier (Theorem 3.21) that A is m-accretive in $L^1(\Omega) \times L^1(\Omega)$. Then, applying the general existence Theorem 2.50, we obtain the following existence result for problem (4.67).

Proposition 4.19 *Let β be a maximal monotone graph in $\mathbb{R}\times\mathbb{R}$ such that $0 \in \beta(0)$. Then, for every $f \in L^1(0,T;L^1(\Omega))$ and every $y_0 \in L^1(\Omega)$, such that $y_0(x) \in \overline{D(\beta)}$, a.e. $x \in \Omega$, the Cauchy problem (equivalently, (4.67))*

$$\begin{cases} \dfrac{dy}{dt}(t) + Ay(t) \ni f(t) \quad in \ (0,T), \\ y(0) = y_0, \end{cases} \tag{4.72}$$

has a unique mild solution $y \in C([0,T];L^1(\Omega))$. Moreover, if $y_0 \geq 0$, a.e. on Ω, then

$$y(t,x) \geq 0, \quad a.e. \ (t,x) \in (0,T) \times \Omega. \tag{4.73}$$

Finally, if $y_0 \in L^\infty(\Omega)$ and $f \equiv 0$, then

$$|y(t,x)| \leq |y_0|_\infty, \quad a.e. \ (t,x) \in (0,T) \times \Omega. \tag{4.74}$$

We note that $\overline{D(A)} = \{y_0 \in L^1(\Omega);\ y_0(x) \in \overline{D(\beta)}, \text{ a.e. } x \in \Omega\}$. Here is the argument. Indeed, $(1+\varepsilon\beta)^{-1}y_0 \to y_0$ in $L^1(\Omega)$ as $\varepsilon \to 0$, if $y_0 \in \overline{D(\beta)}$, a.e. $x \in \Omega$, and therefore $(I+\varepsilon A)^{-1}y_0 \to y_0$ if $j(y_0) \in L^1(\Omega)$. As regards (4.73) and (4.74), they follow by

$$(I+\lambda A)^{-1}u \geq 0, \text{ a.e. in } \Omega \text{ if } u \geq 0, \text{ a.e. in } \Omega,$$
$$\|(I+\lambda A)^{-1}u\|_{L^\infty(\Omega)} \leq \|u\|_{L^\infty(\Omega)}, \ \forall u \in L^\infty(\Omega).$$

Proposition 4.19 amounts to saying that the solution y is given by

$$y(t) = \lim_{\varepsilon\to 0} y_\varepsilon(t) \quad \text{in } L^1(\Omega), \text{ uniformly on } [0,T],$$

where y_ε is the solution to the finite difference scheme

$$\begin{cases} \dfrac{1}{\varepsilon}(y_\varepsilon(t,x) - y_\varepsilon(t-\varepsilon),x) - \Delta v_\varepsilon(t,x) = f_\varepsilon(t,x) & \text{in } (0,T)\times\Omega, \\ v_\varepsilon(t,x) \in \beta(y_\varepsilon(t,x)), & \text{a.e. in } (0,T)\times\Omega, \\ v_\varepsilon = 0 & \text{on } (0,T)\times\partial\Omega, \\ y_\varepsilon(t,x) = y_0(x) & \text{for } t \leq \varepsilon,\ x \in \Omega. \end{cases} \tag{4.75}$$

The function $t \to y_\varepsilon(t) \in W_0^{1,1}(\Omega)$ is piecewise constant on $[0,T]$, that is, $y_\varepsilon(t) = y_\varepsilon^i$, $\forall t \in [i\varepsilon,(i+1)\varepsilon)$, and $f_\varepsilon(t) = f_i$, $\forall t \in [i\varepsilon,(i+1)\varepsilon]$ is a piecewise constant approximation of $f : [0,T] \to L^1(\Omega)$.

Proposition 4.19 extends to porous media equations with nonlinear source term of the form

$$\begin{aligned} &\frac{\partial y}{\partial t} - \Delta\eta + g(y) \ni f && \text{in } (0,T)\times\Omega, \\ &\eta = 0 && \text{on } (0,T)\times\partial\Omega, \\ &\eta(t,x) \in \beta(y(t,x)), && \text{a.e. } (t,x) \in (0,T)\times\Omega, \\ &y(0,x) = y_0(x), && \text{a.e. } x \in \Omega, \end{aligned} \tag{4.76}$$

which, in the special case $d = 1$, $\Omega = (0,1)$ and $g(r) \equiv \int_0^r h(s)ds$, $h > 0$, is the famous *Nagumo reaction-diffusion equation.* Namely, we have

Proposition 4.20 *Let β be maximal monotone in $\mathbb{R}\times\mathbb{R}$, $0\in\beta(0)$, and $g:\mathbb{R}\to\mathbb{R}$ is continuous, monotonically nondecreasing, $g(0)=0$ and*

$$|g(r)|\le L|r|+C,\quad \forall\, r\in\mathbb{R}.$$

Then, for each $f\in L^1(0,T;L^1(\Omega))$ and $y_0\in L^1(\Omega)$ such that $y_0(x)\in D(\beta)$, a.e. $x\in\Omega$, the Cauchy problem (4.76) has a unique mild solution y.

Proof. We write (4.76) as

$$\frac{dy}{dt}+A_1y\ni f,\ t\in(0,T),$$
$$y(0)=y_0.$$

Here, $A_1=A+B$, where A is the operator (4.71) and $By=g(y)$, $\forall y\in L^1(\Omega)$. It is easily seen that the operator B is continuous and accretive in $L^1(\mathbb{R}^d)$. Then, by Theorem 2.42, A_1 is m-accretive in $L^1(\mathbb{R}^d)$ and so the existence and the uniqueness of a mild solution y follows as above by Theorem 2.69. □

By (4.75), it is readily seen that y is a solution to (4.67) in the sense of Schwartz distributions, that is,

$$\int_0^T\int_\Omega(y\varphi_t+v\Delta\varphi-g(y)\varphi)dx\,dt+\int_\Omega y_0(x)\varphi(0,x)dx=-\int_0^T\int_\Omega f\varphi\,dx\,dt,$$

for all $\varphi\in C_0^\infty([0,T)\times\Omega)$.

In particular, it follows by Proposition 4.19 that the operator A defined by (4.71) generates a semigroup of nonlinear contractions $S(t):\overline{D(A)}\to\overline{D(A)}$. This semigroup is not differentiable in $L^1(\Omega)$, but in some special situations it has regularity properties comparable with those of the semigroup generated by the Laplace operator on $L^2(\Omega)$ under Dirichlet boundary conditions or, more generally, by subpotential operators $A=\partial\varphi$ (see Theorem 2.71). In fact, we have the following smoothing effect of nonlinear semigroup $S(t)$ with respect to the initial data.

Theorem 4.21 *Let $\beta\in C^1(\mathbb{R}\setminus\{0\})\cap C(\mathbb{R})$ be a monotone function satisfying the conditions*

$$\beta(0)=0,\quad \beta'(r)\ge C|r|^{\alpha-1},\quad \forall r\ne 0, \tag{4.77}$$

where $\alpha>0$ if $N\le 2$ and $\alpha>\frac{N-2}{N}$ if $N\ge 3$. Then, $S(t)(L^1(\Omega))\subset L^\infty(\Omega)$ for every $t>0$,

$$|S(t)y_0|_\infty\le Ct^{-\frac{N}{N\alpha+2-N}}|y_0|_1^{\frac{2}{2+N(\alpha-1)}},\quad \forall t>0, \tag{4.78}$$

and $S(t)(L^p(\Omega))\subset L^p(\Omega)$ for all $t>0$ and $1\le p<\infty$.

Proof. First, we assume that $p>1$ and establish first the estimates

$$|(I+\lambda A)^{-1}f|_p+C\lambda\left(\int_\Omega|(I+\lambda A)^{-1}f|^{\frac{(p+\alpha-1)N}{N-2}}dx\right)^{\frac{N-2}{N}}\le|f|_p^p, \tag{4.79}$$

$\forall f \in L^p(\Omega)$, $\lambda > 0$, for $N > 2$, and

$$|(I+\lambda A)^{-1}f|_p + C\lambda\left(\int_\Omega |(I+\lambda A)^{-1}f|^{(p+1-\alpha)q}dx\right)^{\frac{1}{q}} \le \int_\Omega |f|^p dx, \tag{4.80}$$

$\forall q>1$, if $N = 2$. Here C is independent of $p \ge 1$, and A is the operator defined by (4.71). For simplicity, we denote by $|\cdot|_q$ the norm $\|\cdot\|_{L^q(\Omega)}$ for $1 \le q \le \infty$.

We set $u = (I+\lambda A)^{-1}f$, that is,

$$\begin{cases} u - \lambda\Delta\beta(u) = f & \text{in } \Omega, \\ \beta(u) = 0 & \text{on } \partial\Omega. \end{cases} \tag{4.81}$$

We recall that $\beta(u) \in W_0^{1,q}(\Omega)$, where $1 < q < \frac{N}{N-2}$ (see Corollary 2.51).

Multiplying equation (4.81) by $|u|^{p-1}\operatorname{sign}(u)$ and integrating on Ω, we get

$$\int_\Omega |u|^p dx + \lambda p(p-1)\int_\Omega \beta'(u)|u|^{p-2}|\nabla u|^2 dx \le \int_\Omega |f|^p dx.$$

Now, using the identity

$$|u|^{p+\alpha-3}|\nabla u|^2 = \frac{4}{(p+\alpha-1)^2}\left|\nabla |u|^{\frac{p+\alpha-1}{2}}\right|^2, \quad \text{a.e. in } \Omega$$

and condition (4.77), we get

$$\int_\Omega |u|^p dx + \frac{4\lambda p(p-1)}{(p+\alpha-1)^2}\int_\Omega \left|\nabla |u|^{\frac{p+\alpha-1}{2}}\right|^2 dx \le C\int_\Omega |f|^p dx. \tag{4.82}$$

On the other hand, by the Sobolev embedding theorem

$$\int_\Omega \left|\nabla |u|^{\frac{p+\alpha-1}{2}}\right|^2 dx \le C\left(\int_\Omega |u|^{\frac{(p+\alpha-1)N}{N-2}}dx\right)^{\frac{N-2}{N}} \quad \text{if } N > 2,$$

and

$$\int_\Omega \left|\nabla |u|^{\frac{p+\alpha-1}{2}}\right|^2 dx \le C\left(\int_\Omega |u|^{\frac{p+\alpha-1}{q}}dx\right)^{\frac{1}{q}}, \quad \forall q > 1,$$

for $N = 2$. Then, substituting these inequalities into (4.82), we get (4.79) and (4.80), respectively.

We set $J_\lambda = (I+\lambda A)^{-1}$ and

$$\varphi(u) \equiv |u|_p^p, \quad \psi(u) = C|u|^{p+\alpha-1}_{(p+\alpha-1)\frac{N}{N-2}}.$$

Then, inequality (4.79) can be written as

$$\varphi(J_\lambda f) + \lambda\psi(J_\lambda f) \le \varphi(f), \quad \forall f \in L^p(\Omega).$$

This yields

$$\varphi(J_\lambda^k f) + \lambda\psi(J_\lambda^k f) = \varphi(J_\lambda^{k-1}), \quad \forall k.$$

Summing these equations from $k = 1$ to $k = n$, and taking $\lambda = \frac{t}{n}$, yields

$$\varphi(J_{\frac{t}{n}}^n f) + \sum_{k=1}^n \frac{1}{n}\,\psi(J_{\frac{t}{n}}^k f) = \varphi(f).$$

Recalling that, by Theorem 2.52, $J^n_{\frac{t}{n}} f \to S(t)$ for $n \to \infty$, the latter equation implies that

$$\varphi(S(t)f) + \int_0^t \psi(S(\tau)f)d\tau = \varphi(f), \quad \forall t \geq 0. \tag{4.83}$$

In particular, it follows that the function $t \to \varphi(S(t)f)$ is decreasing and so is $t \to \psi(S(t)f)$. Then, by (4.83), we see that $\varphi(S(t)f) + t\psi(S(t)f) \leq \varphi(f)$, $\forall t > 0$; that is,

$$|S(t)f|_p^p + Ct|S(t)f|^{p+\alpha-1}_{\frac{(p+\alpha-1)N}{N-2}} \leq |f|_p^p, \quad \forall t > 0, \tag{4.84}$$

where C is independent of p and f.

Let p_n be inductively defined by

$$p_{n+1} = (p_n + \alpha - 1)\frac{N}{N-2}.$$

Then, by (4.84), we see that

$$|S(t_{n+1})f|_{p_{n+1}}^{\frac{N}{N-2}p_{n+1}} \leq \frac{|S(t_n)f|_{p_n}^{p_n}}{C(t_{n+1} - t_n)},$$

where $t_0 = 0$ and $t_{n+1} > t_n$. Choosing $t_{n+1} - t_n = \frac{t}{2^{n+1}}$, we get after some calculation that

$$\limsup_{n\to\infty} |S(t)f|_{p_{n+1}}^{\frac{N-2}{N} n p_{n+1}} \leq C|f|_{p_0}\left(\frac{2}{t}\right)^{\mu}, \quad \forall t > 0,$$

where $\mu = \frac{N}{2}$, because p_n is given by

$$p_n = \left(\frac{N}{N-2}\right)^n p_0 + \frac{N\alpha}{2(N-2)}\left(\left(\frac{N}{N-2}\right)^n - 1\right)$$

(here, we have used the fact that $\alpha > \frac{N-2}{N}$), we get the final estimate

$$|S(t)f|_\infty \leq C|f|_{p_0}^{\frac{2p_0}{2p_0+N(\alpha-1)}}\, t^{-\frac{N}{2p_0+N(\alpha-1)}}, \quad \forall p_0 > 1, \tag{4.85}$$

as claimed.

The latter extends to $p_0 = 1$ by the following argument we briefly sketch below. The first step is the estimate

$$|S(t)y_0|_{p_0} \leq C_{p_0}\, t^{-\frac{p_0-\gamma}{(\gamma+\alpha-1)p_0}}\, |y_0|_1, \ \forall t > 0, \tag{4.86}$$

for $kp_0 < C_\alpha$ and

$$\gamma = \frac{2p_0 + (\alpha-1)N}{(p_0+\alpha-2)N+2} \in (0,1).$$

To get (4.86), we recall that

$$S(t)y_0 = \lim_{h\to n} y_h(t),$$

where

$$y_h(t) = y_h^i,\ t \in [ih,(i+1)h) \text{ and } y_h^{i+1} - h\Delta\beta(y_h^{i+1}) = y_h^i \text{ in } \Omega,\ u_h^0 = y_0.$$

Then, arguing as above, we get

$$t|y_h(t)|_{p_0}^{p_0} + C\int_0^t S|y_h(s)|^{p_0+\alpha-1}_{\frac{(p_0+\alpha-1)N}{N-2}}\,ds \le \int_0^t |y_h(s)|_{p_0}^{p_0}ds + h|y_0|_{p_0}^{p_0},$$

and, therefore,

$$t|y_h(t)|_{p_0}^{p_0} + C\int_0^t s|y_h(s)|^{p_0+\alpha-1}_{\frac{(p_0+\alpha-1)N}{N-2}}\,ds \le C|y_0|_1^{\frac{\gamma(p_0+\alpha-1)}{\gamma+\alpha-1}}\, t^{\frac{2\gamma+\alpha-p_0-1}{\gamma+\alpha-1}} + h|y_0|_{p_0}^{p_0}.$$

This yields

$$|S(t)y_0|_{p_0} \le C_{p_0}|y_0|_1^{\frac{\gamma(p_0+\alpha-1)}{p_0(\gamma+\alpha-1)}}\, t^{\frac{\gamma-p_0}{p_0(\gamma+\alpha-1)}},\ \forall t>0. \tag{4.87}$$

On the other hand, taking into account that, by the semigroup property,

$$S(t)y_0 = S\left(\frac{t}{2}\right)S\left(\frac{t}{2}\right)y_0,\ \forall t>0,$$

we get by (4.86), (4.87),

$$\begin{aligned}|S(t)y_0|_\infty &\le C_{p_0}\left(\frac{t}{2}\right)^{-\frac{N}{2p_0+(\alpha-1)N}}\left|S\left(\frac{t}{2}\right)y_0\right|_1^{\frac{2p_0}{2p_0+N(\alpha-1)}}\\ &\le C_{p_0}\, t^{-\frac{N}{2+(\alpha-1)N}}\,|y_0|_1^{\frac{2}{2+(\alpha-1)N}},\ \forall t>0,\end{aligned}$$

which yields (4.78), as claimed.

The case $N=2$ follows similarly. This completes the proof of Theorem 4.21. □

The asymptotic behavior

We come back to the porous media equation (4.67) (or (4.66)) and note that, since $A=\partial\varphi$ where φ is defined by (4.66), we have by Theorem 2.71

Corollary 4.22 *Under conditions of Theorem* 4.17, *where* $f\equiv 0$, *we have*

$$y(t)\to y_\infty \quad \textit{weakly in } H^{-1}(\Omega) \quad \textit{as } t\to\infty, \tag{4.88}$$

where y_∞ *is a solution to the equation*

$$-\Delta\beta(y_\infty)\ni 0 \textit{ in } \Omega;\quad \beta(y_\infty)\in H_0^1(\Omega).$$

However, it should be said that, if $j(y_0)\in L^1(\Omega)$ where $\partial j=\beta$, then as easily seen the solution y to (4.67) satisfies

$$\int_\Omega j(y(t,x))dx + \int_0^t\int_\Omega |\nabla\beta(y(s,x))|^2 dsdx = \int_\Omega j(y_0(x))dx < \infty,\ \forall t\ge 0,$$

and, therefore, if

$$j(r)\ge \alpha_1|r|^{\frac{2N}{N+2}}+\alpha_2,\quad \forall r\in\mathbb{R},$$

where $\alpha_1>0$, then we have

$$|y(t)|_{\frac{2N}{N+2}}\le C,\quad \forall t\ge 0,$$

and since, by Theorem 1.12, $L^{\frac{2N}{N+2}}(\Omega)$ is compactly embedded in $H^{-1}(\Omega)$, we infer that (4.88) is strengthen to

$$y(t) \to y_\infty \text{ strongly in } H^{-1}(\Omega) \text{ and weakly in } L^{\frac{2N}{N+2}}(\Omega), \quad \text{as } t \to \infty.$$

Consider now the equation (see (4.76))

$$\begin{aligned} &\frac{\partial y}{\partial t} - \Delta\beta(y) + g(y) = 0 \ \text{ in } (0,\infty) \times \Omega, \\ &\beta(y) = 0 \qquad\qquad\qquad\qquad \text{on } (0,\infty) \times \partial\Omega, \\ &y(0) = y_0, \end{aligned} \tag{4.89}$$

where $\beta : \mathbb{R} \to \mathbb{R}$ and $g : \mathbb{R} \to \mathbb{R}$ are monotonically nonincreasing and

$$|g(r)| \le L|r| + C, \quad \forall\, r \in \mathbb{R}, \quad g(0) = 0.$$

As seen earlier in Proposition 4.20, for each $y_0 \in L^1(\Omega)$ there is a unique mild solution $y \in C([0,\infty); L^1)$ to (4.89).

Since the operator $A_1 y = -\Delta\beta(y) + g(y)$ is not a subpotential operator in any Hilbert space (that is, of the form $\partial\varphi$), Theorem 2.71 cannot be applied here. However, the longtime behavior of $y(t)$ can be studied via the invariance principle described in Theorem 2.72. Namely, we define the omega-limit set

$$\omega(y_0) = \left\{\zeta = \lim_{t_n \to \infty} y(t_n) \ \text{ in } L^1(\Omega)\right\}$$

and get

Theorem 4.23 *Assume that $\beta'(r) > 0$, $\forall\, r \neq 0$, and*

$$\lim_{|r|\to\infty} \left|\frac{\beta(r)}{r}\right| \ge \alpha > 0. \tag{4.90}$$

Then, $\omega(y_0)$ is a nonempty set and, if y^* is a solution (if any) to the equation

$$-\Delta\beta(y^*) + g(y^*) = 0 \ \text{in } \Omega; \beta(y^*) = 0 \ \text{on } \partial\Omega, \tag{4.91}$$

then

$$\omega(y_0) \subset \{z \in L^1(\Omega);\ |y^* - z|_1 = r\}, \tag{4.92}$$

where $0 < r < |y_0 - y^*|_1$.

Proof. By Theorem 2.73, it suffices to show that the operator $(I + \lambda A_1)^{-1}$, where $\lambda > 0$, is compact in $L^1(\Omega)$. Indeed, if $\{f_n\}$ is bounded in $L^1(\Omega)$ and $u_n = (I + \lambda A_1)^{-1} f_n$, that is,

$$\begin{aligned} &u_n - \lambda\Delta\beta(u_n) + \lambda g(u_n) = f_n \ \text{ in } \Omega, \\ &\beta(u_n) = 0 \qquad\qquad\qquad\qquad \text{on } \partial\Omega, \end{aligned} \tag{4.93}$$

then it is easily seen that $|u_n|_1 \le C$, $\forall n$, and so, by Theorem 3.17, it follows that

$$\|\beta(u_n)\|_{W_0^{1,q}(\Omega)} \le C, \quad \forall n,$$

where $1 < q < \frac{N}{N-1}$. This implies that $\{\beta(u_n)\}$ is compact in $L^1(\Omega)$ and so, on a subsequence, we have

$$\beta(u_n) \to \zeta, \quad \text{a.e. in } \Omega.$$

Since β is strictly monotone, this implies that $\{u_n\}$ is a.e. convergent to $u = \beta^{-1}(\zeta)$. Inasmuch as by (4.90), for each $R > 0$, $|u_n| \le \alpha^{-1}|\beta(u_n)|$ for $|u_n| \ge R$, we infer that $u_n \to u$ in $L^1(\Omega)$ and so $(I + \lambda A_1)^{-1}$ is compact, as claimed. □

In particular, it follows by Theorem 4.23 that, if equation (4.91) has a unique solution y^*, then $\omega(y_0) = y^*$ and so, $\lim_{t\to\infty} y(t) = y^*$ in $L^1(\Omega)$. This happens, for instance, if β^{-1} is continuous. Indeed, we may rewrite (4.91) as

$$-\Delta z + g(\beta^{-1}(z)) = 0 \ \text{ in } \Omega; \quad z = 0 \ \text{ on } \partial\Omega,$$

and, since the function $h(z) \equiv g(\beta^{-1}(z))$ is continuous and monotone, it follows that the operator $A_0 z = -\Delta z + h(z)$, $\forall z \in D(A_0) = H_0^1(\Omega) \cap H^2(\Omega)$ is maximal monotone and coercive in $L^2(\Omega)$, and so surjective.

The porous media equation in $\mathbb{R}^N$

Consider now equation (4.55) in $\Omega = \mathbb{R}^N$, for $N = 1, 2, 3$,

$$\begin{cases} \dfrac{\partial y}{\partial t} - \Delta\beta(y) \ni f & \text{in } \mathbb{R}^N \times (0,T), \\ y(0,x) = y_0(x), & x \in \mathbb{R}^N, \\ \beta(y(t)), y(t) \in L^1(\mathbb{R}^n), & \forall t \in [0,T] \end{cases} \tag{4.94}$$

where $\frac{\partial}{\partial t}$ and Δ are taken in the sense of distributions on $(0,T) \times \mathbb{R}^N$ (see (4.56)). We may rewrite equation (4.94) in the form (2.96) on the space $X = L^1(\mathbb{R}^N)$, where

$$Ay = \{-\Delta w;\ w(x) \in \beta(y(x)), \text{ a.e. } x \in \Omega,\ w, \Delta w \in L^1(\mathbb{R}^N)\},\ \forall y \in D(A),$$

$$D(A) = \{y \in L^1(\mathbb{R}^N); \exists w \in L^1(\mathbb{R}^N), \Delta w \in L^1(\mathbb{R}^N),\ w(x) \in \beta(y(x)), \text{a.e. } x \in \mathbb{R}^N\},$$

where Δw is taken in the sense of distributions. Here β is a maximal monotone graph in $\mathbb{R} \times \mathbb{R}$ such that $0 \in \beta(0)$ and $0 \in \text{int}(D(\beta))$ if $N = 1, 2$. Then, as shown earlier in Theorem 3.24, A is m-accretive in $L^1(\mathbb{R}^N) \times \mathbb{R}^N$ and so, by Theorem 2.49, we obtain the following.

Proposition 4.24 *Assume that $f \in L^1(0,T; L^1(\mathbb{R}^N))$ and $y_0 \in L^1(\mathbb{R}^N)$ is such that $\exists w \in L^1(\mathbb{R}^N)$, $\Delta w \in L^1(\mathbb{R}^N)$, $w(x) \in \beta(y_0(x))$, a.e. $x \in \mathbb{R}^N$. Then, problem* (4.94) *has a unique mild solution $y \in C([0,T]; L^1(\mathbb{R}^N))$. This extends to all of $y_0 \in \overline{D(A)}$. The mild solution y is, in this case too, given as* (4.75), *where $\Omega = \mathbb{R}^N$.*

Remark 4.25 The continuity of solutions to (4.94) with respect to φ, that is, the structural stabilization follows by Theorem 2.76 (see Bénilan and Crandall [34] and Alikakos and Rostamian [3].)

It turns out that, in this case also, the mild solution y given by Proposition 4.24 is a solution to (4.106) in the sense of Schwartz distributions on $(0,T)\times\mathbb{R}^N$, that is,

$$\int_0^T\int_{\mathbb{R}^N}(y(t,x)\,\frac{\partial\varphi}{\partial t}\,(t,x)+\beta(y(t,x))\Delta\varphi(t,x))dx\,dt+\int_{\mathbb{R}^N}y_0(x)\varphi(0,x)dx$$
$$=-\int_0^T\int_{\mathbb{R}^N}\varphi(t,x)f(t,x)dx\,dt,\ \forall\varphi\in C_0^\infty([0,T)\times\mathbb{R}^N).$$

If β is monotonically increasing and continuous it turns out that y is the unique distributional solution to (4.67) such that $y\in L^\infty((0,T)\times\mathbb{R}^N)$ (see Brezis and Crandall [45]).

Localization of solutions to porous media equations

A nice feature of solutions to the porous media equation are finite time extinction in the fast diffusion case (i.e., $\beta(y)=y^\alpha$, $0<\alpha<1$), and the propagation with finite velocity for the low diffusion porous media equation (i.e., $1<\alpha<\infty$). We refer the reader to the work of Pazy [90]. (See also the Vasquez monograph [102] for a detailed study of the localization of solutions to porous media equations.) Here we shall present some particular results.

Proposition 4.26 *Let $y\in C([0,\infty);L^1(\Omega)\cap H^{-1}(\Omega))$ be the solution to equation*

$$\frac{\partial y}{\partial t}-\mu\Delta(|y|^\alpha\operatorname{sign}y)=0\quad in\ \ \Omega\times(0,\infty),\tag{4.95}$$

where $y_0\in H^{-1}(\Omega)\cap L^1(\Omega)$, $\mu>0$, $0<\alpha<1$ if $N=1,2$ and $\frac{1}{5}\le\alpha<1$ if $N=3$. Then,

$$y(t,x)=0\quad for\ \ t\ge T(y_0),$$

where

$$T(y_0)=\frac{|y_0|_{-1}^{1-\alpha}}{\mu\gamma^{1+\alpha}}\,.$$

If $\alpha=0$ and $N=1$, then $y(t,x)=0$ for $t\ge\frac{|y_0|_{-1}}{\mu\gamma}$.

Proof. Assume first that $N>1$. As seen earlier, the equation has a unique smooth solution $y\in W^{1,2}([0,T];H^{-1}(\Omega))$ for each $T>0$. Multiplying scalarly in $H^{-1}(\Omega)$ equation (4.95) by y and integrating on $(0,T)$, we obtain

$$\frac{1}{2}\,\frac{d}{dt}\,|y(t)|_{-1}^2+\mu\int_\Omega|y(s,x)|^{\alpha+1}dx=0,\quad\forall t\ge 0.$$

Now, by the Sobolev embedding theorem (see Theorem 1.11), we have

$$\gamma|y(s)|_{-1}\le|y(s)|_{\alpha+1}\quad\text{for all }\alpha>0\text{ if }N=1,2\text{ and for }\alpha\ge\frac{N-2}{N+2}\text{ if }N\ge 3.$$

(Here, $|\cdot|_{-1}$ is the $H^{-1}(\Omega)$ norm and $|\cdot|_p$ is the $L^p(\Omega)$-norm.) This yields

$$\frac{d}{dt}\,|y(t)|_{-1}^2+2\mu\gamma^{\alpha+1}|y(t)|_{-1}^{\alpha+1}\le 0,\quad\forall t\ge 0,$$

and, therefore,

$$\frac{d}{dt}\,|y(t)|_{-1}^{1-\alpha} + \mu\gamma^{1+\alpha} \le 0, \quad \text{a.e. } t > 0.$$

Hence,

$$|y(t)|_{-1} = 0 \quad \text{for } t \ge \frac{|y_0|_{-1}^{1-\alpha}}{\mu\gamma^{1+\alpha}}.$$

If $N = 1$, then, multiplying scalarly in $H^{-1}(\Omega)$ equation (4.95) by $y(t)$, we get

$$\frac{1}{2}\,\frac{d}{dt}\,|y(t)|_{-1}^2 + \mu|y(t)|_1 \le 0, \quad \text{a.e. } t > 0.$$

This yields (we have $|y|_1 \ge \gamma\,|y_0|_{-1}$)

$$|y(t)|_{-1} + \mu\gamma t \le |y_0|_{-1}, \quad \forall t \ge 0$$

and, therefore,

$$|y(t)|_{-1} = 0 \quad \text{for } t \ge \frac{|y_0|_{-1}}{\mu\gamma}. \qquad \square$$

Remark 4.27 The extinction in finite time is a significant nonlinear behavior of solutions to fast diffusion porous media equations and this implies that the diffusion process reaches its critical state (which is zero in this case) in finite time. This dynamic resembles that described in Theorem 2.74, but the case considered here is not, however, covered by that theorem.

If $\alpha > 1$, that is, in a slow diffusion case, equation (4.66) has the finite speed propagation property, that is, if *support* $y_0 \subset \Omega_0 \subset \Omega$, then *support* $y(t) = S(t)y_0 \subset \Omega_t$, $\forall t \ge 0$, where $\{\Omega_t\}$ is an increasing family of domains $\Omega_t \subset \Omega$. Namely, we have

Proposition 4.28 *Let $y_0 \in L^\infty(\Omega)$ be such that $y_0 \ge 0$ in Ω and*

$$y_0(x) = 0, \quad \text{for } |x| \le r_0 < \infty. \tag{4.96}$$

Then, there is a decreasing function $r : [0,T] \to (0, r_0)$ such that

$$y(t,x) = 0, \quad \text{for } |x| \le r(t), \quad \forall t \in [0,T]. \tag{4.97}$$

Proof. We note first that, as seen earlier, we have $y \ge 0$, a.e. on $(0,T) \times \Omega$ and $y \in L^\infty((0,T) \times \Omega)$. Moreover, by Theorem 4.17 we know that

$$y \in C([0,T]; H^{-1}(\Omega)) \cap C([0,T]; L^1(\Omega)), \quad y^\alpha \in L^2(0,T; H_0^1(\Omega)),$$
$$\frac{dy}{dt} \in L^2(0,T; H^{-1}(\Omega)).$$

Then, by (4.95) it follows via Green's formula

$$\begin{aligned}\frac{1}{\alpha+1}\int_\Omega y^{\alpha+1}(t,x)\psi(x)dx + \int_0^t\int_\Omega \nabla(y^\alpha(s,x)) \cdot \nabla\psi(x)dx \\ = \frac{1}{\alpha+1}\int_\Omega y_0(x)\psi(x)dx, \quad \forall \psi \in C_0^\infty(\Omega).\end{aligned} \tag{4.98}$$

Now, we introduce the *energy function*

$$\Phi(t,r) = \int_0^t \int_{B_r} |\nabla y(s,x)|^2 dx\, ds, \quad \forall t \in [0,T],\ r > 0, \tag{4.99}$$

where $B_r = \{x \in \mathbb{R}^N;\ |x| < r\}$.

To prove (4.97), it suffices to show that

$$\frac{\partial \Phi}{\partial r}(t,r) \geq Ct^{\theta-1}(\Phi(t,r))^{\delta}, \quad \forall t \in (0,T),\ r \in (0,r_0), \tag{4.100}$$

for some $\theta, \delta \in (0.1)$. Indeed, by (4.96) and (4.100), it follows that $\varphi(t,r) \equiv (\Phi(t,r))^{1-\delta}$ satisfies the inequality

$$\frac{\partial \varphi}{\partial r}(t,r) \geq C(1-\delta)t^{\theta-1}, \quad \forall t \in (0,T),\ r \in (0,r_0).$$

We set $r(t) = \inf\{r \geq 0;\ \Phi(t,r) > 0\} \wedge r_0$, $\forall t \in [0,T]$.

Clearly, $\varphi(t,r(t)) = \Phi(t,r(t)) = 0$ and, therefore,

$$\varphi(t,r_0) \geq C(1-\delta)t^{\theta-1}(r_0 - r(t)), \quad \forall t \in [0,T].$$

This implies that $r(t) \geq r_0 - C(1-\delta)t^{\theta-1}$ and, therefore, there is $t_0 \in (0,T)$ sufficiently small such that $r(t) > 0$. Hence, $\Phi(t_0,r) = 0$ for $r \leq r(t_0)$ and, since $r \to \Phi(t,r)$ as well as $t \to \Phi(t,r)$ are decreasing, we infer that $\Phi(t,r) = 0$ for $r \leq r(t)$ and so, by (4.99), we get (4.97), as claimed.

Proof of (4.100). We set $B_r^\varepsilon = B_{r+2\varepsilon} \setminus B_{r+\varepsilon}$ and in (4.98) we choose $\psi = y^\alpha \psi_\varepsilon$, where $\psi_\varepsilon(x) \equiv \rho_\varepsilon(|x|)$ and $\rho \in C^\infty(\mathbb{R}^+)$ be a cut-off function such that $\rho_\varepsilon(s) = 1$ for $0 \leq s \leq r+\varepsilon$, $\rho_\varepsilon(s) = 0$ for $s \geq r+2\varepsilon$ and, for $\mathcal{X}_\varepsilon = \mathbf{1}_{[r+\varepsilon, r+2\varepsilon]}$,

$$\lim_{\varepsilon \to 0} \left| \rho'_\varepsilon(s) + \frac{1}{\varepsilon} \right| \mathcal{X}_\varepsilon(s) = 0,$$

uniformly in $s \in [0,\infty)$. We get

$$\begin{aligned}
&\frac{1}{\alpha+1} \int_{B_{r+2\varepsilon}} y^{\alpha+1}(t,x)\psi_\varepsilon(x)dx + \int_0^t ds \int_{B_{r+2\varepsilon}} \psi_\varepsilon(x)|\nabla y^\alpha(s,x)|^2 dx\, ds \qquad (4.101)\\
&\quad = \frac{1}{\alpha+1} \int_{B_{r+2\varepsilon}} \psi_\varepsilon(x) y^\alpha(x)dx + \int_0^t \int_{B_{r+2\varepsilon}} y^\alpha(s,x)\nabla y^\alpha(s,x) \cdot \nabla \psi_\varepsilon(x)dx\, ds\\
&\quad = \frac{1}{\alpha+1} \int_{B_{r+2\varepsilon}} \psi_\varepsilon(x) y_0^\alpha(x)dx + \int_0^t \int_{B_r^\varepsilon} y^\alpha(s,x)\left(\nabla y^\alpha(s,x) \cdot \frac{x}{|x|}\right)\rho'_\varepsilon(|x|)dx\, ds\\
&\quad \leq \frac{1}{\alpha+1} \int_{B_{r+2\varepsilon}} \psi_\varepsilon(x) y_0^\alpha(x)dx + \left(\int_0^t \int_{B_r^\varepsilon} |\rho'_\varepsilon|\, |\nabla y(s,x)|^2\right)^{\frac{1}{2}}\\
&\qquad \times \left(\int_0^t \int_{B_r^\varepsilon} |\rho'_\varepsilon| y^{2\alpha}(s,x)dx\, ds\right)^{\frac{1}{2}}.
\end{aligned}$$

On the other hand, as easily seen, we have

$$\lim_{\varepsilon \to 0} \int_0^t \int_{B_r^\varepsilon} |\rho'_\varepsilon|\, |\nabla y^\alpha(s,x)|dx\, ds = \frac{\partial \Phi}{\partial r}(t,r),$$

and so, letting $\varepsilon \to 0$ in (4.101) and taking into account (4.96), we obtain that

$$H(t,r) + \Phi(t,r) \le \left(\frac{\partial \Phi}{\partial r}(t,r)\right)^{\frac{1}{2}} \left(\int_0^t \int_{\Sigma_r} y^{2\alpha}(s,x)dx\right)^{\frac{1}{2}}, \ \forall r \in [0, r_0],$$

where $\Sigma_r = \{x;\ |x| = r\}$ and

$$H(t,r) = \frac{1}{\alpha+1} \int_{B_r} y^{\alpha+1}(t,x)dx.$$

Then, by Proposition 1.15, we get via Hölder's inequality

$$H(t,r) + \Phi(t,r) \le C\left(\frac{\partial \Phi}{\partial t}(t,r)\right)^{\frac{1}{2}} H^{\frac{\alpha(1-\theta)}{\alpha+1}} t^{\frac{1-\theta}{2}} \left(((\Phi(t,r))^{\frac{1}{2}} + H^{\frac{\alpha}{\alpha+1}}(t,r))^{\frac{1}{2}}\right),$$
$$\forall t \in (0,T),\ r \in [0, r_0].$$

This implies (4.100) by a standard computation and the proof of Proposition 4.28 is complete. □

Remark 4.29 The propagation with finite speed of the fast diffusion porous media flow $S(t)$ is a property which resembles the hyperbolic flows. In the case $\Omega = \mathbb{R}^N$, a simple proof of this result can be given by comparing the solution $y(t) = S(t)y_0$ with the *Barenblatt source solution*

$$U(t,x) = t^{-\frac{N}{(\alpha-1)N+2}} \left[C - \frac{\alpha-1}{2\alpha(\alpha-1)N+2} \frac{|x|^2}{t^{(\alpha-1)N+2}}\right]_+^{\frac{1}{\alpha-1}}$$

which has the support in $\{(t,x) \in (0,T) \times \mathbb{R}^N;\ |x|^2 \le C_1 t^{\frac{2}{(\alpha-1)N+2}}\}$.

4.3 The Phase Field Transition with Mushy Region

It should be said that the Stefan model of phase transition discussed in the previous section is quite inappropriate to describe the melting process because the real physical process exhibits superheating which leads to unstability of the phase change surface $\{x;\ \theta(t,x) = 0\}$. In fact, during a melting (or a freezing) process, a *mushy zone* arises, where θ equals to the melting temperature, while the substance is neither pure solid, nor pure liquid and so the interface is not a smooth region. This phenomenon leads to new mathematical models which try handle such a situation.

Such a model is discussed in [11], where the two-phase Stefan problem (4.61) is replaced by the equation

$$\begin{aligned} &\frac{\partial y}{\partial t} - \operatorname{div}(H_\lambda(y)\nabla y) = 0 \quad \text{in } Q = (0,T) \times \Omega, \\ &H_\lambda(y)(\nabla y \cdot u) = 0 \qquad\quad \text{on } \Sigma = (0,T) \times \partial\Omega, \\ &y(0,x) = y_0(x) \qquad\qquad\ \ x \in \Omega, \end{aligned} \tag{4.102}$$

where H_λ is a smooth approximation for $\lambda \to 0$ to the function β which keeps in system more of the existence of mushy region. (In fact, by construction, $H_\lambda(y) = \beta'(y)$ outside some region $|y| \le \delta_\lambda$.)

Here, we shall treat another phase-transition model with mushy region, due to C. Caginalp [51]. Namely,

$$\begin{cases} \dfrac{\partial}{\partial t}\,\theta(t,x) + l\,\dfrac{\partial \varphi}{\partial t}\,(t,x) - k\Delta\theta(t,x) = f_1(t,x), & \text{in } \ Q = (0,T)\times\Omega, \\ \dfrac{\partial}{\partial t}\,\varphi(t,x) - \alpha\Delta\varphi(t,x) - \kappa(\varphi(t,x) - \varphi^3(t,x)) & \\ \qquad\qquad\qquad\qquad +\delta\theta(t,x) = f_2(t,x), & \text{in } \ Q, \\ \theta(0,x) = \theta_0(x), \ \ \varphi(0,x) = \varphi_0(x), & x\in\Omega, \\ \theta = 0, \ \ \varphi = 0, & \text{on } \ (0,T)\times\partial\Omega, \end{cases} \tag{4.103}$$

where $l, k, \alpha, \kappa, \delta$ are positive constants. This system, called in the literature the *phase-field system*, was introduced as an improved model of the melting and solidification phenomena on the lines discussed above. In this latter case, $\theta = \theta(t,x)$ is the temperature, whereas φ is the phase-field transition function. The two-phase Stefan problem presented above can be viewed as a particular limit case of this model. In fact, it can be obtained from the two-phase Stefan model of phase transition by the following heuristic argument.

As seen earlier, the classical two-phase Stefan problem (4.61) (take $K(\theta) \equiv K\theta$) can be rewritten as

$$\frac{\partial}{\partial t}\,\gamma(\theta) - k\Delta\theta = f \quad \text{in} \ \ \mathcal{D}'((0,T)\times\Omega),$$

where γ is the multivalued graph (4.58), that is, $\gamma = C + \rho H$. Equivalently,

$$\frac{\partial}{\partial t}\,\varphi(\theta) - k\Delta\theta = f \quad \text{in} \ \ \mathcal{D}'((0,T)\times\Omega), \tag{4.104}$$

where $\varphi : \mathbb{R} \to \mathbb{R}$ is given by

$$\varphi(\theta) = \begin{cases} C_1\theta & \text{if } \ \theta < 0, \\ [0,\rho] & \text{if } \ \theta = 0, \\ C_2\theta + \rho & \text{if } \ \theta > 0. \end{cases} \tag{4.105}$$

In this model, the presence of a multivalued graph φ describes a sharp phase transition from solid to liquid which, as mentioned above, is quite unrealistic. The idea behind Caginalp's model of phase transition is to replace the multivalued graph φ by a smooth function $\varphi = \varphi(t,x)$, called the *phase function* and equation (4.104) by

$$\frac{\partial\theta}{\partial t} + l\,\frac{\partial\varphi}{\partial t} - k\Delta\theta = f. \tag{4.106}$$

The phase function φ should be interpreted as a measure of phase transition and more precisely as the proportion related to the first phase and the second one. For instance, in the case of liquid–solid transition, one has, formally, $\varphi \geq 1$ in the liquid zone $\{(t,x);\ \theta(t,x) > 0\}$ and $\varphi < 0$ in the solid zone $\{(t,x);\ \theta(t,x) < 0\}$. Formally, $\{(t,x);\ 0 < \varphi(t,x) < 1\}$ is the mushy zone of the phase transition. In general, however, φ remains in an interval $[\varphi_*, \varphi^*]$ which is determined by the specific physical model. This is the reason why φ is taken as the solution to a parabolic equation of the Ginzburg–Landau type

$$\frac{\partial\varphi}{\partial t} - \alpha\Delta\varphi - \kappa(\varphi - \varphi^3) + \delta\theta = f_2, \tag{4.107}$$

which is the basic mathematical model of phase transition. Equations (4.106) and (4.107) lead, after further simplifications, to system (4.103).

As regards the existence in problem (4.103), we have the following.

Theorem 4.30 *Assume that* $\varphi_0, \theta_0 \in H_0^1(\Omega) \cap H^2(\Omega)$, $\Omega \subset \mathbb{R}^N$, $N = 1, 2, 3$, *and that* $f_1, f_2 \in W^{1,2}([0,T]; L^2(\Omega))$. *Then, there is a unique solution* (θ, φ) *to system* (4.103) *satisfying*

$$(\theta, \varphi) \in (W^{1,\infty}([0,T]; L^2(\Omega)))^2 \cap (L^\infty(0,T; H_0^1(\Omega) \cap H^2(\Omega)))^2. \tag{4.108}$$

Proof. We set $y = \theta + l\varphi$ and reduce system (4.103) to

$$\begin{cases} \dfrac{\partial}{\partial t} y - k\Delta y + kl\Delta\varphi = f_1 & \text{in } Q, \\ \dfrac{\partial}{\partial t}\varphi - \alpha\Delta\varphi - \kappa(\varphi - \varphi^3) + \delta(y - l\varphi) = f_2 & \text{in } Q, \\ y(0) = y_0 = \theta_0 + l\varphi_0,\ \varphi(0) = \varphi_0 \text{ in } \Omega,\ y = \varphi = 0 & \text{on } \Sigma. \end{cases} \tag{4.109}$$

In the space $X = L^2(\Omega) \times L^2(\Omega)$ consider the operator $A : X \to X$,

$$A\begin{pmatrix} y \\ \varphi \end{pmatrix} = \begin{pmatrix} -k\Delta y + kl\Delta\varphi \\ -\alpha\Delta\varphi - \kappa(\varphi - \varphi^3) + \delta(y - l\varphi) \end{pmatrix}$$

with the domain $D(A) = \{(y, \varphi) \in (H^2(\Omega) \cap H_0^1(\Omega))^2;\ \varphi \in L^6(\Omega)\}$. Then, system (4.109) can be written as

$$\begin{cases} \dfrac{d}{dt}\begin{pmatrix} y \\ \varphi \end{pmatrix} + A\begin{pmatrix} y \\ \varphi \end{pmatrix} = \begin{pmatrix} f_1 \\ f_2 \end{pmatrix}, & t \in (0,T), \\ \begin{pmatrix} y \\ \varphi \end{pmatrix}(0) = \begin{pmatrix} y_0 \\ \varphi_0 \end{pmatrix}. \end{cases} \tag{4.110}$$

In order to apply Theorem 2.60 to (4.110), we check that A is quasi-m-accretive in X. To this aim we endow the space $X = L^2(\Omega) \times L^2(\Omega)$ with an equivalent Hilbertian norm provided by the scalar product

$$\left\langle \begin{pmatrix} y \\ \varphi \end{pmatrix}, \begin{pmatrix} \widetilde{y} \\ \widetilde{\varphi} \end{pmatrix} \right\rangle = a(y, \widetilde{y})_{L^2(\Omega)} + (\varphi, \widetilde{\varphi})_{L^2(\Omega)},$$

where $a = \frac{\alpha}{kl^2}$. Then, as easily seen, we have

$$\left\langle A\begin{pmatrix} y \\ \varphi \end{pmatrix} - A\begin{pmatrix} y^* \\ \varphi^* \end{pmatrix}, \begin{pmatrix} y \\ \varphi \end{pmatrix} - \begin{pmatrix} y^* \\ \varphi^* \end{pmatrix} \right\rangle$$

$$\geq \eta(\|\nabla(y - y^*)\|^2_{L^2(\Omega)} + \|\nabla(\varphi - \varphi^*)\|^2_{L^2(\Omega)}) - \omega(\|y - y^*\|^2_{L^2(\Omega)} + \|\varphi - \varphi^*\|^2_{L^2(\Omega)}),$$

for some $\omega, \eta > 0$. Clearly, this implies that A is quasi-accretive and, more precisely, that is, $A + \omega I$ is accretive.

Now, consider for given $g_1, g_2 \in L^2(\Omega)$ the equation

$$\lambda \begin{pmatrix} y \\ \varphi \end{pmatrix} + A \begin{pmatrix} y \\ \varphi \end{pmatrix} = \begin{pmatrix} g_1 \\ g_2 \end{pmatrix}. \tag{4.111}$$

Equivalently,

$$\begin{cases} \lambda y - k\Delta y + kl\Delta\varphi = g_1 \quad \text{in } \Omega, \\ \lambda\varphi - \alpha\Delta\varphi - \kappa(\varphi - \varphi^3) + \delta(y - l\varphi) = g_2 \text{ in } \Omega, \\ y = \varphi = 0 \quad \text{on } \partial\Omega. \end{cases} \tag{4.112}$$

System (4.112) can be equivalently rewritten as

$$\begin{pmatrix} \lambda y \\ (\lambda - \kappa - l\delta)\varphi + \delta y \end{pmatrix} + A_0 \begin{pmatrix} y \\ \varphi \end{pmatrix} + F \begin{pmatrix} y \\ \varphi \end{pmatrix} = \begin{pmatrix} q_1 \\ q_2 \end{pmatrix}, \tag{4.113}$$

where the operators F and $A_0 : L^2(\Omega) \times L^2(\Omega) \to L^2(\Omega) \times L^2(\Omega)$ are given by

$$A_0 \begin{pmatrix} y \\ \varphi \end{pmatrix} = \begin{pmatrix} -k\Delta y + kl\Delta\varphi \\ -\alpha\Delta\varphi \end{pmatrix}, \; D(A_0) = (H^2(\Omega) \times H_0^1(\Omega))^2$$

$$F \begin{pmatrix} y \\ \varphi \end{pmatrix} = \begin{pmatrix} 0 \\ \kappa\varphi^3 \end{pmatrix}, \qquad D(F) = L^2(\Omega) \times L^6(\Omega).$$

By the Lax–Milgram lemma, it is easily seen that A_0 is m-accretive and coercive in $X = L^2(\Omega) \times L^2(\Omega)$. On the other hand, F is quasi-m-accretive in X and, as easily seen,

$$\left\langle A_0 \begin{pmatrix} y \\ \varphi \end{pmatrix}, F_\lambda \begin{pmatrix} y \\ \varphi \end{pmatrix} \right\rangle \geq 0, \;\; \forall \begin{pmatrix} y \\ \varphi \end{pmatrix} \in D(A_0), \; \forall \lambda > 0,$$

where F_λ is the Yosida approximation of F. Hence, by Proposition 2.43, the operator $A_0 + F$ is quasi-m-accretive and this implies that (4.113) has a solution for λ sufficiently large. □

Remark 4.31 The liquid and solid regions in the case of a melting solidification problem are those that remain invariant by the flow $t \to (\theta(t), \varphi(t))$. This is one way of determining in specific physical models the range interval $[\varphi_*, \varphi^*]$ of phase-field function φ. In the literature on phase transition there is a large variety of such phase-field models which can be treated in a similar way. Such a model was studied recently by P.L. Colli et al. in [56]. Note also that the exact controllability of phase-field systems of the form (4.103) with the distributed input controller u in the first equation was studied in the work [20]. In particular, the existence of a sliding mode controller u for this system was proven.

In a similar way, it can be treated the reaction-diffusion system

$$\begin{aligned}
&\frac{\partial y}{\partial t} - D(x)\Delta y + B(x)\cdot\nabla y + C(x,y) && \text{in } (0,T)\times\Omega,\\
&y(0) = y_0 && \text{in } \Omega, \\
&y(t,x) = 0 && \text{on } (0,T)\times\Omega,
\end{aligned} \tag{4.114}$$

where $y = \text{col}(y_i)_{i=1}^N$, $D(x) = \text{diag}(D_1(x), D_2(x), ..., D_N(x))$, $B(x) = (B_1(x), ..., B_N(x))$, $D_j \in C^2(\overline{\Omega})$, $D_j \geq \gamma > 0$, $B \in C^1(\overline{\Omega}), R^N)$ and $C \in C(\overline{\Omega}\times\mathbb{R}^N) \to \mathbb{R}^N$ is such that

$$(C(x,y) - C(x,\bar{y}))\cdot(y-\bar{y}) \geq -\gamma|y-\bar{y}|_N^2, \quad \forall y,\bar{y}\in\mathbb{R}^N, x\in\overline{\Omega}.$$

Then, we can write (4.114) as

$$\begin{aligned}
&\frac{dy}{dt} + Ay = 0, \ t\in(0,T),\\
&y(0) = y_0,
\end{aligned}$$

where $A : (L^2(\Omega))^N \to (L^2(\Omega))^N$ is defined by

$$\begin{aligned}
Ay &= D(x)\Delta y + B(x)\cdot\nabla y + C(x,y),\\
D(A) &= (H_0^1(\Omega))^N \cap (H^2(\Omega))^N.
\end{aligned}$$

It is easily seen that A is quasi-m-accretive and so the general existence Theorem 2.59 applies as well.

4.4 The Self-organized Criticality Equation

The *self-organized criticality* is the property of dynamical systems to emerge spontaneously to an equilibrium state. One mathematical model in 2-D (the sandpile cellular model) we briefly describe below was introduced by Bak, Tang and Wiesenfeld [9] as a model of evolution but in the last decades the self-organized criticality has become a new theory for the explanation of a large category of dynamics in nature and society. The self-organized means that the evolution to equilibrium occurs spontaneously and apparently is no outside agent to influence it. Likewise other models in statistical mechanics, this system describes the transient evolution far from equilibrium, via a transport mechanics to be made explicit below (see, e.g., [52]).

Schematically, the evolution process is described on an $N\times N$ square lattice, representing a discrete region $\Omega = \{(i,j)\}_{i,j=1}^N$. To each site (i,j) is assigned at time t an integer height variable h_{ij}. The system is perturbed externally at the site (i,j) until the height h_{ij} exceeds a threshold (critical) value h_{ij}^c. Then, a toppling (avalanche) event occurs. The toppling at the *activated* site (i,j) is described by the following dynamics

$$h_{ij}^{t+1} \to h_{ij}^t - M_{ij}^{kl} \quad \text{for } (k,l)\in\Gamma_{ij},$$

where $\Gamma_{ij} = \{(i+1), j), (i, j+1), (i-1), (i, j-1)\}$ is the set of all 4-nearest neighbors of (i, j) and $M_{ij}^{k,l} = 4$ if $i = k$, $j = l$, $M_{ij}^{k,l} = -1$ if $(k, l) \in \Gamma_{ij}$, $M_{ij}^{kl} = 0$ if $(k, l) \in \Gamma_{ij}$. Then, the transition from h^t to h^{t+1} can be expressed by

$$h_{ij}^{t+1} - h_{ij}^t = -M_{ij} H(h_{ij}^t - h_{ij}^c), \quad i, j = 1, ..., N, \tag{4.115}$$

where H is the Heaviside function

$$H(r) = \begin{cases} 1 & \text{if } r > 0, \\ 0 & \text{if } r < 0. \end{cases}$$

This toppling dynamics is iterated at each supercritical site (i, j) until each site is below the threshold (critical) state $\rho_c = \{h_{ij}^c\}_{i,j=1}^N$. If we view h_{ij} as the discrete version of the continuous height-density function $\rho = \rho(t, x)$, $x \in \Omega \subset \mathbb{R}^2$, we may replace equation (4.115) by its continuous version

$$\frac{\partial \rho}{\partial t}(t, x) - \Delta H(\rho(t, x) - \rho_c(x)) = 0, \quad x \in \Omega, \ t \geq 0. \tag{4.116}$$

The system is initialized by taking in equation (4.116), the condition $\rho(0, x) = \rho_0(x)$, $x \in \Omega$. Also, the Dirichlet boundary value conditions $H(\rho(t, x) - \rho_c(x)) = 0$, $x \in \partial\Omega$, $t \geq 0$, will be imposed. It should be noticed that, because the Heaviside function H is discontinuous in origin, the standard existence theory for nonlinear diffusion equations is not applicable, unless we *fill* the jump by replacing the function H by its multivalued version

$$\widetilde{H}(r) = \begin{cases} 1 & \text{for } r > 0, \\ [0, 1] & \text{for } r = 0, \\ 0 & \text{for } r < 0, \end{cases} \tag{4.117}$$

which has the advantage to be a maximal monotone graph in $\mathbb{R} \times \mathbb{R}$. Thus, instead (4.116) we consider the multivalued nonlinear diffusion equation

$$\begin{aligned} &\frac{\partial \rho}{\partial t} - \Delta \widetilde{H}(\rho - \rho_c) \ni 0 \ \text{ in } (0, \infty) \times \Omega, \\ &\rho(0, x) = \rho_0(x), \ x \in \Omega, \\ &0 \in \widetilde{H}(\rho(t, x) - \rho_c(x)), \ \ t \geq 0, \ x \in \partial\Omega. \end{aligned} \tag{4.118}$$

Setting $\rho - \rho_c = y$, we may rewrite (4.118) as

$$\begin{aligned} &\frac{\partial y}{\partial t} - \Delta \widetilde{H}(y) \ni 0, \ \ t \geq 0, \ x \in \Omega, \\ &y(0, x) = y_0(x) = \rho_0(x) - \rho_c(x), \ \ x \in \Omega, \\ &\widetilde{H}(y) \ni 0 \ \text{ on } (0, \infty) \times \partial\Omega. \end{aligned} \tag{4.119}$$

This is an equation of the form (4.55) where $\beta = \widetilde{H}$.

Then, by Theorem 4.17 (see also Proposition 4.19), we have

Proposition 4.32 *Let $\rho_0 \in H^{-1}(\Omega) \cap L^1(\Omega)$. Then there is a unique solution $y \in C([0,T]; H^{-1}(\Omega) \cap L^1(\Omega)) \cap W^{1,2}([0,T]; H^{-1}(\Omega))$ to (4.115) such that*

$$\begin{aligned} &\frac{\partial y}{\partial t} - \Delta\eta(t) = 0, \quad a.e.\ t > 0,\ x \in \Omega, \\ &y(0) = y_0, \quad \eta(t,x) \in \widetilde{H}(y(t,x)), \quad a.e.\ (t,x) \in (0,T), \\ &\eta(t) \in H_0^1(\Omega), \quad \forall\, t \in (0,T). \end{aligned} \tag{4.120}$$

If $y_0 \in L^1(\Omega)$, then (4.119) has a unique mild solution $y \in C([0,T]; L^1(\Omega))$. If $y_0 \in L^\infty(\Omega)$ and $y_0 \geq 0$, a.e. in Ω, then

$$0 \leq y(t,x) \leq |y_0|_\infty, \quad a.e.\ (t,x) \in (0,T) \times \Omega. \tag{4.121}$$

Moreover, the semiflow $t \to y(t,x)$ is a continuous semigroup of contractions in $L^1(\Omega)$.

(Here, we shall denote by $|\cdot|_p$, $1 \leq p \leq \infty$, the norm in $L^p(\Omega)$.)

Equation (4.120) is hard to solve in explicit terms, but it can be approximated however by the finite difference scheme (4.75) which, as seen earlier, is convergent to y.

At each time t, we divide the domain Ω in the following regions

$$\begin{aligned} \Omega_0^t &= \{x \in \Omega; \rho(t,x) = \rho_c(x)\} \quad \text{(critical region)} \\ \Omega_-^t &= \{x \in \Omega; \rho(t,x) < \rho_c(x)\} \quad \text{(subcritical region)} \\ \Omega_+^t &= \{x \in \Omega; \rho(t,x) > \rho_c(x)\} \quad \text{(supercritical region).} \end{aligned}$$

(Here, ρ is the solution to equation (4.118).)

A distinctive feature of the *self-organized criticality* process described by the semiflow $S(t)\rho_0 \equiv \rho(t)$ is that the *subcritical and supercritical regions are unstable and are absorbed in the finite time T by the critical region Ω_0^T*. This is one of the ways the system organizes itself by emerging in finite time toward a stable critical state.

More precisely, we have

Theorem 4.33 *Assume that $\rho_0, \rho_c \in L^\infty(\Omega)$ and that $\rho_0(x) \geq \rho_c(x)$, a.e. $x \in \Omega$. Then,*

$$\rho(t,x) = \rho_c(x), \quad a.e.\ x \in \Omega,\ \forall t \in T^*, \tag{4.122}$$

where

$$T^* = \frac{p^*}{p^*-2}\gamma^2 |\rho_0 - \rho_c|_\infty^{\frac{2}{p^*}} |\rho_0 - \rho_c|_1^{1-\frac{2}{p^*}}.$$

Here, $p^* = \frac{2N}{N-2}$ for $N \geq 3$; $p^* > 2$ for $N = 1, 2$, and

$$\gamma = \sup\{|u|_{p^*} \|u\|_{H_0^1(\Omega)}^{-1}\}.$$

Proof. By the substitution $y = \rho - \rho_c$, we reduce as above equation (4.118) to (4.119) and (4.122) to the extinction infinite time T^* of the solution y to (4.119).

It should be said, however, that here this property, though similar, is not under the incidence of Proposition 4.26 and so a separate treatment is necessary.

To begin with, we recall that the *mild solution* y to (4.119) is given by

$$y(t)=\lim_{\varepsilon\to 0} y_\varepsilon(t) \text{ strongly in } L^1(\Omega),\ \forall t\in[0,T],$$

where

$$\begin{aligned} &y_\varepsilon(t)=y_i^\varepsilon \text{ for } t\in[i\varepsilon,(i+1)\varepsilon),\ i=0,1,...,N,\ N=\left[\tfrac{T}{\varepsilon}\right],\\ &y_{i+1}^\varepsilon-\varepsilon\Delta\eta_{i+1}^\varepsilon=y_i^\varepsilon,\ \eta_{i+1}^\varepsilon\in H_0^1(\Omega),\ \eta_{i+1}^\varepsilon\in\widetilde{H}(y_{i+1}^\varepsilon),\ \forall i. \end{aligned} \tag{4.123}$$

Taking into account that $y_i^\varepsilon\ge 0$, a.e. in Ω, and that

$$\int_\Omega \eta_{i+1}^\varepsilon(y_{i+1}^\varepsilon-y_i^\varepsilon)dx\ge\int_\Omega(y_{i+1}^\varepsilon-y_i^\varepsilon)dx,\quad\forall i,$$

we get by (4.123) that

$$\int_\Omega y_{i+1}^\varepsilon dx+\varepsilon\int_\Omega|\nabla\eta_{i+1}^\varepsilon|^2dx\le\int_\Omega y_i^\varepsilon dx,\quad\forall i=0,1,... \tag{4.124}$$

On the other hand, by the Sobolev embedding theorem (see Theorem 1.11), we have $H_0^1(\Omega)\subset L^{p^*}(\Omega)$ for $1\le p^*\le\frac{2N}{N-2}$, $H_0^1(\Omega)\subset\bigcap_{p\ge 1}L^p(\Omega)$ for $N=2$, $H_0^1(\Omega)\subset L^\infty(\Omega)$ for $N=1$. This yields

$$\int_\Omega|\eta_{i+1}^\varepsilon|^{p^*}dx\le\gamma\int_\Omega|\nabla|\eta_{i+1}^\varepsilon|^2dx,\quad\forall i=0,1,...,$$

and so, by (4.124), we get

$$\int_\Omega y_{i+1}^\varepsilon dx+\varepsilon\gamma^{-2}\left(\int_\Omega|\eta_{i+1}^\varepsilon|^{p^*}dx\right)^{\frac{2}{p^*}}\le\int_\Omega y_i^\varepsilon dx,\ i=0,1,... \tag{4.125}$$

On the other hand, since $\eta_{i+1}^\varepsilon\in\widetilde{H}(y_{i+1}^\varepsilon)$, a.e. in Ω, we have

$$\left(\int_\Omega|\eta_{i+1}^\varepsilon|^{p^*}dx\right)^{\frac{2}{p^*}}\ge m\left(x\in\Omega;y_{i+1}^\varepsilon(x)>0\right)^{\frac{2}{p^*}},$$

where m is the Lebesgue measure. Taking into account that, as easily follows by (4.123),

$$0\le y_{i+1}^\varepsilon(x)\le|y_0|_\infty,\quad\forall i,$$

we get

$$|y_0|_\infty m\left(x\in\Omega;y_{i+1}^\varepsilon(x)>0\right)\ge\int_\Omega y_{i+1}^\varepsilon(x)dx,$$

and so, (4.125) yields

$$\int_\Omega y_{i+1}^\varepsilon dx+\varepsilon\gamma^{-2}\sum_{j=k}^{i+1}\int_\Omega|\eta_j^\varepsilon|^{p^*}dx\le\int_\Omega y_k^\varepsilon dx,\quad\forall i\ge k,$$

and, therefore,

$$\int_\Omega y_{i+1}^\varepsilon dx+\varepsilon|y_0|_\infty^{-\frac{2}{p^*}}\sum_{j=k}^{i+1}\left(\int_\Omega y_{j+1}dx\right)^{\frac{2}{p^*}}\le\int_\Omega y_k^\varepsilon dx,\ \forall k<i.$$

(Here, $|\cdot|_\infty=\|\cdot\|_{L^\infty(\Omega)}$.)

Letting $\varepsilon \to 0$, we get, for all $0 \le s \le t < \infty$,

$$\int_\Omega y(t,x)dx + \gamma^{-2}|y_0|_\infty^{-\frac{2}{p^*}} \int_s^t \left(\int_\Omega y(\tau,x)dx\right)^{\frac{2}{p^*}} d\tau \le \int_\Omega y(s,x)dx. \tag{4.126}$$

Now, we set

$$\varphi(t) = \int_\Omega y(t,x)dx, \quad t \ge 0,$$

and rewrite (4.126) as

$$\varphi(t) + \gamma^{-2}|y_0|_\infty^{-\frac{2}{p^*}} \int_0^t (\varphi(\tau))^{\frac{2}{p^*}} d\tau \le \varphi(s), \quad 0 < s \le t < \infty.$$

This yields

$$\varphi'(t) + \gamma^{-2}|y_0|_\infty^{-\frac{2}{p^*}} (\varphi(t))^{\frac{2}{p^*}} \le 0, \quad \text{a.e. } t > 0,$$

and, therefore,

$$\frac{p^*}{p^*-2}\,\frac{d}{dt}\,(\varphi(t))^{1-\frac{2}{p^*}} + \gamma^{-2}|y_0|_\infty^{-\frac{2}{p^*}} \le 0, \quad \text{a.e } t > 0.$$

Integrating on $(0,t)$, we get

$$(\varphi(t))^{1-\frac{2}{p^*}} + \frac{p^*-2}{p^*\gamma^2}\,|y_0|_\infty^{-\frac{2}{p^*}}\, t \le |y_0|_1^{1-\frac{2}{p^*}},$$

and so,

$$y(t) \equiv 0 \ \text{ for } t \ge T^* = |y_0|_1^{1-\frac{2}{p^*}} p^*\gamma^2 |y_0|_\infty^{\frac{2}{p^*}} (p^*-1)^{-1},$$

as claimed. □

Remark 4.34 The conclusion of Theorem 4.33 can be obtained formally by multiplying (4.120) by $\operatorname{sign} y$ and integrating on $(0,t) \times \Omega$. However, since as happens in other situations as well, the mild solution $y(t)$ to (4.120) is not differentiable in t, such an argument is not rigorous and so one should work as above with the finite difference scheme (4.123). (See also Remark 2.58.)

4.5 The Conservation Law Equation

We consider here the Cauchy problem

$$\begin{cases} \dfrac{\partial y}{\partial t} + \displaystyle\sum_{i=1}^N \frac{\partial}{\partial x_i} a_i(y) = 0 & \text{in } \mathbb{R}^N \times \mathbb{R}^+, \\ y(0,x) = y_0(x), & x \in \mathbb{R}^N, \end{cases} \tag{4.127}$$

where $a = (a_1, ..., a_N)$ is a continuous map from $\mathbb{R}$ to $\mathbb{R}^N$ satisfying the condition

$$\limsup_{|r|\to 0} \frac{\|a(r)\|}{|r|} < \infty,$$

and $y_0 \in L^1(\mathbb{R}^N)$. This equation models the movement of a fluid in $\mathbb{R}^N$ in the absence of sources. If the speed v is a function of density y, that is, $v = a(y)$, then by the continuity equation

$$\frac{\partial y}{\partial t} + \operatorname{div}(v) = 0$$

one gets (4.127). Perhaps the most famous example is the Hopf's equation

$$\frac{\partial y}{\partial t} + \frac{1}{2}\,\frac{\partial}{\partial y}\,y^2 = 0.$$

Though it is a hyperbolic type equation, it can be treated as a nonlinear Cauchy problem in the space $X = L^1(\mathbb{R}^N)$ by the same semigroups techniques as used for the treatment of parabolic equations in $L^1(\mathbb{R}^N)$. In fact, we have seen earlier in Theorem 3.26 that the first-order differential operator $y \to \sum_{i=1}^N \left(\frac{\partial}{\partial x_i}\right) a_i(y)$ admits an m-accretive extension $A \subset L^1(\mathbb{R}^N) \times L^1(\mathbb{R}^N)$ defined as the closure in $L^1(\mathbb{R}^N) \times L^1(\mathbb{R}^N)$ of the operator A_0 given by Definition 3.25. Then, by Theorem 2.52, the Cauchy problem

$$\begin{cases} \dfrac{dy}{dt} + Ay \ni 0 \quad \text{in } (0, +\infty), \\ y(0) = y_0, \end{cases}$$

has for every $y_0 \in \overline{D(A)}$ a unique mild solution $y(t) = S(t)y_0$ given by the exponential formula (2.112) or, equivalently (see (2.128)),

$$y(t) = \lim_{h\to 0} y_h(t) \quad \text{uniformly on compact intervals,}$$

where y_h is the solution to difference equation

$$\begin{aligned} h^{-1}(y_h(t) - y_h(t-h)) + Ay_h(t) = 0 \quad &\text{for } t > h, \\ y_h(t) = y_0 \quad &\text{for } t < 0. \end{aligned} \tag{4.128}$$

We call such a function $y(t) = S(t)y_0$ a *semigroup solution* or *mild solution* to the Cauchy problem (4.127).

We see in Theorem 4.35 below that this solution is in fact an entropy solution to the equation of conservation laws.

Theorem 4.35 *Let $y = S(t)y_0 \in C([0,T]; L^1(\mathbb{R}^N))$ be the semigroup solution to problem* (4.127)*. Then,*

(i) *$S(t)L^p(\mathbb{R}^N) \subset L^p(\mathbb{R}^N)$ for all $1 \le p < \infty$ and*

$$|S(t)y_0|_p \le |y_0|_p, \quad \forall y_0 \in \overline{D(A)} \cap L^p(\mathbb{R}^N). \tag{4.129}$$

(ii) *If $y_0 \in \overline{D(A)} \cap L^\infty(\mathbb{R}^N)$, then*

$$\begin{aligned} \int_0^T \int_{\mathbb{R}^N} &(|y(t,x) - k|\varphi_t(t,x) \\ &+ \operatorname{sign}_0(y(t,x) - k)(a(y(t,x)) - a(k)) \cdot \varphi_x(t,x))\,dx\,dt \ge 0 \end{aligned} \tag{4.130}$$

for every $\varphi \in C_0^\infty(\mathbb{R}^N \times (0,T))$ such that $\varphi \ge 0$, and all $k \in \mathbb{R}^N$ and $T > 0$.

Here $\varphi_t = \frac{\partial\varphi}{\partial t}$, $\varphi_x = \nabla_x \varphi$, and $|\cdot|_p$, $1 \le p \le \infty$, is the norm in $L^p(\Omega)$.

Inequality (4.130) is Kruzkhov's definition of entropy solution to the Cauchy problem (4.127) (see [75]) and its exact significance will be discussed later on.

Proof of Theorem 4.35. Because, as seen in the proof of Theorem 3.26, $(I+\lambda A)^{-1}$ maps $L^p(\mathbb{R}^N)$ into itself and

$$|(I+\lambda A)^{-1}u|_p \le |u|_p, \quad \forall \lambda > 0,\ u \in L^p(\mathbb{R}^N) \text{ for } 1 \le p \le \infty,$$

we deduce (i) by the exponential formula (2.112).

To prove inequality (4.129), consider the solution y to equation (4.128), where $y_0 \in L^1(\mathbb{R}^N) \cap L^\infty(\mathbb{R}^N)$ and $A_0 = A$. (Recall that $L^1(\mathbb{R}^N) \cap L^\infty(\mathbb{R}^N) \subset R(I + \lambda A)^{-1}$ for all $\lambda > 0$.) Then, $\|y_h(t)\|_{L^p(\mathbb{R}^N)} \le \|y_0\|_{L^p(\mathbb{R}^N)}$ for $p = 1, \infty$ and so, by Definition 3.25 and by (4.128), we have

$$\begin{aligned}\int_{\mathbb{R}^N} &(\text{sign}_0(y_h(t,x)-k)(a(y_h(t,x))-a(k)))\cdot\varphi_x(t,x)\\ &+h(y_\varepsilon(x,t-h)-y_h(t,x))\,\text{sign}_0(y_h(t,x)-k)\varphi(t,x))dx \ge 0,\\ &\forall k\in\mathbb{R},\ \varphi\in C_0^\infty(\mathbb{R}^N\times(0,T)),\ \varphi\ge 0,\ t\in(0,T).\end{aligned} \tag{4.131}$$

On the other hand, we have

$$\begin{aligned}&(y_h(x,t-h)-y_h(t,x))\text{sign}_0(y_h(t,x)-k)\\ &\quad= (y_h(x,t-h)-k)\text{sign}_0(y_h(t,x)-k)-(y_h(t,x)-k)\text{sign}_0(y_h(t,x)-k)\\ &\quad\le z_h(x,t-h)-z_h(t,x),\end{aligned}$$

where $z_h(t,x) = |y_h(t,x)-k|$.

Substituting the latter into (4.131) and integrating on $\mathbb{R}^N \times [0,T]$, we get

$$\begin{aligned}\int_0^T\int_{\mathbb{R}^N} &(\text{sign}_0(y_h(t,x)-k)(a(y_h(t,x))-a(k))\cdot\varphi_x(t,x)\\ &+h^{-1}(z_h(x,t-h)-z_h(t,x))\varphi(t,x))dx\,dt \ge 0.\end{aligned}$$

This yields

$$\begin{aligned}\int_0^T\int_{\mathbb{R}^N} &(\text{sign}_0(y_h(t,x)-k)(a(y_e(t,x))-a(k))\cdot\varphi_x(t,x))dx\,dt\\ &-h^{-1}\int_0^h\int_{\mathbb{R}^N}|y_h(t,x)-k|\varphi(t,x)dxdt+h^{-1}\int_0^T\int_{\mathbb{R}^N}z_h(t,x)\varphi(t,x)dxdt\\ &+h^{-1}\int_{T-h}^T\int_{\mathbb{R}^N}z_h(t,x)(\varphi(x,t+h)-\varphi(t,x))dxdt \ge 0.\end{aligned}$$

Now, letting h tend to zero, we get (4.130) because $y_h(t) \to y(t)$ uniformly on $[0,T]$ in $L^1(\mathbb{R}^N)$ and $h^{-1}(z_h(t-h,x)-z_h(t,x)) \to |y(t,x)-k|$. This completes the proof of Theorem 4.30. $\square$

We note that, if y is an entropy solution to (4.116), then it also satisfies (4.116) in the sense of distributions, that is,

$$\int_0^T\int_{\mathbb{R}^N}\left(y(t,x)\frac{\partial\varphi}{\partial t}(t,x)+a(y(t,x))\cdot\nabla_x\varphi(t,x)\right)dx\,dt = 0,\ \forall\varphi\in C_0^\infty((0,T)\times\mathbb{R}^N),$$

while the initial datum $y(0,\cdot) = y_0$ is taken in the sense of $L^1(\mathbb{R}^N)$, that is,

$$\lim_{t\to 0} \int_{\mathbb{R}^N} |y(t,x) - y_0(x)|dx = 0.$$

Such a function y can be viewed as a generalized (or distributional) solution to equation (4.116), but one main limitation of this notion (besides nonuniqueness) is that it does not include a condition for the entropy growth in discontinuity points which characterizes the irreversibility of the process. It should be said that Kruzkhov's entropy solution defined above includes the general concept of entropy to the present case, a function $\eta : \mathbb{R} \to \mathbb{R}$ is called an *entropy* of system (4.127) if there is a function $q : \mathbb{R} \to \mathbb{R}^N$ (the *entropy flux* associated with entropy η) such that

$$\nabla q_j(y) = \nabla \eta(y) \cdot \nabla a_j(y), \quad \forall y \in \mathbb{R}^N,\ j = 1, ..., N.$$

Such a pair (η, q) is called an *entropy pair.*

The bounded measurable function $y : [0,T] \times \mathbb{R}^N \to \mathbb{R}$ is called an *entropy solution* to (4.127) if, for all convex entropy pairs (η, q),

$$\frac{\partial}{\partial t}\,\eta(y(t,x)) + \operatorname{div} q(y(t,x)) \le 0 \quad \text{in } \mathcal{D}'((0,T) \times \mathbb{R}^N), \tag{4.132}$$

that is,

$$\int_0^T \int_{\mathbb{R}^N} (\eta(y(t,x))\varphi_t(t,x) + q(y(t,x)) \cdot \varphi_x(t,x))dtdx \ge 0$$

for all $\varphi \in C_0^\infty((0,T) \times \mathbb{R}^N)$, $\varphi \ge 0$.

If take $\eta(y) \equiv |y - k|$ and $q(y) \equiv \operatorname{sign}_0(y-k)(a(y) - a(k))$, we see that y satisfies equation (4.130). The existence and uniqueness of the entropy solution were proven by S. Kruzkhov [75].

Recalling that the resolvent $(I + \lambda A)^{-1}$ of the operator A can be approximated by the family of approximating equation (3.149), one might deduce via the Trotter–Kato Theorem 2.76 that the entropy solution y can also be obtained as the limit for $\varepsilon \to 0$ to solutions y_ε to the parabolic nonlinear equation

$$\frac{\partial y}{\partial t} - \varepsilon\Delta y + (a(y))_x = 0, \quad t \ge 0,\ x \in \mathbb{R}^N,$$

which is related to the vanishing viscosity solution approach discussed in Section 3.3. (See Lemma 3.28.) For this reason, the solution $u = S(t)u_0$ is also called the viscosity solution to (4.127).

4.6 The Fokker–Planck Equation

We shall study here the nonlinear parabolic equation

$$\begin{aligned} &\frac{\partial y}{\partial t} - \Delta\beta(y) + \operatorname{div}(D(x)b(y)y) = 0, \text{ in } (0,\infty) \times \mathbb{R}^N, \\ &y(0,x) = y_0(x), \qquad\qquad\qquad\qquad x \in \mathbb{R}^N, \end{aligned} \tag{4.133}$$

where $N \geq 1$ and the functions $\beta : \mathbb{R} \to \mathbb{R}$, $D : \mathbb{R}^N \to \mathbb{R}^N$, $b : \mathbb{R} \to \mathbb{R}$ satisfy hypotheses (i)–(iv) of Theorem 3.31.

This equation, called in literature the Fokker–Planck (Kolmogorov) equation, is relevant in statistical mechanics as well as in the mean field theory where describes the dynamic of a set of interacting particles in nonequilibrium many body systems (anomalous diffusion) (see, e.g., [73]). In physical models, the solution y to the Fokker–Planck equation (4.133) is a probability density, that is, $y(t) \in \mathcal{P}_0(\mathbb{R}^N)$, $\forall t \geq 0$, where

$$\mathcal{P}_0(\mathbb{R}^N) = \left\{ u \in L^1(\mathbb{R}^N),\ u \geq 0, \text{ a.e. in } \mathbb{R}^N,\ \int_{\mathbb{R}^N} u(x)dx = 1 \right\}.$$

In fact, it turns out (see, e.g., [22,23]) that the solution y to (4.133) is the probability density of the law $\mathcal{L}_{X(t)}$ of the solution X to the stochastic differential equation (McKean–Vlasov equation)

$$dX(t) = D(X(t))b(y(t,X(t)))dt + \frac{1}{\sqrt{2}} \left(\frac{\beta(y(t,X(t)))}{y(t,X(t))} \right)^{\frac{1}{2}} dW(t), \tag{4.134}$$

where $W(t)$ is an N-dimensional Brownian motion in a probability space $\{\Omega, \mathcal{F}, \mathbb{P}\}$. Roughly speaking, this means that, for each Borelian set $B \subset \mathbb{R}^N$,

$$\mathbb{P}[X(t) \in B] = \int_B y(t,x)dx, \quad \forall t \geq 0.$$

We note that, in the special case $\beta(y) \equiv y$ and $b \equiv 1$, equation (4.134) reduces to the classical Einstein–Langevin equation, $dX = D(X)dt + dW$ which was firstly introduced by A. Einstein in the case $D \equiv 0$ to explain the Brownian motion via the kinetic theory. Equation (4.133) for $b(y) \equiv y$ is also known as the mean field Smoluchowski equation and models the dynamic of density of Brownian particles $y = y(t,x)$ in a field of forces, where the drift D describes the effect of the fluctuation of the particle. This allows the reconstruction of stochastics dynamics $X(t)$ from a deterministic parabolic equation of the form (4.132). Of course, this does not give as much information as equation (4.133) because it does not provide the sample path of the stochastic process $X(t)$ but it is, however, useful to model the evolution of its probability density ρ to avoid tedious simulation of a stochastic trajectory. Other significant examples are $\beta(r) \equiv a(\operatorname{sign} r)\log(1+|r|)$ and $\beta(r) \equiv a|r|^{m-1}r$ which lead to the boson equation and to the Plastino model, respectively.

The equivalence of (4.133) and (4.134) taken in a weak probabilistic sense is true for more general equations and it is known in literature of the *superposition* principle (see, e.g., [22]). In the case $b \equiv 0$, (4.132) reduces to the porous media equation (4.55).

Another formal way to derive the nonlinear Fokker–Planck equation (4.133) is to start from an generalized entropy functional

$$S^*[p] = \int_{\mathbb{R}^N} \widetilde{S}(p(x))dx,$$

where p is a probability density and $\widetilde{S} \in C((0,\infty))$ is a strictly concave function such that $\lim_{z\to 0} \widetilde{S}'(z) = \infty$. Then the corresponding Fokker–Planck equation is given by virtue of the so called canonical ensemble principle as (see [72])

$$\frac{\partial p}{\partial t}(t,x) = -\mathrm{div}_x(g(p(t,x))(h(x)p(t,x)) - \nabla_x L(\widetilde{S}(p(t,x))),$$

where $\widetilde{L}(u) \equiv F(u) - xu'(x)$, $\forall u \in C^\infty(\mathbb{R})$.

For instance, in our case, the functional S is defined by

$$S^*[u] = -\int_{\mathbb{R}^N} \eta(u(x))dx - 1, \quad \eta(r) = -\int_0^r d\tau \int_\tau^1 \frac{\beta'(s)}{sb(s)}\,ds \tag{4.135}$$

which, in the special case $\beta(y) \equiv y$, $b \equiv 1$, $S([u])$ reduces to the classical Boltzman–Gibbs–Shannon entropy,

$$S^*[u] = -\int_{\mathbb{R}^N} u(x)\log u(x)dx, \quad \forall\, u \in \mathcal{P}_0(\mathbb{R}^N).$$

We shall see later on that S^* is, in fact, an entropy function in the sense of (4.132). The Fokker–Planck equation can also be viewed as an open or a transient system far from the thermodynamic equilibrium and can be represented as well as the continuity equation

$$\frac{\partial \rho}{\partial t} = \mathrm{div}\, J,$$

where $J(y) \equiv \nabla\beta(y) - Db(y)y)$ is the so called the *probability current.*

As regards the existence for equation (4.133) in agreement with Definition 2.47, we shall look here for a *mild solution* corresponding to the operator A defined in Section 3.4. Such a solution will be also called *generalized solution.*

Definition 4.36 A continuous function $u : [0,\infty] \to L^1$ is said to be a *generalized solution* (or *mild solution*) to equation (4.133) on $(0,\infty)$ if, for each $T > 0$,

$$y(t) = \lim_{h\to 0} y_h(t) \text{ in } L^1 \text{ uniformly on each interval } [0,T],$$

where $y_h : (0,T] \to L^1$ is the step function defined by

$$\begin{aligned} &y_h(t) = y_h^i,\ \forall t \in (ih,(i+1)h],\ i = 0,1,...,n-1,\\ &y_h^0 = y_0,\ \beta(y_h^i) \in L^1_{\mathrm{loc}},\ y_h^i \in L^1,\ \forall i = 1,2,...,n,\ nh = T,\\ &y_h^{i+1} - h\Delta(\beta(y_h^{i+1})) + h\,\mathrm{div}(Db(y_h^{i+1})y_h^{i+1}) = y_h^i,\ \text{in } \mathcal{D}'(\mathbb{R}^N), \end{aligned} \tag{4.136}$$

for all $i = 0,1,....,n-1$.

The main existence result is

Theorem 4.37 *Under hypotheses* (i)–(iv) *of Theorem* 3.31, *for each* $u_0 \in L^1(\mathbb{R}^N)$, *there is a unique generalized solution* $u = u(t,u_0)$ *to equation* (4.133).

If $y_0 \in \mathcal{P}_0(\mathbb{R}^N)$, *then*

$$y(t) \in \mathcal{P}_0(\mathbb{R}^N),\ \forall t \geq 0. \tag{4.137}$$

Moreover, $t \to S(t)y_0 = y(t,y_0)$ *is a semigroup of nonlinear contractions on* L^1.

If $y_0 \in L^1 \cap L^\infty$, then $y \in L^\infty((0,T) \times \mathbb{R}^N)$ and it is a solution to (4.133) in the sense of Schwartz distributions on $(0,\infty) \times \mathbb{R}^N$, that is,

$$\int_0^\infty \int_{\mathbb{R}^N} (y(\varphi_t + b(y)D \cdot \nabla\varphi) + \beta(y)\Delta\varphi)dt\,dx + \int_0^\infty y_0(x)\varphi(0,x)dx = 0, \quad \forall \varphi \in C_0^\infty([0,\infty) \times \mathbb{R}^N). \tag{4.138}$$

Proof. We write (4.133) as the Cauchy problem in the space $L^1(\mathbb{R}^N)$

$$\begin{aligned} &\frac{dy}{dt} + Ay = 0,\ t \in [0,T], \\ &y(0) = y_0, \end{aligned} \tag{4.139}$$

where $A : D(A) \subset L^1(\mathbb{R}^N) \to L^1(\mathbb{R}^N)$ is the m-accretive operator defined by (3.178) (see Theorem 3.32). Then, by Theorem 2.52, for each $y_0 \in \overline{D(A)} = L^1(\mathbb{R}^N)$ and each $T > 0$, there is a unique mild solution $y \in C([0,T]; L^1(\mathbb{R}^N))$ to (4.133), that is, $y(t) = \lim_{h\to 0} y_h(t)$ in $L^1(\mathbb{R}^N)$, where y_h is given as in Definition 4.36, that is (see (2.128)),

$$\begin{aligned} &y_h(t) = y_h^1, \quad \text{for } t \in [ih,(i+1)h),\ i = 1,...,N-1, \\ &y_h^{i+1} + hAy_h^{i+1} = y_h^i,\ i = 0,1,...,N,\ Nh = T, \\ &y_h^0 = y_0. \end{aligned} \tag{4.140}$$

Taking into account the definition of A and that $(I + \lambda A)^{-1}f \in (I + \lambda A_0)^{-1}f$, $\forall \lambda > 0$, $f \in L^1$ (see (3.179)), we infer that the mild solution y to the Cauchy problem (4.139) is just the generalized solution y arising in Definition 4.36. Invoking Remark 3.18 and the definition of the operator A, one might also call y a *viscosity solution* to (4.133).

As regards (4.137), it follows by (4.140) and (3.171), (3.172). It is also easily seen by (4.140) that the mild solution y is a distributional solution, that is, (4.138) holds. In fact, equation (4.136) written as (we take $y_h(s,x) \equiv y_0(x)$ for $s \in (-h,0]$)

$$\int_h^\infty \int_{\mathbb{R}^N} \frac{1}{h}(y_h(t,x) - y_h(t-h,x))\varphi(t,x) + \text{div}(Db(y_h(t,x))y_h(t,x))\varphi(t,x))dx\,dt - \int_h^\infty \Delta\beta(y_h(t))(\varphi)dt = 0,\ \forall \varphi \in C([0,\infty) \times \mathbb{R}^N),$$

leads to

$$-\int_0^\infty \int_{\mathbb{R}^N} \Big(\frac{1}{h} y_h(t,x)(\varphi(t+h,x) - \varphi(t,x) + \beta(y_h(t,x))\Delta\varphi(t,x) + b(y_h(t,x))y_h(t,x)D \cdot \nabla\varphi(t,x))\Big)dx\,dt - \int_0^h \int_{\mathbb{R}^N} y_0(x)\varphi(t,x)dx\,dt,$$

and so, for $h \to 0$, we get (4.138). □

To resume, Theorem 4.37 amounts to saying that under the above hypotheses there is a nonlinear semiflow (semigroup) $S(t)$ defined by equation (4.132) which is called the *Fokker–Planck flow*. One should emphasize that it is not differentiable in t but satisfies the equation in the sense of the finite difference scheme (4.136) and it is also a distributional solution to this equation. Though $y(t) = S(t)y_0 : [0,\infty) \to L^1(\mathbb{R}^N)$ is not differentiable, by Proposition 2.66 it follows that, for each $y_0 \in \hat{D}(A) \supset D(A)$ the function $t \to y(t)$ is Lipschitz on bounded intervals.

As mentioned earlier, the uniqueness of a generalized solution y given by Theorem 4.37 is relative to the m-accretive section A of the operator A_0 and refers strictly to the class of functions $y \in C([0,T];L^1)$ defined by (4.140). In principle, for two different m-accretive sections A might correspond two generalized (mild) solutions y. However, if $(I+\lambda A_0)^{-1}$ is single valued, then $A = A_0$ and so one has the global uniqueness given by (4.136).

Taking into account that every generalized (mild) solution is a distributional solution, that is one for which (4.138) holds, the uniqueness of distributional solutions also implies the global uniqueness of mild solutions. This is a delicate problem and so for only a partial result is known (see the work [26]). Based on this result, it follows the uniqueness under our hypotheses in the class of functions $y \in L^\infty([0,T]\times\mathbb{R}^d)\cap C([0,T];L^1)$ and so, in particular, for initial data $y_0 \in L^1\cap L^\infty$.

Remark 4.38 Contrary to the porous media equations with the nondegenerate diffusion term β, the Fokker–Planck equation (4.133) has a nonstationary (equilibrium) solution. As happens in many other physical situations, the solution $y = y(t,x)$ to (4.133) is, in fact, a transient state which links the initial nonequilibrium state y_0 with a final equilibrium state and the convergence to equilibrium state is called in literature the H-theorem. In the work [25], it is proven the H-theorem for (4.133) under additional assumptions on β, D and b. More precisely, it is shown that, if $D = -\nabla\Phi$ (this means that the system is conservative) and if β is nondegenerate, that is, $0 < \gamma_1 \le \beta' \le \gamma_2 < \infty$, $b \ge b_0 > 0$, while the potential Φ satisfies a certain concavity condition, then, for $t \to \infty$,

$$\lim_{t\to\infty} y(t) = y_\infty \quad \text{in } L^1(\mathbb{R}^N),$$

where $y_\infty \in L^1(\mathbb{R}^N)$ is the solution to the stationary Fokker–Planck equation

$$-\Delta\beta(y_\infty) + \operatorname{div}(Db(y_\infty)y_\infty) = 0, \text{ in } \mathcal{D}'(\mathbb{R}^N), \quad y_\infty \in \mathcal{P}_0.$$

In the special case of classical Fokker–Planck equations, that is, $\beta(r) \equiv r$, $b \equiv 1$, the equilibrium solution y_∞ is just the Gibbs distribution $\exp(-\Phi)\|\exp(-\Phi)\|_{L^1}^{-1}$.

It should be emphasized that $L^1(\mathbb{R}^N)$ is the natural space for a Fokker–Planck flow $S(t)$ and this not only for the physical significance of the solution $y(t)$ to (4.133) as a probability distribution of particles, but also because only in this space it is a semigroup of contractions $S(t)$, and so evolves as is natural as a dissipative flow for $t \to \infty$. This longtime behavior is specific to nonlinear diffusion dynamic systems and happens in $L^1(\mathbb{R}^N)$ only because in this space only the generator of the flow $S(t)$ is dissipative. This is the main distinctive feature compared with the porous media equation which, as seen earlier, is also well posed in the Sobolev space $H^{-1}(\mathbb{R}^N)$ too. However, under additional conditions on b, one could show (see the proof of Theorem 3.32) that the operator $Ay = -\Delta\beta(y) + \text{div}(Db(y)y)$ with the domain $\{y \in L^2(\mathbb{R}^N), \beta(y) \in H^1(\mathbb{R}^N)\}$ is quasi m-accretive in $H^{-1}(\mathbb{R}^N)$ and so, one could show that there is a unique solution $y \in C([0,T]; H^{-1}(\mathbb{R}^N))$ to (4.133).

If $\mathbb{R}^N$ is replaced by a subdomain Ω of $\mathbb{R}^N$, the treatment of equation (4.133) is similar but it should be said, however, that equation (4.133) on a bounded open domain $\Omega \subset \mathbb{R}^d$ with usual Dirichlet boundary conditions cannot be interpreted as a Fokker–Planck equation associated with a certain stochastic process and so it is seen as a porous media equation perturbed by the nonlinear drift term $y \to \text{div}(Db(y)y)$.

Remark 4.39 Taking into account that, as seen in the proof of Theorem 3.31 (see (3.198)),

$$(I + \lambda A)^{-1} f = \lim_{\varepsilon \to 0} (I + \lambda A_\varepsilon)^{-1} f, \ \forall f \in L^1(\mathbb{R}^N),$$

where $A_\varepsilon u = -\Delta(\beta(u) + \varepsilon u) + \varepsilon\beta(u) + \text{div}(Db_\varepsilon(u_\varepsilon)u_\varepsilon)$ with the domain $u \in H^1(\mathbb{R}^N) \cap L^1(\mathbb{R}^N)$, we derive by the Trotter–Kato theorem for nonlinear semigroups (Theorem 2.76) that the solution y to (4.133) is given by

$$y = \lim_{\varepsilon \to 0} y_\varepsilon \ \text{ in } (C[0,T]; L^1(\mathbb{R}^N)), \tag{4.141}$$

where y_ε is the solution to the regular equation

$$\begin{aligned} &(y_\varepsilon)_t - \Delta(\beta_\varepsilon(y_\varepsilon) + \varepsilon y_\varepsilon) + \varepsilon\beta(y_\varepsilon) + \text{div}(Db_\varepsilon(y_\varepsilon)y_\varepsilon) = 0 \text{ in } \mathcal{D}'(\mathbb{R}^N), \\ &y_\varepsilon(0) = y_0. \end{aligned} \tag{4.142}$$

This result shows that the generalized solution y given by Theorem 4.37 can be viewed as a *viscosity solution* to equation (4.133).

The entropy function. Assume that $b \geq 0$ on $[0,\infty)$. If $\eta \equiv \eta(r)$ is the function arising in (4.135) and y is the solution to (4.133), then we have

$$\begin{aligned} &\int_{\mathbb{R}^N} \frac{d}{dt}\eta(y(t,x))dx \\ &= -\int_{\mathbb{R}^N} \int_0^{y(t,x)} \frac{\beta'(s)}{sb(s)}\, ds(\Delta\beta(y(t,x)) - \text{div}(D(x)b(y(t,x))(t,x)))dx \\ &= -\int_{\mathbb{R}^N} \frac{\beta'(y(t,x))}{y(t,x)b(y(t,x))}\, \nabla y(t,x) \cdot (\beta'(y(t,x))\nabla y(t,x) - D(x)b(y(t,x))y(t,x))dx \\ &\leq \int_{\mathbb{R}^N} D(x) \cdot \nabla\beta(y(t,x))dx. \end{aligned}$$

(The above computation becomes rigorous and extends to all the generalized solutions $\mathbb{R}^N$ if we use the approximation equation (4.142).) This yields

$$\frac{\partial}{\partial t}\, \eta(y) + \text{div}(D\beta(y)) \leq 0 \ \text{ in } \mathcal{D}'(0,\infty) \times \mathbb{R}^d)$$

and hence η is an entropy function for system (4.133) with the flux $q \equiv D\beta(y)$ (see (4.132)).

Remark 4.40 Theorem 4.37 extends, by a similar method, to more general nonlinear Fokker–Planck equations of the form

$$\frac{\partial y}{\partial t} - \sum_{i,j=1}^{N} (a_{ij}(x,y)y)_{x_i x_j} + \operatorname{div}(b(x,y)y) = 0 \text{ in } (0,\infty)\times\mathbb{R}^d,$$
$$y(0,x) = y_0(x),$$

under appropriate regularity assumptions on a_{ij} and $b = (b_j)_{j=1}^d$ and the ellipticity hypothesis

$$\sum_{i,j=1}^{N} (a_{ij}(x,y) + (a_{ij}(x,y)))_y u\xi_i\xi_j \geq \gamma(\xi)^2, \quad \forall \xi \in \mathbb{R}^N,\ y \in \mathbb{R}.$$

We refer to [22] (see also [23]) for the treatment in this general case.

4.7 A Fokker–Planck Like Parabolic Equation in $\mathbb{R}^N$

Consider here the equation

$$\begin{aligned} &\frac{\partial y}{\partial t} - \sum_{i,j=1}^{N} a_{ij} D_{ij}^2(\beta(y)) + \operatorname{div}(Db(y)y) + F(y) = 0 \text{ in } (0,\infty)\times\mathbb{R}^N, \\ &y(0,x) = y_0(x),\ x \in \mathbb{R}^N, \end{aligned} \tag{4.143}$$

where the functions $\beta : \mathbb{R} \to \mathbb{R}$, $D : \mathbb{R}^N \to \mathbb{R}^N$ and $b : \mathbb{R} \to \mathbb{R}$ satisfy hypotheses (i)–(iii) of Theorem 3.31,

$$\sum_{i,j=1}^{N} a_{ij}\xi_i\xi_j \geq \alpha|\xi|^2,\ \forall \xi \in \mathbb{R}^d, \quad a_{ij} \equiv a_{ji},\ \forall i,j = 1,...,N, \tag{4.144}$$

for some $\alpha > 0$ and $F : L^1(\mathbb{R}^N) \to L^1(\mathbb{R}^N)$ is a continuous and quasi-accretive operator, that is

$$|u_1 - u_2 + \lambda(F(u_1) - F(u_2))|_1 \geq (1 - \lambda\omega)|u_1 - u_2|_1,\ \forall u_1, u_2 \in L^1,\ \lambda \in (0, \omega^{-1}),$$

for some $\omega > 0$ (see (2.71)). Here $|\cdot|_1$ is the norm of $L^1 = L^1(\mathbb{R}^N)$.

The existence of a mild solution to (4.143) follows exactly as in the case of Theorem 4.37 by writing it as a Cauchy problem in $L^1(\mathbb{R}^N)$,

$$\frac{dy}{dt} + Ay + Fy = 0,\ t \geq 0; \quad y(0) = y_0, \tag{4.145}$$

where $A : D(A) \subset L^1(\mathbb{R}^N) \to L^1(\mathbb{R}^N)$ is the m-accretive operator defined as Section 3.4, but with $A_0 y = -\sum_{i,j=1}^{N} a_{ij} D_{ij}^2(\beta(y)) + \operatorname{div}(Db(y)y) + F(y)$ instead of $-\Delta\beta(y) + \operatorname{div}(Db(y)y)$ (see Remark 3.33). Then, by Theorem 2.42, the operator $A + F$ is quasi-m-accretive in $L^1(\mathbb{R}^N)$ and so the Cauchy problem (4.145) has a unique mild solution $y \in C([0,\infty); L^1(\mathbb{R}^N))$. We have, therefore,

Theorem 4.41 *Under the above assumptions, equation* (4.143) *has a unique mild solution* $y \in C([0,T]; L^1(\mathbb{R}^N))$. *Moreover, the semiflow* $y_0 \to S(t)y_0 = y(t)$ *is a quasi-contractive continuous semigroup on* $L^1(\mathbb{R}^N)$.

A special case is that where F is the Nemytskii operator

$$(F(y))(x) = f(x, y(x)), \quad \text{a.e. } x \in \mathbb{R}^N, \ y \in L^1(\mathbb{R}^N),$$

where $f : \mathbb{R}^N \times \mathbb{R} \to \mathbb{R}$ is continuous and quasi-monotone in y, Lebesgue measurable in x and satisfies

$$|f(x,r)| \leq C|r| + g(x), \quad \text{a.e. } x \in \mathbb{R}^N, \ r \in \mathbb{R}^N, \ g \in L^1(\mathbb{R}^N).$$

Theorem 4.41 applies as well to the case where $F : L^1(\mathbb{R}^N) \to L^1(\mathbb{R}^N)$ is Lipschitz and, in particular, is an integral operator of the form

$$F(y)(x) = \int_{\mathbb{R}^N} K(x, \bar{x}) g(y) dx, \quad \forall\, y \in L^1(\mathbb{R}^N),$$

where $K \in L^\infty(\mathbb{R}^N \times \mathbb{R}^N)$ and $g \in \mathrm{Lip}(\mathbb{R})$.

It should be emphasized that (4.143) is not, however, a Fokker–Planck equation in the classical sense because the semiflow $S(t) : L^1(\mathbb{R}^N) \to L^1(\mathbb{R}^N)$ does not conserve the mass and so $y(t) = S(t)y_0$ is not a probability density except the case where $F \equiv 0$. However, this equation is relevant in other situations as well, and one of them will be briefly described below.

The stochastic dynamic programming equation

For a stochastic optimal control problem, the dynamic programming equation is usually expressed as a nonlinear parabolic equation (the Kolmogorov equation) which, in general, is well posed in a generalized sense (viscosity solution) only. (See, e.g., [70, 71].) However, as seen below, one can develop a simple mechanism to reduce such an equation to a Fokker–Planck equation of the form (4.143). We shall illustrate here this procedure on a particular example of a stochastic optimal control problem arising in dynamics of stock price $X(t)$ on $(0,T)$. (See [18, 19].)

Namely, consider the problem

$$\text{(P)} \ \textit{Minimize} \left\{ \mathbb{E} \int_0^T \left(g(X(t)) + \frac{1}{2}\, |u(t)|^2\right) dt + \mathbb{E} g_0(X(T)) \right\}$$

subject to $u \in \mathcal{U}$ *and to the stochastic differential equation*

$$dX = f(X)dt + \sqrt{u}\,\sigma(X)dW, \ t \in (0,T),$$
$$X(0) = X_0.$$

Here

$$\sigma(X)dW = \left\{ \sum_{j=1}^m \sigma_{ij} d\beta_j \right\}_{i=1}^d, \quad X = \{X_j\}_{j=1}^d,$$

where $\{\beta_j\}_{j=1}^m$ is an independent system of Brownian motions in a probability space $(\Omega, \mathcal{F}, \mathbb{P})$, and $\mathcal{U}$ is the class of $(\mathcal{F}_t)_{t\geq 0}$-adapted processes $u : [0,T] \times \Omega \to [0, +\infty)$. Here, $(\mathcal{F}_t)_{t\geq 0}$ is the natural filtration generated by $\{\beta_j\}_{j=1}^m$.

We assume the following hypotheses which though are not optimal, simplify the treatment of the problem and provide a better insight on the present approach.

(j) $g, g_0 \in C^2(\mathbb{R}^N)$, $D^2_{ij}g, D^2_{ij}g_0 \in L^1(\mathbb{R}^N)$, $\forall i,j = 1,...,N$, $f \in C^2(\mathbb{R}^N)$, *support* f *is compact.*

(jj) $\sum\limits_{i,j=1}^N a_{ij}\xi_i\xi_j \geq \alpha_2|\xi|_N^2$, $\forall \xi \in \mathbb{R}^N$, *where* $\alpha_2 > 0$, $a_{ij} = \sum\limits_{k=1}^N \sigma_{ik}\sigma_{jk}$, $\forall i,j{=}1,...,N$.

Consider the function $h : \mathbb{R} \to]-\infty, +\infty]$

$$h(u) = \begin{cases} \frac{1}{2}|u|^2 & \text{if } u \geq 0, \\ +\infty & \text{otherwise,} \end{cases}$$

and denote by h^* its conjugate, that is,

$$h^*(p) = \sup\{pu - h(u);\ u \in \mathbb{R}\} = \frac{1}{2}(p^+)^2.$$

The dynamic programming equation corresponding to problem (P) reads as (see, e.g., [71])

$$\begin{aligned} &\varphi_t(t,x) + \min_u \left\{ \frac{u}{2} \sum_{i,j=1}^N a_{ij}\varphi_{x_ix_j}(t,x) + h(u) \right\} + f(x)\cdot\varphi_x(t,x) + g(x) = 0, \\ &\hspace{22em} \forall t \in [0,T],\ x \in \mathbb{R}^N, \\ &\varphi(T,x) = g_0(x),\ x \in \mathbb{R}^N. \end{aligned}$$

Equivalently,

$$\begin{aligned} &\varphi_t(t,x) - h^*\left(-\frac{1}{2} \sum_{i,j=1}^N a_{ij}\varphi_{x_ix_j}(t,x) \right) + f(x)\cdot\varphi_x(t,x) + g(x) = 0, \\ &\hspace{22em} \forall t \in [0,T],\ x \in \mathbb{R}^N, \\ &\varphi(T,x) = g_0(x),\ x \in \mathbb{R}^N. \end{aligned} \tag{4.146}$$

If φ is a regular solution to (4.146), then the feedback controller

$$u(t) = \arg\min_u \left\{ \frac{u}{2} \sum_{i,j=1}^N a_{ij}(X(t))\varphi_{x_ix_j}(t,X(t)) + h(u) \right\} = \partial h^*(-L(D)\varphi) \tag{4.147}$$

is optimal in problem (P). Here, $\partial h^*(r) \equiv r^+$ is the subdifferential of h^* and

$$\varphi_t = \frac{\partial}{\partial t}\varphi,\ \varphi_{x_i} = \frac{\partial\varphi}{\partial x_i},\ i = 1,...,N,$$

$$\varphi_{x_ix_j} = D^2_{ij}\varphi = \frac{\partial^2\varphi}{\partial x_i \partial x_j},\ x = \{x_j\}_{j=1}^N,$$

$$L(D)\varphi = \frac{1}{2}\sum_{i,j=1}^N a_{ij}\varphi_{x_ix_j},\ \nabla\varphi = \varphi_x = \{\varphi_{x_i}\}_{i=1}^N,$$

are taken in the sense of the Schwartz distributions, i.e., in $\mathcal{D}'((0,\infty)\times\mathbb{R}^N)$.

By the transformation

$$y(t,x) = -L(D)(\varphi(T-t,x)), \tag{4.148}$$

we reduce (4.146) to

$$\begin{aligned} &y_t(t,x) - \frac{1}{2}L(D)(y^+(t,x))^2 + L(D)(f(x)\cdot\varphi_x(T-t,x)) + L(D)(g) = 0, \\ &\qquad\qquad\qquad\qquad\qquad\qquad\qquad\qquad\qquad \text{in } (0,T)\in\mathbb{R}^N, \\ &y(0,x) = L(D)g_0)(x), \quad x\in\mathbb{R}^N. \end{aligned} \tag{4.149}$$

On the other hand, we have

$$L(D)(f\cdot\varphi_x(T-t)) = L(D)(f)\varphi_x(T-t) - f\cdot y_x(t) + \sum_{j=1}^N \widetilde{\nabla} f_j \cdot \widetilde{\nabla}\varphi_{x_j}(T-t),$$

where $\widetilde{\nabla} y = \left\{\sum_{j=1}^N a_{ij}(y)_{x_j}\right\}_{i=1}^N$ and $f = \{f_j\}_{j=1}^N$.

For each $\psi\in L^1(\mathbb{R}^N)$, we set $L^{-1}(D)\psi = z$, where $L(D)z = \psi$ in $\mathcal{D}'(\mathbb{R}^N)$. By Proposition 1.28, we have for $N\geq 3$

$$\|\nabla(L^{-1}(D)(\psi)))\|_{M^{\frac{N}{N-1}}} \leq C\|\psi\|_{L^1(\mathbb{R}^N)}, \quad \forall\psi\in L^1(\mathbb{R}^N).$$

Then, by (4.148) and assumption (jj), we get

$$|L(D)(f)_x|_1 \leq C|y|_1$$

$$\left|\sum_{j=1}^N \widetilde{\nabla} f_j\cdot\nabla\varphi_{x_j}\right|_1 \leq C|y|_1$$

where $|\cdot|_1$ is usually the norm in $L^1(\mathbb{R}^N)$. Then, (4.149) yields

$$\begin{aligned} &y_t - \frac{1}{2}L(D)(y^+)^2 + \operatorname{div}(fy) + F(y) = 0 \text{ in } (0,T)\times\mathbb{R}^N, \\ &y(0) = y_0 = L_0(g_0)\in L^1(\mathbb{R}^N) \ \text{ in } \mathbb{R}^N, \end{aligned} \tag{4.150}$$

where $F: L^1(\mathbb{R}^N)\to L^1(\mathbb{R}^N)$ is a linear continuous operator.

Now, one applies Theorem 4.41, where $\beta(r)\equiv\frac{1}{2}(r^+)^2$, $D\equiv f$, $b\equiv 1$. We get

Theorem 4.42 *Under assumptions* (j), (jj), *there is a unique mild solution* y *to equation* (4.149). *Then,*

$$\varphi(t) = L^{-1}(D)(-y(t)), \quad t\in(0,T),$$

is the solution to the dynamic programming equation (4.146) *and the optimal feedback controller* u *is given by*

$$u(t) = \partial h^*(-L(D)\varphi(X(t))) = y^+(t,X(t)), \quad t\in(0,T). \tag{4.151}$$

Remark 4.43 The approach developed above in a special case, which is applicable in a more general case, can be described in few words as replacing the classical Hamilton–Jacobi equation as dynamic program equation by a Fokker–Planck equation. The advantage, compared with the Hamilton–Jacobi equation, is that the latter is well posed in $L^1(\mathbb{R}^N)$ and, therefore, can be solved via the finite difference scheme, and so can be used to derive a simple formula for the stochastic optimal control. This is apparent by Theorem 4.42, where the stochastic optimal feedback control u for problem (P) has the simple representation (4.151).

4.8 Generalized Fokker–Planck Equation

In literature, a *generalized Fokker–Planck equation* is an equation of the form (4.13), where the Nemytskii drift-transport term $Db(y)y$ is replaced by $K(y)y$, namely,

$$\begin{aligned} &\frac{\partial y}{\partial t} - \Delta\beta(y) + \operatorname{div}(K(y)y) = 0 \ \text{ in } (0,\infty)\times\mathbb{R}^N, \\ &y(0,x) = y_0(x), \end{aligned} \tag{4.152}$$

where $\beta : \mathbb{R} \to \mathbb{R}$ is as in Theorem 4.37, while $K : L^1(\mathbb{R}^N) \to (L^1_{\text{loc}}(\mathbb{R}^N))^N$ is a nonlinear operator. Such an equation arises, for instance, when the diffusion-transport equation

$$\begin{aligned} &\frac{\partial y}{\partial t} - \Delta\beta(y) + \operatorname{div}(uy) = 0 \ \text{ in } (0,\infty)\times\mathbb{R}^N, \\ &y(0,x) = y_0(x), \end{aligned}$$

is coupled with a stationary equation of the form $L(u) = y$ where L is an elliptic differential operator in $\mathbb{R}^N$. This happens in the description of chemostatic models, fluid dynamics and in statistical mechanics with long-range interactions. For instance, if the interactions are described by inverse forces in 2-D, then K is of the form

$$K(y)(x) = \int_{\mathbb{R}^N} \frac{g(x,x')(x-x')}{|x-x'|^3}\,dx'.$$

Formally, equation (4.152) can be written as

$$\begin{aligned} &\frac{dy}{dt} + Ay = 0, \ t \geq 0, \\ &y(0) = y_0, \end{aligned} \tag{4.153}$$

where the operator $Ay \equiv -\Delta\beta(y) + \operatorname{div}(K(y)y)$ is defined in a suitable space $\mathcal{X}$ of functions on $\mathbb{R}^N$. As seen earlier, the best choice for $\mathcal{X}$ would be either $L^1(\mathbb{R}^N)$ or $H^{-1}(\mathbb{R}^N)$ because only here the nonlinear diffusion operator $y \to -\Delta\beta(y)$ is accretive. In $L^1(\mathbb{R}^N)$, in general, this property fails for $y \to \operatorname{div}(K(y)y)$ if K is not a Nemytskii operator in $L^1(\mathbb{R}^N)$. Then, as mentioned earlier, the best strategy in this case is to work in the space $H^{-1}(\mathbb{R}^N)$. We shall illustrate this approach for an important problem in fluid dynamics related to 2D Navier–Stokes equations.

The vorticity equation

We recall that the Navier–Stokes

$$\begin{aligned} &u_t - \nu\Delta u + (u\cdot\nabla)u = \nabla p \text{ in } (0,\infty)\times\mathbb{R}^N, \\ &\operatorname{div}(u) = 0 \qquad\qquad\qquad\ \text{in } (0,\infty)\times\mathbb{R}^N, \\ &u(0,x) = u_0(x), \qquad\qquad\ x \in \mathbb{R}^N, \end{aligned}$$

where $N = 2, 3$, $\nu > 0$, and

$$(u\cdot\nabla)u = \sum_{i=1}^N u_i \frac{\partial}{\partial x_i} u_j, \quad u = (u_1, \dots, u_N),$$

describes the dynamics of an incompressible fluid velocity u with the viscosity $\nu > 0$. Here, we shall briefly treat the case $N = 2$. In this case, the *vorticity* $\rho(t)$ of the fluid is given by

$$\rho(t) = \operatorname{curl}(u(t)) = \frac{\partial}{\partial x_1}\, u_2(t,x) - \frac{\partial}{\partial x_2}\, u_1(t,x),$$

and is the solution to the equation (*the vorticity equation*)

$$\begin{aligned} &\rho_t - \nu\Delta\rho + \operatorname{div}(u\rho) = 0 \text{ in } (0,\infty)\times\mathbb{R}^2,\\ &\rho(0,x) = y_0(x), \qquad\qquad x \in \mathbb{R}^2. \end{aligned} \tag{4.154}$$

We also note that the fluid velocity u is expressed by the *stream function* ψ as $u = \nabla^\perp\psi$, that is,

$$u_1 = \frac{\partial}{\partial x_2}\,\psi,\quad u_2 = -\frac{\partial}{\partial_{x_1}}\,\psi,\quad \Delta\psi = -\rho \ \text{ in } \mathbb{R}^2.$$

On the other hand, we have (see (1.35)),

$$\psi(t,x) = (E * \rho(t))(x) = \int_{\mathbb{R}^d} E(x-\bar{x})\rho(t,\bar{x})d\bar{x},$$

$$E(x) \equiv -\frac{1}{2\pi}\log|x|,\ x \in \mathbb{R}^2,$$

and, therefore,

$$\rho(t,x) \equiv \nabla^\perp(E * u(t))(x) = -\nabla^\perp(\psi(t))(x).$$

We set

$$K(z) = \nabla^\perp(E * z),\quad \forall\, z \in L^1(\mathbb{R}^2), \tag{4.155}$$

and recall (see (1.38)) that

$$\|K(z)\|_{M^2(\mathbb{R}^2)} \le C|z|_1,\quad \forall\, z \in L^1(\mathbb{R}^2).$$

By substituting u in (4.154), we get the equation

$$\begin{aligned} &\rho_t - \nu\Delta\rho + \nabla\cdot(\rho K(\rho)) = 0 \text{ in } (0,\infty)\times\mathbb{R}^2,\\ &\rho(0,x) = \rho_0(x), \qquad\qquad x \in \mathbb{R}^2, \end{aligned} \tag{4.156}$$

which is a *generalized Fokker–Planck equation* of the form (4.152).

We have

Theorem 4.44 *Let $\rho_0 \in L^1(\mathbb{R}^N)\cap L^\infty(\mathbb{R}^N)$ be such that $K(\rho_0) \in L^2(\mathbb{R}^N)$. Then, there is a unique solution $\rho = \rho(t,\rho_0) \in W^{1,2}([0,\infty);H^{-1}(\mathbb{R}^2))\cap L^2(0,\infty;H^1(\mathbb{R}^2))\cap L^\infty((0,\infty)\times\mathbb{R}^2)$ to (4.156). We have $K(\rho) \in L^\infty(0,T;(L^2(\mathbb{R}^2))^2)$ and*

$$\begin{aligned} &|\rho(t,\rho_0) - \rho(t,\bar{\rho}_0)|_2^2 + \int_0^t |\nabla(\rho(s,\rho_0) - \rho(s,\bar{\rho}_0))|_2^2 ds\\ &\le C(|\rho_0 - \bar{\rho}_0|_2^2 + |\rho_0|_\infty^2(|K(\rho_0 - \bar{\rho}_0)|_2^2 + t^2|K(\bar{\rho}_0)|_2^2)\exp(2|\rho_0|_\infty t),\ \forall t \ge 0. \end{aligned} \tag{4.157}$$

We also have

$$|\rho(t)|_p \le |\rho_0|_p,\ \forall t \ge 0,\ 1 \le p \le \infty, \tag{4.158}$$

$$\frac{1}{2}\,|\rho(t)|_2^2 + \int_0^t |\nabla\rho(s)|_2^2 ds = \frac{1}{2}\,|\rho_0|_2^2. \tag{4.159}$$

If $\rho_0 \ge 0$, a.e. in $\mathbb{R}^2$, then $\rho \ge 0$, a.e. in $(0,\infty)\times\mathbb{R}^2$. Moreover, we have

$$\int_{\mathbb{R}^d} \rho(t,x)dx = \int_{\mathbb{R}^d} \rho_0(x)dx,\quad \forall\, t \ge 0. \tag{4.160}$$

(Here, $|\cdot|_p$, $1 \le p \le \infty$, is the norm of $L^p(\mathbb{R}^2)$.)

Proof. We approximate (4.156) by the equation

$$\begin{aligned} &\frac{du_\varepsilon}{dt} + A^\eta_\varepsilon(u_\varepsilon) = 0, \quad t \geq 0, \\ &u_\varepsilon(0) = \rho_0, \end{aligned} \tag{4.161}$$

where the operator $A^\eta_\varepsilon : L^2(\mathbb{R}^2) \to L^2(\mathbb{R}^2)$ for $\varepsilon, \eta > 0$ is defined by

$$\begin{aligned} A^\eta_\varepsilon(u) &= -\Delta u + \operatorname{div}(K_\varepsilon(u)u), \ \forall u \in D(A^\eta_\varepsilon), \\ D(A^\eta_\varepsilon) &= \{u \in H^2(\mathbb{R}^2); \ |u|_\infty \leq \eta\}. \end{aligned}$$

Here, $K_\varepsilon(u) = \nabla^\perp(\varepsilon I - \Delta)^{-1}u$ and note that

$$\|K_\varepsilon(u)\|_{H^1(\mathbb{R}^2)} \leq \frac{C}{\varepsilon}\ |u|_2, \quad \forall\, u \in L^2(\mathbb{R}^2).$$

We have

Lemma 4.45 *There is $\omega = \omega^\eta_\varepsilon$ such that, for all $0 < \lambda < \omega^{-1}$, we have*

$$D(A^\eta_\varepsilon) = \{u \in L^2; \ |u|_\infty \leq \eta\} \subset R(I + \lambda A^\eta_\varepsilon) \tag{4.162}$$

$$|(I+\lambda A^\eta_\varepsilon)^{-1}f_1 - (I+\lambda A^\eta_\varepsilon)^{-1}f_2|_2 \leq (1-\lambda\omega)^{-1}|f_1 - f_2|_2, \ \forall f_1, \ f_2 \in \overline{D(A^\eta_\varepsilon)}. \tag{4.163}$$

Moreover,

$$|(I + \lambda A^\eta_\varepsilon)^{-1}f|_p \leq |f|_p, \ \forall f \in L^p, \ 0 < \lambda \leq \omega^{-1}, \ 1 \leq p \leq \infty, \tag{4.164}$$

$$(I + \lambda A^\eta_\varepsilon)^{-1}f \geq 0, \ \text{a.e. in } \mathbb{R}^2 \text{ if } f \geq 0, \ \text{a.e. in } \mathbb{R}^2. \tag{4.165}$$

Here, $\overline{D(A^\eta_\varepsilon)}$ is the closure of $D(A^\eta_\varepsilon)$ in $L^2(\mathbb{R}^2)$.

We postpone for the time being the proof of Lemma 4.45 and note that by Theorem 2.52 it implies that for each $\rho_0 \in D(A^\eta_\varepsilon)$ the Cauchy problem (4.161) has a unique solution u_ε such that $\frac{du_\varepsilon}{dt}$, $Au_\varepsilon \in L^\infty(0,T; L^2(\mathbb{R}^2))$. Hence, $u_\varepsilon \in W^{1,\infty}([0,T]; L^2(\mathbb{R}^2)) \cap L^\infty(0,T; H^2(\mathbb{R}^2))$, $\forall \varepsilon > 0$. Moreover, by (4.164)–(4.165) it follows, via the exponential formula (2.112) for the semigroup $S_\varepsilon(t) = \exp(-A^\eta_\varepsilon t)$, that

$$|u_\varepsilon(t)|_p \leq |f|_p, \quad \forall t \geq 0, \ p \in [1, \infty), \tag{4.166}$$

and that $u_\varepsilon \geq 0$, a.e. in $(0,\infty) \times \mathbb{R}^2$, if $\rho_0 \geq 0$, a.e. in $\mathbb{R}^2$.

By (4.161), we also get the following estimates on u_ε,

$$\frac{1}{2}\ \frac{d}{dt}\ |u_\varepsilon(t)|^2_2 + \nu|\nabla u_\varepsilon(t)|^2_2 + \int_{\mathbb{R}^2} K_\varepsilon(u_\varepsilon)u_\varepsilon \cdot \nabla u_\varepsilon\, dx, \ \text{a.e. } t > 0.$$

Taking into account that

$$(K_\varepsilon(u_\varepsilon)u_\varepsilon, \nabla u_\varepsilon)_2 = \frac{1}{2}\int_{\mathbb{R}^2} K_\varepsilon(u_\varepsilon)\dot{\nabla}|u_\varepsilon|^2 dx = 0$$

(because $\nabla \cdot \nabla^\perp v = 0$, $\forall v \in H^1(\mathbb{R}^2)$), this yields

$$|u_\varepsilon(t)|^2_2 + 2\nu \int_0^t |\nabla u_\varepsilon(s)|^2_2 ds = |\rho_0|^2_2, \ \forall t \geq 0. \tag{4.167}$$

Now, multiplying (4.161) by $(\varepsilon I-\Delta)^{-1}u_\varepsilon$ and integrating on $(0,t)\times\mathbb{R}^2$, we get

$$\begin{aligned}&\frac{1}{2}\,((\varepsilon I-\Delta)^{-1}u_\varepsilon(t),u_\varepsilon(t))_2+\nu\int_0^t|u_\varepsilon(t)|_2^2ds\\&\quad=\varepsilon\int_0^t(\varepsilon I-\Delta)^{-1}u_\varepsilon(s),u_\varepsilon(s))_2ds+\frac{1}{2}\,((\varepsilon I-\Delta)^{-1}\rho_0,\rho_0)_2,\end{aligned}$$

where $(\cdot,\cdot)_2$ is the scalar product in $L^2(\mathbb{R}^2)$.

Since

$$(u,(\varepsilon I-\Delta)^{-1}u)_2=\varepsilon|(\varepsilon I-\Delta)^{-1}u|_2^2+|K_\varepsilon(u)|_2^2,\ \forall u\in L^2(\mathbb{R}^2),$$

we get

$$\begin{aligned}&\frac{1}{2}\,|K_\varepsilon(u_\varepsilon)|_2^2+\nu\int_0^t|u_\varepsilon(s)|_2^2ds\\&\quad\le\frac{1}{2}\,|K_\varepsilon(\rho_0)|_2^2+\varepsilon\int_0^t(|K_\varepsilon(u_\varepsilon(s))|^2+|(\varepsilon I-\Delta)^{-1}u_\varepsilon(s)|^2)ds,\end{aligned}\tag{4.168}$$

and, therefore,

$$\limsup_{\varepsilon\to0}\left(|K_\varepsilon(u_\varepsilon(t))|_2^2+2\nu\int_0^t|u_\varepsilon(s)|_2^2ds\right)\le|K(\rho_0)|_2^2.$$

Similarly, for $u_\varepsilon\equiv u_\varepsilon(t,\rho_0)$, $\bar{u}_\varepsilon\equiv u_\varepsilon(t,\bar{\rho}_0)$, we get

$$\begin{aligned}&\frac{1}{2}\,((\varepsilon I-\Delta)^{-1}(u_\varepsilon-\bar{u}_\varepsilon)(t),u_\varepsilon(t)-\bar{u}_\varepsilon(t))_2+\nu\int_0^t|u_\varepsilon(s)-\bar{u}_\varepsilon(s)|_2^2ds\\&\quad=\frac{1}{2}\,((\varepsilon I-\Delta)^{-1}(\rho_0-\bar{\rho}_0),\rho_0-\bar{\rho}_0)_2\\&\qquad+\varepsilon\int_0^t((\varepsilon I-\Delta)^{-1}(u_\varepsilon(s)-\bar{u}_\varepsilon(s)),u_\varepsilon(s)-\bar{u}_\varepsilon(s))ds\\&\qquad+\int_0^t\int_{\mathbb{R}^2}(K_\varepsilon(u_\varepsilon(s)-\bar{u}_\varepsilon(s))\bar{u}_\varepsilon\\&\qquad+K_\varepsilon(\bar{u}_\varepsilon(s))(u_\varepsilon(s)-\bar{u}_\varepsilon(s)))\cdot\nabla(\varepsilon I-\Delta)^{-1}(u_\varepsilon(s)-\bar{u}_\varepsilon(s))dx\,ds\\&\quad\le\frac{1}{2}\,((\varepsilon I-\Delta)^{-1}(\rho_0-\bar{\rho}_0),\rho_0-\bar{\rho}_0)_2\\&\qquad+\varepsilon\int_0^t(|\nabla(u_\varepsilon(s)-\bar{u}_\varepsilon(s))|_2^2+\varepsilon|u_\varepsilon(s)-\bar{u}_\varepsilon(s))|_2^2)ds\\&\qquad+\int_0^t(|\bar{u}_\varepsilon(s)|_\infty|K_\varepsilon(u_\varepsilon(s)-\bar{u}_\varepsilon(s))|_2^2\\&\qquad+|K_\varepsilon(\bar{u}_\varepsilon)|_2|K_\varepsilon(u_\varepsilon(s)-\bar{u}_\varepsilon(s))|_2|(u_\varepsilon(s)-\bar{u}_\varepsilon(s)|_\infty))ds.\end{aligned}$$

This yields

$$\begin{aligned}&\frac{1}{2}\,|K_\varepsilon(u_\varepsilon(t)-\bar{u}_\varepsilon(t))|_2^2+\nu\int_0^t|u_\varepsilon(s)-\bar{u}_\varepsilon(s)|_2^2ds\\&\quad\le\frac{1}{2}\,|K_\varepsilon(\rho_0-\bar{\rho}_0)|_2^2+\int_0^t(|\bar{\rho}_0|_\infty|K_\varepsilon(u_\varepsilon(s)-\bar{u}_\varepsilon(s))|_2^2\\&\qquad+|K_\varepsilon(\bar{\rho}_0)|_2|K_\varepsilon(u_\varepsilon(s)-\bar{u}_\varepsilon(s))|_2)ds+\varphi_\varepsilon(t),\end{aligned}\tag{4.169}$$

where $\varphi_\varepsilon(t) \to 0$ as $\varepsilon \to 0$ for $t \geq 0$. We get, therefore,

$$|K_\varepsilon(u_\varepsilon(t) - \bar{u}_\varepsilon(t))|_2 \leq |K_\varepsilon(\rho_0 - \bar{\rho}_0)|_2$$
$$+|\bar{\rho}_0|_\infty \int_0^t |K_\varepsilon(u_\varepsilon(s) - \bar{u}_\varepsilon(s))|_2 ds + |K_\varepsilon(\bar{\rho}_0)|_2 t + \varphi_\varepsilon(t), \ \forall f \geq 0.$$

Hence,

$$|K_\varepsilon(u_\varepsilon(t) - \bar{u}_\varepsilon(t))|_2 \leq (|K_\varepsilon(\rho_0 - \bar{\rho}_0)|_2 + t|K_\varepsilon(\bar{\rho}_0)_2) \exp(|\bar{\rho}_0|_\infty t) + \eta_\varepsilon(t),$$

and so, by (4.168), we have

$$\begin{aligned}&|K_\varepsilon(u_\varepsilon(t) - \bar{u}_\varepsilon(t))|_2^2 + \int_0^t |u_\varepsilon(s) - \bar{u}_\varepsilon(s)|_2^2 ds \\ &\quad\leq C(|K_\varepsilon(\rho_0 - \bar{\rho}_0)|_2^2 + t^2|K_\varepsilon(\rho_0)|_2^2 \exp(2|\bar{\rho}_0|_\infty t)) + \eta_\varepsilon(t),\end{aligned} \tag{4.170}$$

where $\eta_\varepsilon(t) \to 0$ as $\varepsilon \to 0$. By (4.168), it follows that on a subsequence $\{\varepsilon\} \to 0$,

$$\begin{aligned} u_\varepsilon \to \rho \qquad & \text{weakly in } L^2(0,T;H^1(\mathbb{R}^2)), \\ & \text{strongly in } L^2(0,T;L^2(\mathbb{R}^2)), \\ \frac{du_\varepsilon}{dt} \to \frac{d\rho}{dt} \qquad & \text{weakly in } L^2(0,T;H^{-1}(\mathbb{R}^2)), \\ K_\varepsilon(u_\varepsilon)u_\varepsilon \to K(\rho)\rho \ & \text{weakly in } L^2(0,T;L^2(\mathbb{R}^2)). \end{aligned} \tag{4.171}$$

(We recall that, by (1.38), $K(\rho)\rho$ is well defined and belongs to L^1_{loc}.) Then, letting $\varepsilon \to 0$ in (4.161), it follows that ρ is a solution to (4.156). Moreover, once again by (4.167), (4.168), we get

$$\frac{1}{2}\,|K(\rho(t))|_2^2 + \nu \int_0^t |\rho(s)|_2^2 \leq \frac{1}{2}\,|K(\rho_0)|_2^2, \quad \forall t \geq 0,$$
$$\frac{1}{2}\,|\rho(t)|_2^2 + \nu \int_0^t |\nabla\rho(s)|_2^2 = \frac{1}{2}\,|\rho_0|_2^2, \quad \forall t \geq 0,$$

while (4.166) implies (4.158). By (4.156), we have as for $u_\varepsilon, \bar{u}_\varepsilon$ above,

$$\begin{aligned}&\frac{1}{2}\,|\rho(t) - \bar{\rho}(t)|_2^2 + \nu \int_0^t |\nabla(\rho(s) - \bar{\rho}(s))|_2^2 ds \leq \frac{1}{2}\,|\rho_0 - \bar{\rho}_0|_2^2 \\ &\quad + \int_0^t \left(K(\rho(s) - \bar{\rho}(s)) \cdot \nabla\rho(s)(\rho(s) - \bar{\rho}(s)) + \frac{1}{2}\,K(\bar{\rho}(s)) \cdot \nabla|\rho(s) - \bar{\rho}(s)|^2 \right) ds \\ &= \frac{1}{2}\,|\rho_0 - \bar{\rho}_0|_2^2 + \int_0^t \int_{\mathbb{R}^2} (K(\rho(s) - \bar{\rho}(s))\rho(s) \cdot \nabla(\rho(s) - \bar{\rho}(s))\,dx\,ds \\ &\leq \frac{1}{2}\,|\rho_0 - \bar{\rho}_0|_2^2 + \|\rho\|_{L^\infty((0,\infty)\times\mathbb{R}^2)} \int_0^t |K(\rho(s) - \bar{\rho}(s))|_2 |\nabla(\rho(s) - \bar{\rho}(s))|_2 ds.\end{aligned}$$

Taking into account that, by (4.170),

$$|K(\rho(t) - \bar{\rho}(t))|_2 \leq (|K(\rho_0 - \bar{\rho}_0)|_2 + t|K(\bar{\rho}_0)|_2) \exp(|\bar{\rho}_0|_\infty t), \ \forall t \geq 0,$$

we get

$$|\rho(t)-\bar{\rho}(t)|_2^2+\int_0^t |\nabla(\rho(s)-\bar{\rho}(s))|_2^2 ds$$
$$\le |\rho_0-\bar{\rho}_0|_2^2+C(|\rho_0|_\infty^2(|K(\rho_0-\bar{\rho}_0|_2^2+t^2|K(\rho_0)|_2^2\exp(2|\bar{\rho}_0|_\infty t))$$

and so (4.157) follows.

As regards (4.161), it follows by integrating (4.156) on $\mathbb{R}^2$.

Proof of Lemma 4.45. We fix $f\in L^2$ and consider the equation $u+\lambda(A_\varepsilon^\eta)u=f$, that is,

$$u-\lambda\Delta u+\lambda\nabla\cdot(K_\varepsilon(u)u)=f \ \text{ in } \mathcal{D}'(\mathbb{R}^2). \tag{4.172}$$

We shall prove first that, for each $k>0$, there is a solution $u_0^\varepsilon\in H_0^1(B_k)$ to the equation

$$u_k-\lambda\Delta u_k+\lambda\nabla\cdot(K_\varepsilon(u_k)u_k)=f \text{ in } B_k=\{x\in\mathbb{R}^2; |x|<k\}. \tag{4.173}$$

Equivalently,

$$u_k=F(u_k)+(1-\lambda\Delta)^{-1}f, \tag{4.174}$$

where $F(u)=-\lambda(I-\lambda\Delta)^{-1}(\nabla\cdot(K_\varepsilon(u)u))$.

It is easily seen that the operator F is continuous and bounded from $H_0^1(B_k)$ to $H^2(B_k)$ and, therefore, continuous and compact in $H_0^1(B_k)$. Consider the set

$$\Sigma_k=\{u\in H_0^1(B_k), \|u\|_{L^2(B_k)}^2+\lambda\|\nabla u\|_{L^2(B_k)}^2<R^2\},$$

where $R=\|f\|_{L^2(B_k)}+1$. We have

$$F(f)\overline{\in}(I-\sigma F)(\partial\Sigma_k),\ \ \forall\sigma\in[0,1]. \tag{4.175}$$

Indeed, otherwise there is

$$v\in\partial\Sigma_k=\{v\in H_0^1(B_k); \|v\|_{L^2(B_k)}^2+\lambda\|\nabla u\|_{L^2(B_k)}^2=R^2\}$$

such that

$$v-\lambda\Delta v+\lambda\sigma\nabla\cdot(K_\varepsilon(v)v)=f \ \text{ in } B_k.$$

If we multiply the latter by v and integrate on B_k, we get

$$\|v\|_{L^2(B_k)}^2+\lambda\sigma\|\nabla v\|_{L^2(B_k)}^2=\int_{B_k} f\,v_k\,dx\le\frac{1}{2}\|f\|_{L^2(B_k)}^2+\frac{1}{2}\|v\|_{L^2(B_k)}^2$$

because

$$\int_{B_k}\nabla|v|^2\cdot K_\varepsilon(v)dx=-\int_{B_k}|v|^2\nabla\cdot(\nabla^\perp(\varepsilon I-\Delta)^{-1}v)dx=0.$$

This yields

$$\|v\|_{L^2(B_k)}^2+\lambda\sigma\|\nabla v\|_{L^2(B_k)}^2\le\|f\|_{L^2(B_k)}^2\le(R-1)^2<R, \tag{4.176}$$

and so $v\overline{\in}\partial\Sigma_k$, contrary to hypothesis (4.173).

Denote by $d(I-\sigma F,\Sigma_k,(I-\lambda\Delta)^{-1}f)$ the Leray–Schauder degree of the map $I-\sigma F$ relative to the set Σ_k at $(I-\lambda\Delta)^{-1}f$. By (4.175) and the invariance of the Leray–Schauder degree, it follows that

$$d(I-\sigma F,\Sigma_k,(I-\lambda\Delta)^{-1}f)=d(I,\Sigma_k,(I-\lambda\Delta)^{-1}f)=1,\ \forall\sigma\in[0,1],$$

because $(I-\lambda\Delta)^{-1}f\in\Sigma_k$. Hence, equation (4.173), equivalently (4.174), has at least one solution $u=u_k^\varepsilon\in\Sigma_k$, as claimed.

Moreover, by (4.176) we have the estimate

$$\|u_k^\varepsilon\|_{L^2(B_k)}+\lambda\|\nabla u_k^\varepsilon\|_{L^2(B_k)}\le\|f\|_{L^2(B_k)},\quad\forall k. \tag{4.177}$$

Then, on a subsequence $\{k\}\to\infty$, we have

$$u_k^\varepsilon\to u_\varepsilon \text{ strongly in } L^2_{\rm loc}(\mathbb{R}^2) \text{ and weakly in } H^1_{\rm loc}(\mathbb{R}^2),$$

where $u_\varepsilon\in H^1(\mathbb{R}^2)$ is a solution to equation (4.172), that is,

$$u_\varepsilon-\lambda\Delta u_\varepsilon+\lambda\nabla\cdot(K_\varepsilon(u_\varepsilon)u_\varepsilon)=f \text{ in } \mathcal{D}'(\mathbb{R}^2). \tag{4.178}$$

By (4.177) we also have the estimate

$$|u_\varepsilon|_2+\lambda|\nabla u_\varepsilon|_2\le|f|_2, \tag{4.179}$$

for all $\lambda>0$ and $f\in L^2(\mathbb{R}^2)$.

To prove (4.164) for $p=\infty$, we take $f\in L^\infty$ and write (4.178) as

$$(u_\varepsilon-M)-\lambda\Delta(u_\varepsilon-M)+\lambda K_\varepsilon(u_\varepsilon)\cdot\nabla(u_\varepsilon-M)=f-M$$

$M=|f|_\infty$, and multiply by $(u_\varepsilon-M)^+$. After integration, we get $(u_\varepsilon-M)^+=0$, a.e. in $\mathbb{R}^2$. Similarly, it follows $u_t\ge -M$, a.e. in $\mathbb{R}^2$. Hence, $u_\varepsilon\in D(A_\varepsilon^\eta)$ if $f\in\overline{D(A_\varepsilon^\eta)}$.

Next, we multiply (4.178) by $\mathcal{X}_\delta(u_\varepsilon)$ and set $G_\varepsilon=\lambda\Delta u_\varepsilon-\lambda\nabla\cdot(K_\varepsilon(u_\varepsilon)u_\varepsilon)$. This yields

$$\begin{aligned}\int_{\mathbb{R}^2}u_\varepsilon\mathcal{X}_\delta(u_\varepsilon)dx&=-\int_{\mathbb{R}^2}G_\varepsilon\cdot\nabla\mathcal{X}_\delta(u_\varepsilon)dx+\int_{\mathbb{R}}f\mathcal{X}_\delta(u_\varepsilon)dx\\&=-\int_{\mathbb{R}^2}(G_\varepsilon\cdot\nabla)\mathcal{X}_\delta'(u_\varepsilon)dx+\int_{\mathbb{R}^2}f\mathcal{X}_\delta(u_\varepsilon)dx.\end{aligned} \tag{4.180}$$

It follows that

$$\lim_{\delta\to0}\frac{1}{\delta}\int_{[|u|\le\delta]}(|K_\varepsilon(u_\varepsilon)u_\varepsilon|)|\nabla u_\varepsilon|dx\le C_\varepsilon\lim_{\delta\to0}\left(\int_{[|u_\varepsilon|\le\delta]}|\nabla u_\varepsilon|^2dx\right)^{\frac{1}{2}}=0.$$

This yields (see Proposition 1.16)

$$\lim_{\delta\to0}\int_{\mathbb{R}^2}u_\varepsilon\mathcal{X}_\delta(u_\varepsilon)dx\le\int_{\mathbb{R}^2}|f|\,dx$$

and so (4.164) follows for $p=1$.

For $1<p<\infty$, we multiply (4.180) where $f\in L^1\cap L^p$ by $|u_\varepsilon|^{p-2}u_\varepsilon$ and integrate on $\mathbb{R}^d$. We get

$$|u_\varepsilon|_p^p\le|f|_p|u_\varepsilon|_p^{p-1}-\frac{\lambda}{p}\int_{\mathbb{R}^2}K_\varepsilon(u_\varepsilon)\cdot\nabla(|u_\varepsilon|^p)dx=|f|_p|u_\varepsilon|_p^{p-1}$$

and this implies (4.167), as claimed.

For $f_1, f_2 \in L^2 \cap L^\infty$, $|f_i|_\infty \le \eta$, $i = 1, 2$, we set $u_\varepsilon = u_\varepsilon(\lambda, f_1) - u_\varepsilon(\lambda, f_2)$. By (4.172), we have

$$u_\varepsilon - \lambda\Delta u_\varepsilon + \lambda K_\varepsilon(u_\varepsilon)\cdot\nabla u_\varepsilon(\lambda, f_1) + \lambda K_\varepsilon(u_\varepsilon(\lambda, f_2))\cdot\nabla u_\varepsilon = f_1 - f_2,$$

because $\operatorname{div}(\nabla^\perp u) = 0$. If we multiply the latter by u_ε and integrate, we get

$$|u_\varepsilon|_2 + \lambda|\nabla u_\varepsilon|_2^2 + \lambda\int_{\mathbb{R}^2} K_\varepsilon(u_\varepsilon)\cdot\nabla u_\varepsilon(\lambda, f_1)u_\varepsilon\,dx \le |f_1 - f_2, u_\varepsilon|_2.$$

This yields

$$\begin{aligned}|u_\varepsilon|_2^2 + \lambda|\nabla u_\varepsilon|_2^2 &\le \lambda\int_{\mathbb{R}^2}(K_\varepsilon(u_\varepsilon)\cdot\nabla u_\varepsilon(\lambda, f_1))u_\varepsilon\,dx + |f_1 - f_2|_2|u_\varepsilon|_2\\ &\le \lambda\eta|\nabla u_\varepsilon|_2|K_\varepsilon(u_\varepsilon)|_2 + |f_1 - f_2|_2|u_\varepsilon|_2\\ &\le \frac{C}{\varepsilon}\,\lambda\eta|\nabla u_\varepsilon|_2|u_\varepsilon|_2 + |f_1 - f_2|_2|u_\varepsilon|_2,\end{aligned}$$

and, therefore, (4.163) holds for $\omega = \left(\frac{C\eta}{\varepsilon}\right)^2$. □

Remark 4.46 It follows by Theorem 4.30 that, if ρ_0 is a probability density, then so is $\rho(t)$ for all $t \ge 0$. This implies, as for the Fokker–Planck equation (4.133), that the vorticity $\rho(t)$ can be represented as the density probability of a solution $X(t)$ to the McKean–Vlasov stochastic differential equation

$$\begin{aligned}dX(t) &= K(\rho(t, X(t))dt + \frac{1}{2}\,dW(t),\\ X(0) &= X_0,\end{aligned}$$

where $\mathcal{L}_{X_0} = \rho_0$.

Taking into account (4.159), it follows that the solution ρ to (4.156) can be extended to a solution (eventually not unique) $\rho \in L^2(0, T; V)\cap W^{1,2}((0, T); V')$ for $\rho_0 \in L^2(\mathbb{R}^2)$. However, a sharper result can be obtained for initial data in $W^{1,1}(\mathbb{R}^2)$.

Theorem 4.47 *Let $\rho_0 \in W^{1,1}(\mathbb{R}^2)$ and let $T > 0$ be arbitrary but fixed. Then, if $|\nabla\rho_0|_1 T^{\frac{1}{2}}$ is sufficiently small, there is a solution ρ to (4.156) such that*

$$\rho \in L^2(0, T; H^1(\mathbb{R}^2)) \cap L^\infty(0, T; W^{1,1}(\mathbb{R}^2)) \cap W^{1,2}([0, T]; H^{-1}(\mathbb{R}^2)). \tag{4.181}$$

Proof. For $M \in \mathbb{N}$, consider the set

$$\mathcal{K}_M = \{z \in L^\infty(0, T; W^{1,1}(\mathbb{R}^2)); \|z\|_{L^\infty(0,T;W^{1,1}(\mathbb{R}^2))} \le M\}.$$

For $z \in \mathcal{K}_M$ we consider the equation

$$\begin{aligned}&\frac{dy}{dt} - \nu\Delta y + \operatorname{div}(K(z)y) = 0, \text{ a.e. } t \in (0, T)\\ &y(0) = \rho_0,\end{aligned} \tag{4.182}$$

which can be written as

$$\begin{aligned}&\frac{dy}{dt} + Ay = 0, \quad \text{a.e. } t \in (0, T),\\ &y(0) = y_0,\end{aligned} \tag{4.183}$$

where $A : H^1(\mathbb{R}^2) \to H^{-1}(\mathbb{R}^2)$ is given by

$$(Ay, \varphi) = \int_{\mathbb{R}^2}(\nu\nabla y - K(z)y)\cdot\nabla\varphi\,dx = \nu\int_{\mathbb{R}^2}\nabla y\cdot\nabla\varphi\,dx, \ \forall\varphi \in H^1(\mathbb{R}^2).$$

Then, by Theorem 2.77, it follows that (4.183) has a unique solution $y = \Phi(z) \in W^{1,2}([0,T];H^{-1}(\mathbb{R}^2)) \cap L^2(0,T;H^1(\mathbb{R}^2)) \subset C([0,T];L^2(\mathbb{R}^2))$. We also note that

$$\|\Phi(z)\|_{W^{1,2}([0,T];H^{-1}(\mathbb{R}^2))} + \|z\|_{L^2(0,T;H^1(\mathbb{R}^2))} \le C, \ \forall z \in \mathcal{K},$$

which, by Theorem 1.35, implies that the set $\Phi(\mathcal{K})$ is *relatively compact in the space* $L^2(0,T;L^2_{\rm loc}(\mathbb{R}^2))$ endowed with the standard local convex topology. We also note that the mapping $\Phi : \mathcal{K}_M \to L^2(0,T;L^2_{\rm loc}(\mathbb{R}^2))$ is continuous because if $z_n \to z$ in $L^2(0,T;L^2(\Sigma_k))$, $\Sigma_k = \{x \in \mathbb{R}^2; |x| \le k\}$, then clearly ${\rm div}(K(z_n)y_n) \to {\rm div}(K(z)y)$.

Let us prove now that $\Phi(\mathcal{K}_M) \subset \mathcal{K}_M$ for a suitable M. To this purpose, we fix $z \in \mathcal{K}_M$, set $y = \Phi(z)$ and set $w_i = \frac{\partial}{\partial x_i} z$, $\mu_i = \frac{\partial y}{\partial x_i}$, $i = 1,2$. By (4.182), we see that

$$(\mu_i)_t - \nu\Delta\mu_i + {\rm div}(K(w_i)y + K(z)\mu_i) = 0 \text{ in } (0,T) \times \mathbb{R}^2.$$

If we multiply the latter by $\mathcal{X}_\delta(\mu_i)$ and integrate on $\mathbb{R}^2$, we get for $j_\delta(r) \equiv \int_0^r \mathcal{X}_\delta(s)ds$,

$$\begin{aligned}\frac{d}{dt}\int_{\mathbb{R}^2} j_\delta(\mu_i(t,x))dx + \nu\int_{\mathbb{R}^2} |\nabla\mu_i(t,x)|^2\mathcal{X}'_\delta(\mu_i(t,x))dx \\ = -\int_{\mathbb{R}^2} K(w_i(t,x)) \cdot \nabla y(t,x)\mathcal{X}_\delta(\mu_i(t,x))dx,\end{aligned}$$

because ${\rm div}\, K(z) \equiv 0$, and so

$$\int_{\mathbb{R}^2} K(z) \cdot \nabla\mathcal{X}_\delta(\mu_i)\mu_i\, dx = \int_{\mathbb{R}^2} K(z) \cdot \nabla j_\delta(\mu_i)dx = -\int_{\mathbb{R}^2} j_\delta(\mu_i){\rm div}\, K(z)dx = 0.$$

On the other hand, we have

$$\begin{aligned}\left|\int_{\mathbb{R}^2} K(w_i(t,x)) \cdot \nabla y(t,x)\mathcal{X}_\delta(\mu_i(t,x))dx\right| \le |K(w_i(t))|_2|\nabla y(t)|_2 \\ = \left|\frac{\partial}{\partial x_i} K(z(t))\right|_2 |\nabla y(t)|_2 \le C|\nabla z|_1|\nabla y|_2,\end{aligned}$$

because (see, e.g., Bourgain and Brezis [37])

$$|\nabla(-\Delta)^{-1}\nabla^\perp u|_2 \le C|\nabla u|_1, \ \ \forall u \in W^{1,1}(\mathbb{R}^2).$$

This yields

$$\int_\Omega j_\delta(y_1(t,x))dx \le \int_\Omega \left|\frac{\partial}{\partial x_i}\rho_0(x)\right| dx + CMt^{\frac{1}{2}}\left(\int_0^t |\nabla y(s)|_2^2 ds\right)^{\frac{1}{2}}, \ \forall t \in (0,T),$$

and, for $\delta \to 0$, we get

$$|\mu_i(t)|_1 \le \left|\frac{\partial}{\partial x_i}\rho_0\right|_1 + 2CMt^{\frac{1}{2}}|\rho_0|_2, \ i = 1,2, \tag{4.184}$$

where C is independent of M and ρ_0. Taking into account that $W^{1,1}(\mathbb{R}^2) \subset L^2(\mathbb{R}^2)$, $|\rho_0|_2 \le C_1|\nabla\rho_0|_1$, we see by (4.184) that

$$|\nabla y(t)|_1 \le |\nabla\rho_0|_1(1 + 4CMt^{\frac{1}{2}}).$$

Since $|y(t)|_1 \le |\rho_0|_1$, we infer that

$$|y(t)\|_{W^{1,1}(\mathbb{R}^2)} \le (1 + 4CMt^{\frac{1}{2}})|\nabla\rho_0|_1 + |\rho_0|_1, \ \ \forall t \ge 0,$$

and so, for $t^{\frac{1}{2}}|\nabla\rho_0|_1$ sufficiently small, there is $M > 0$ such that $\Phi(\mathcal{K}_M) \subset \mathcal{K}_M$. Then, by the Schauder–Tihonov theorem (see, e.g., [67], p. 138), in the space $L^2(0,T;L^2_{\rm loc}(\mathbb{R}^N))$ it follows that Φ has a fixed point $y \in \mathcal{K}_M$ and so equation (4.156) has a solution y satisfying (4.181).

Comments. There is an extensive literature on semilinear parabolic equations, parabolic variational inequalities, and the Stefan problem (see Lions [78] for significant results and complete references on this subject). Here, we were primarily interested in the existence results that arise as direct consequences of the general theory developed previously, and we tried to put in perspective those models of free boundary problems that can be formulated as nonlinear differential equations of accretive type. The L^1-space semigroup approach to the nonlinear diffusion equation was initiated by Bénilan [31] and the $H^{-1}(\Omega)$ approach is due to Brezis [40]. The smoothing effect of the semigroup generated by the semilinear elliptic operator in $L^1(\Omega)$ (Proposition 4.5) is due to Evans [68]. The analogous result for the nonlinear diffusion operator in $L^1(\Omega)$ (Theorem 4.21) was first established by Bénilan [31], and Véron [103], but the proof given here is from [23] and follows the one due to Pazy [90] with a small modification to correct a gap. An extension to nonlinear Fokker–Planck equations is given in [23] by an essential modification of the original proof given in [90]. For other related contributions to the existence and regularity of solutions to the porous medium equation, we refer to Bénilan, Crandall, and Pierre [35], and Brezis and Crandall [45]. There is a large literature devoted to applications of porous media type equations biseds that mentioned in Section 4.2, including image restoring (see, e.g., [28]). Theorem 4.33 was proven in [17]. We also mention on these lines the author's book [13] for applications of this result to the exact controllability of nonlinear parabolic equations, and also to D. Mosco [89] for a discrete model of self-organized behavior. The semigroup approach to the conservation law equation (Theorem 4.35) is due to Crandall [58]. As regards Proposition 4.28, the idea of the proof is essentially due to S.N. Antontsev (see, e.g., [6]). The semigroup approach to nonlinear Fokker–Planck equations in $L^1(\mathbb{R}^N)$ was initiated by V. Barbu and M. Röckner and the results presented here (Theorem 4.17) was firstly established in the work [22] and in [25] in a more general setting. We mention also the work [55] of Chen and Perthame. The results of Section 4.7 are related to the author works [18,19] but that in Section 4.8 are new in this context. However, there is an extensive literature on chemostatic equations (see [84] for some recent results). The semigroup approach of vorticity equations is new but there is an extensive literature on this equation we mention the results in [54,81], which are closer of that in Theorem 4.47.

Bibliography

[1] D. Adams, *Sobolev Spaces*, Academic Press, San Diego, 1975.

[2] S. Agmon, A. Douglis, L. Nirenberg, Estimates near the boundary for solutions of elliptic partial differential equations satisfying general boundary conditions, *Comm. Pure Appl. Math.*, **12** (1959), pp. 623–727.

[3] N. Alikakos, R. Rostamian, Large time behaviour of solutions of Neumann boundary value problems for the porous medium equations, *Indiana Univ. Math. J.*, **39** (1981), pp. 749–785.

[4] L. Ambrosio, N. Fusco, D. Pallara, *Functions of Bounded Variations and Free Discontinuous Processes*, Oxford University Press, Oxford, UK, 2000.

[5] F. Andreu, C. Ballester, V. Caselles, J.M. Mazón, Minimizing total variation flow, *Differential and Integral Equations*, **14** (3) (2001), pp. 321–360.

[6] A.N. Antontsev, J.I. Diaz, S. Shmarev, *Energy Methods for Free Boundary Problems*, Birkhäuser, Basel, 2002.

[7] E. Asplund, Average norms, *Israel J. Math.*, **5** (1967), pp. 227–233.

[8] J.P. Aubin, Un théorème de compacité, *C.R. Acad. Sci.*, **256** (1963), pp. 5042–5044.

[9] P. Bak, C. Tang, K. Wiesenfeld, Self-organized criticality; an explanation of $1/f$ noise, *Physical Review Letters*, **59** (1987), pp. 381–394.

[10] V. Barbu, *Analysis and Control of Nonlinear Infinite Dimensional Systems*, Academic Press, Boston, 1993.

[11] V. Barbu, Boundary controllability of phase-transition region of a two-phase Stefan problem, *System & Control Letters*, **150** (2021), pp. 1–7.

[12] V. Barbu, Continuous perturbation of nonlinear m-accretive operators in Banach spaces, *Boll. Unione Mat. Ital.*, **6** (1972), pp. 270–278.

[13] V. Barbu, *Controllability and Stabilization of Parabolic Equations*, Birkhäuser, 2018.

[14] V. Barbu, *Nonlinear Semigroups and Differential Equations in Banach Spaces*, Noordhoff, Leyden, 1976.

[15] V. Barbu, *Nonlinear Differential Equations of Monotone Type in Banach Spaces*, Springer, 2010.

[16] V. Barbu, *Partial Differential Equations and Boundary Value Problems*, Kluwer, Dordrecht, 1998.

[17] V. Barbu, Self-organized criticality of cellular automata model; absorbtion infinite time of supercritical region into the critical one, *Mathematical Methods in Applied Siences*, **36** (13) (2013).

[18] V. Barbu, The dynamic programming equation for a stochastic volatility optimal control problem, *Automatica*, **107** (2019), pp. 119–124.
[19] V. Barbu, C. Benazzoli, L. Di Persio, Mild solutions to the dynamic programming equation for stochastic optimal control problems, *Automatica*, **83** (2018), pp. 920–226.
[20] V. Barbu, P.L. Colli, G. Gilardi, G. Marinoschi, E. Rocca, Sliding mode control for a nonlinear phase-field system, *SIAM J. Control. Optim.*, vol. **95** (2017), pp. 2108–2133.
[21] V. Barbu, T. Precupanu, *Convexity and Optimization in Banach Spaces*, D. Reidel, Dordrecht, 1986.
[22] V. Barbu, M. Röckner, From Fokker–Planck equations to solutions of distribution dependent SDE, *Annals of Probability*, **48** (4) (2020), pp. 1902–1920.
[23] V. Barbu, M. Röckner, Solutions for nonlinear Fokker–Planck equations with measures as initial data and McKean–Vlasov equations, *J. Funct. Anal.*, **280** (7) 2021.
[24] V. Barbu, M. Röckner, Stochastic variational inequalities and applications to the total variation flow perturbed by linear multiplicative noise, *Archives Rat. Mech. Anal.*, **209** (2013), 797–834.
[25] V. Barbu, M. Röckner, The evolution to equilibrium of solutions to nonlinear Fokker–Planck equations, *Indiana Univ. Math. J.*
[26] V. Barbu, M. Röckner, Uniqueness for Fokker–Planck equations and weak uniqueness for McKean–Vlasov SDE, *Stochastic PDEs, Anal. Comput.*, (2020).
[27] T. Barbu, A PDE based model for sonar image and video denoising, *Anal. St. Univ. Ovidius, Constanta*, **19** (2011), pp. 51–58.
[28] T. Barbu, Nonlinear PDE model for image restoration using second-order hyperbolic equations, *Numerical Functional Analysis and Optimization*, **38** (11) (2015), pp. 1375–1387.
[29] T. Barbu, V. Barbu, V. Biga, D. Coca, A PDE variational approach to image denoising and restoration, *Nonlinear Anal. Real World Appl.*, **10** (1009), pp. 1351–1361.
[30] Ph. Bénilan, *Equations d'évolution dans un espace de Banach quelconque et applications*, Thèse, Orsay, 1972.
[31] Ph. Bénilan, Opérateurs accétifs et semigroupes dans les espaces L^p, $1 \leq p \leq \infty$, *Functional Analysis and Numerical Analysis*, pp. 15–51, T. Fuzita (ed.), Japan Soc., Tokyo, 1978.
[32] Ph. Benilan, H. Brezis, M.G. Crandall, A semilinear equation in $L^1(\mathbb{R}^N)$, *Annali di Scuola Normale Superiore di Pisa*, **4** (1975), pp. 523–555.
[33] Ph. Bénilan, H. Brezis, Solutions faibles d'équations d'évolution dans les espaces de Hilbert, *Ann. Inst. Fourier*, **22** (1972), pp. 311–329.
[34] Ph. Bénilan, M.G. Crandall, The continuous dependence on φ of solutions of $u_t - \Delta\varphi(u) = 0$, *Indiana Univ. Math. J.*, **30** (1981), pp. 161–177.
[35] Ph. Benilan, M.G. Crandall, M. Pierre, Solutions of the porous medium equations in $\mathbb{R}^N$ under optimal conditions on initial values, *Indiana Univ. Math. J.*, **33** (1984), pp. 51–87.
[36] Ph. Benilan, N. Kruzkhov, Conservation laws with continuous flux conditions, *Nonlinear Differential Eqs. Appl.*, **3** (1996), pp. 395–419.

[37] J. Bourgain, H. Brezis, New estimates for elliptic equations and Hodge type system, *J. Eur. Math. Soc.*, **9** (2007), pp. 277–315.

[38] H. Brezis, *Functional Analysis, Sobolev Spaces and Partial Differential Equations*, Springer, New York, 2011.

[39] H. Brezis, Intégrales convexes dans les espaces de Sobolev, *Israel J. Math.*, **13** (1972), pp. 9–23.

[40] H. Brezis, Monotonicity methods in Hilbert spaces and some applications to nonlinear partial differential equations, *Contributions to Nonlinear Functional Analysis*, E. Zarantonello (ed.), Academic Press, New York, 1971.

[41] H. Brezis, *Opérateurs Maximaux Monotones et Semigroupes de Contractions dans un Espace de Hilbert*, North-Holland, Amsterdam, 1973.

[42] H. Brezis, Problemes unilatéraux, *J. Math. Pures Appl.*, **51** (1972), pp. 1–168.

[43] H. Brezis, Propriétés régularisantes de certaines semi-groupes nonlinéaires, *Israel J. Math.*, **9** (1971), pp. 513–514.

[44] H. Brezis, F. Browder, Some properties of higher order Sobolev spaces, *J. Math. Pures Appl.*, **61** (1982), pp. 245–259.

[45] H. Brezis, M.G. Crandall, Uniqueness of solutions of the initial-value problem for $u_t - \Delta\varphi(u) = 0$, *J. Math. Pures Appl.*, **58** (1979), pp. 153–163.

[46] H. Brezis, A. Friedman, Nonlinear parabolic equations involving measures as initial conditions, *J. Math. Pures Appl.*, **62** (1983), pp. 73–97.

[47] H. Brezis, A. Pazy, Convergence and approximation of semigroups of nonlinear operators in Banach spaces, *J. Funct. Anal.*, **9** (1971), pp. 63–74.

[48] H. Brezis, G. Stampacchia, Sur la régularité de la solution d'inéquations elliptiques, *Bull. Soc. Math. France*, **95** (1968), pp. 153–180.

[49] H. Brezis, W. Strauss, Semilinear elliptic equations in L^1, *J. Math. Soc. Japan*, **25** (1973), pp. 565–590.

[50] F. Browder, *Problèmes Nonlinéaires*, Les Presses de l'Université de Montréal, 1966.

[51] G. Caginalp, An analysis of a phase field model of a free boundary, *Arch. Ration. Mech. Anal.*, **92** (1996), pp. 206–245.

[52] J.M. Carlson, G.H. Swindle, Self-organized criticality: sandpiles, singularities and scaling, *Proc. Nat. Acad. Sci. USA*, **92** (1995), 6712–6719.

[53] A. Chamballe, P.L. Lions, Image recovery via total viariation minimization and related problems, *Numer. Math.*, **76** (1997), pp. 167–182.

[54] S. Chanillo, J. Van Schaftingen, P.-L. Young, The incompressible Navier–Stokes flow in two dimensions with prescribed vorticity. In: Chanillo S., Franchi B., Lu G., Perez C., Sawyer E. (eds.) Harmonic Analysis, Partial Differential Equations and Applications. Applied and Numerical Harmonic Analysis. Birkhäuser, Cham. https://doi.org/10.1007/978-3-319-52742-0_2. Springer, 2017, pp. 19–25.

[55] G.-Q. Chen, B. Perthame, Well-posedness for non-isotropic degenerate parabolic-hyperbolic equations, *Ann. Inst. H. Poincaré (C) Nonlin. Anal.*, **20** (4) (2003), pp. 645–668.

[56] P.L. Colli, G. Gilardi, G. Marinoschi, E. Rocca, Optimal control for a phase field system with a possibly singular potential, *Math. Control Relat. Fields*, **6** (2016), pp. 95–112.

[57] M.G. Crandall, Nonlinear semigroups and evolutions generated by accretive operators, *Nonlinear Functional Analysis and Its Applications*, pp. 305–338, F. Browder (ed.), American Mathematical Society, Providence, RI, 1986.

[58] M.G. Crandall, The semigroup approach to the first order quasilinear equations in several space variables, *Israel J. Math.*, **12** (1972), pp. 108–132.

[59] M.G. Crandall, L.C. Evans, On the relation of the operator $\partial/\partial s + \partial/\partial t$ to evolution governed by accretive operators, *Israel J. Math.*, **21** (1975), pp. 261–278.

[60] M.G. Crandall, T.M. Liggett, Generation of semigroups of nonlinear transformations in general Banach spaces, *Amer. J. Math.*, **93** (1971), pp. 265–298.

[61] M.G. Crandall, A. Pazy, Nonlinear evolution equations in Banach spaces, *Israel J. Math.*, **11** (1972), pp. 57–94.

[62] M.G. Crandall, A. Pazy, On accretive sets in Banach spaces, *J. Funct. Anal.*, **5** (1970), pp. 204–217.

[63] M.G. Crandall, A. Pazy, Semigroups of nonlinear contractions and dissipative sets, *J. Funct. Anal.*, **3** (1969), pp. 376–418.

[64] C. Dafermos, M. Slemrod, Asymptotic behaviour of nonlinear contraction semigroups, *J. Funct. Anal.*, **12** (1973), pp. 96–106.

[65] G. Dinca, J. Mawhin, *Brower Degree*, Birkhäuser, 2021.

[66] J.L. Diaz, L. Veron, Local vanishsing properties of solutions of elliptic and parabolic quasilinear equations, *Trans. Amer. Math. Soc.*, **290** (1985), pp. 787–814.

[67] N. Dunford, T. Schwartz, *Linear Operators.* Part I: *General Theory*, Interscience Publishers, New York, London, 1958.

[68] L.C. Evans, Applications of nonlinear semigroup theory to certain partial differential equations, *Proc. Symp. Nonlinear Evolution Equations*, M.G. Crandall (ed.), Academic Press, New York (1978), pp. 163–188.

[69] R.E. Edwards, *Functional Analysis*, Holt, Rinehart and Winston, New York, 1965.

[70] G. Fabbri, F. Gozzi, A. Święch, *Stochastic Optimal Control in Infinite Dimension. Dynamic Programming and HJB Equations*, Springer International Publishing, 2017.

[71] W.H. Flemming, R.W. Rishel, *Deterministic and Stochastic Optimal Control*, Springer Science & Business Media, New York, 2012.

[72] T.D. Frank, Generalized Fokker-Planck equations derived from generalized linear nonequilibrium thermodynamics, *Physica*, A, **310** (2002), pp. 397–412.

[73] T.D. Frank, *Nonlinear Fokker–Planck Equations*, Springer, 2005.

[74] L.I. Hedberg, Two approximation problems in function spaces, *Ark. Mat.*, **16** (1978), pp. 51–81.

[75] S.N. Kružkov, First-order quasilinear equations in several independent variables, *Math. Sbornik*, **10** (1970), pp. 217–236.

[76] O.A. Ladyzenskaya, V.A. Solonnikov, N.N. Ural'ceva, *Linear and Quasilinear Equations of Parabolic Type*, American Mathematical Society Transl., American Mathematical Society, Providence, RI, 1968.

[77] P.L. Lions, Axiomatic derivation of image processing models, *Math. Models in Applied Sciences*, 4 (1994), 467–475.

[78] J.L. Lions, *Quelques Méthodes de Résolution des Problèmes aux Limites Nonlinéaires*, Dunod, Gauthier–Villars, Paris, 1969.

[79] T. Kato, Accretive operators and nonlinear evolution equations in Banach spaces, *Nonlinear Functional Analysis*, Proc. Symp. Pure Math., vol. **13**, F. Browder (ed.), American Mathematical Society, (1970), pp. 138–161.

[80] T. Kato, Nonlinear semigroups and evolution equations, *J. Math. Soc. Japan*, **19** (1967), pp. 508–520.

[81] T. Kato, The Navier–Stokes equations for an incompressible fluid in $\mathbb{R}^2$ with a measure as the initial vorticity, *Diff. Integral Eqns.*, **7** (1994), pp. 949–966.

[82] Y. Kōmura, Nonlinear semigroups in Hilbert spaces, *J. Math. Soc. Japan*, **19** (1967), pp. 508–520.

[83] G. Marinoschi, *Functional Approach to Nonlinear Models of Water Flows in Soils*, Springer, New York, 2006.

[84] G. Marinoschi, Semigroup approach to a reaction-diffusion equation with cross diffusion (to appear).

[85] G. Minty, Monotone (nonlinear) operators in Hilbert spaces, *Duke Math. J.*, **29** (1962), pp. 341–346.

[86] G. Minty, On the generalization of a direct method of the calculus of variations, *Bull. Amer. Math. Soc.*, **73** (1967), pp. 315–321.

[87] J.J. Moreau, *Fonctionnelles Convexes*, Seminaire sur les équations aux dérivées partielles, Collège de France, Paris, 1966–1967.

[88] J.J. Moreau, Proximité et dualité dans un espace hilbertien, *Bull. Soc. Math. France*, **93** (1965), pp. 273–299.

[89] U. Mosco, Finite-time self-organized criticality on synchronized infite grids, *SIAM J. Math. Anal.*, **50** (3) (2018), pp. 2469–2446.

[90] A. Pazy, The Lyapunov method for semigroups of nonlinear contractions in Banach spaces, *Journal d'Analyse Mathématiques*, **40** (1982), pp. 239–262.

[91] P. Perona, J. Malik, Scale space and edge detection using anisotropic diffusion, *IEEE Transactions Pattern Anal. Mach. Intell.*, 12 (1990), 629–639.

[92] A. Pazy, *Semigroups of Linear Operators and Applications to Partial Differential Equations*, Springer-Verlag, New York. Berlin, Heidelberg. Tokyo, 1983.

[93] A. Pazy, The Lyapunov method for semigroups of nonlinear contractions in Banach spaces, *Journal Analyse Math.*, **40** (1981), pp. 239–262.

[94] R.T. Rockafellar, *Convex Analysis*, Princeton University Press, Princeton, NJ, 1969.

[95] R.T. Rockafellar, Monotone operators and the proximal point algorithm, *SIAM J. Control Optim.*, **14** (1976), pp. 877–898.

[96] R.T. Rockafellar, On the maximal monotonicity of subdifferential mappings, *Pacific J. Math.*, **33** (1970), pp. 209–216.

[97] R.T. Rockafellar, On the maximality of sums of nonlinear operatorsw, *Trans. Amer. Math. Soc.*, **149** (1970), pp. 75–88.

[98] R.T. Rockafellar, Integral functionals, normal integrands and measurable selections, *Nonlinear Operators and the Calculus of Variations*, J.P. Gossez, E. Dozo, J. Mawhin, L. Waelbroeck (eds.), Lecture Notes in Mathematics, Springer-Verlag, New York, 1976, pp. 157–205.

[99] L.I. Rudin, S. Osher, E. Fatemi, Nonlinear total variation based noise removal algorithms, *Physica*, D, **60** (1992), pp. 259–260.

[100] J. Simon, Compactness sets in the space $L^{p_0}(0,T;B)$, *Annali di Matematica Pura ed Applicata*, **146** (1986), pp. 65–96.

[101] G. Stampacchia, *Equations Elliptiques du Second Ordre à Coefficients Discontinues*, Les Presses de l'Université de Montréal, Montréal, 1966.

[102] J.L. Vasquez, *The Porous Medium Equation*, Oxford University Press, Oxford, UK, 2006.

[103] V. Veron, Effects régularisants de semigroupes nonlinéaire dans les espaces de Banach, *Annales Faculté Sciences Toulouse*, **1** (1979), pp. 171–200.

[104] K. Yosida, *Functional Analysis*, Springer-Verlag, New York, 1980.

Index